SCIENCE, CULTURE, AND SOCIETY:
A WILEY-INTERSCIENCE SERIES

BERNARD BARBER, Editor

Syntony and Spark—The Origins of Radio
Hugh G. J. Aitken

Astronomy Transformed: The Emergence of Radio
Astronomy in Britain
David O. Edge and Michael J. Mulkay

The Spaceflight Revolution: A Sociological Study
William Sims Bainbridge

Toward a Metric of Science: The Advent of
Science Indicators
Edited by Yehuda Elkana, Joshua Lederberg,
Robert K. Merton, Arnold Thackray, and
Harriet Zuckerman

A 16th century measuring rod, for measuring pastures, fields, vineyards, meadows, and fruit gardens. "To find the length of a measuring rod the right way and as it is common in the craft . . . Take sixteen men, short men and tall ones as they leave church and let each of them put one shoe after the other and the length thus obtained shall be a just and common measuring rod to survey the land with." Jacob Köbel, Geometrei, von Künstlichen Messen und absehen . . . (1536), Dii–Diii.

3 3001 00533 9196

TOWARD A METRIC OF SCIENCE:

The Advent of Science Indicators

Edited by

Yehuda Elkana
Joshua Lederberg
Robert K. Merton
Arnold Thackray
Harriet Zuckerman

Based upon a conference sponsored by the Center for Advanced Study in the Behavioral Sciences and the Social Science Research Council.

A Wiley-Interscience Publication

JOHN WILEY & SONS, New York · Chichester · Brisbane · Toronto

The preparation of this volume was supported in part by grants to the Center for Advanced Study in the Behavioral Sciences and the Social Science Research Council from the Division of Social Sciences, National Science Foundation. The views expressed in this volume are not necessarily those of the Foundation.

Q
1ุ2
ᴛ6 8

Copyright © 1978 by John Wiley & Sons, Inc.

All rights reserved. Published simultaneously in Canada.

Reproduction or translation of any part of this work beyond that permitted by Sections 107 or 108 of the 1976 United States Copyright Act without the permission of the copyright owner is unlawful. Requests for permission or further information should be addressed to the Permissions Department, John Wiley & Sons, Inc.

Library of Congress Cataloging in Publication Data:

Main entry under title:

Toward a metric of science.

(Science, culture, and society)
"A Wiley-Interscience publication."
"Based upon a conference sponsored by the Center for Advanced Study in the Behavioral Sciences and the Social Science Research Council."
Includes bibliographical references and indexes.
1. Science indicators—Congresses. I. Elkana, Yehuda, 1934-   II. Stanford, Calif. Center for Advanced Study in the Behavioral Sciences. III.   Social Science Research Council.

Q172.T68   301.5   77-24513
ISBN 0-471-98435-3

Printed in the United States of America

10 9 8 7 6 5 4 3 2 1

Paul F. Lazarsfeld, 1901-76

Master
of
Quantitative and Qualitative Social Research

298043

Foreword

Toward A Metric of Science: The Advent of Science Indicators is indicative of two happy recent developments in the social study of science. Along with consideration of the significant qualitative material, every effort is made to identify tenable quantitative measures of scientific values, activities, rewards, and problems. And this volume is interdisciplinary. The editors and authors are from history, sociology, economics, statistics, philosophy of science, and political science. Not all interdisciplinary efforts are successful. This one is, partly because, as so rarely happens, a group of outstanding people from several disciplines have genuinely worked together.

The interdisciplinary character of these authors is also related to their internationality. But the internationality also bespeaks the worldwide concern with the uses of science and science indicators for social welfare, and displays the character of the widespread community of scholars now working on scientific measurement.

The recency of the systematic use of quantitative measurement in the social studies of science should be no surprise. We forget how quantitative work first appeared in other areas of science, for example, in biology. Some of the scientists who resisted Mendel felt that his ratios were taking us back to Paracelsian number magic. As late as the early twentieth century, Karl Pearson had to establish a separate journal, *Bio-Metrika,* to publish and legitimatize quantitative work in biology.

This discussion of quantitative measures in the social study of science is, then, offered as a step along the way. It is an important step. The editors have provided what one of them, Robert K. Merton, has called "disciplined eclecticism." While this book vastly illuminates its subject, it offers no final theory of science or of the methodology of measurement. It is a fine-grained critique of social indicators and social measurement in general, a critique that is essential for continued improvement in this area. It will make an excellent volume for social scientists and their students who want to know more about the nature and problems of measurement in their field.

What I have just said about measurement should alert the reader to the fact that there is much general significance in many of the specific discussions in this volume. The discussion of the political context of science indicators provides another example where a specific problem is treated in a generalized way. That discussion is an excellent generalizable statement about the complex and mutually interactive relations between all kinds of knowledge and social policy.

In brief, there is much to ponder and to enjoy in this volume. The social study of science has made great progress recently. This volume establishes a further mark along the way.

BERNARD BARBER

Preface

Because the aims of this volume are discussed in the Introduction, we report here only how it came into being.

As one of the pioneers in the development of social indicators, Dr. Eleanor Bernert Sheldon, President of the Social Science Research Council, had immediate interest in the first report on science indicators by the National Science Board. In 1974, soon after its publication, Dr. Sheldon asked a small group of Fellows at the Center for Advanced Study in the Behavioral Sciences at Stanford, California to examine the field of science indicators in general and this first report in particular. That group was already committed to informal cooperation, after the traditions of the Center, in the historical sociology of scientific knowledge. But we were far from expert either in social indicators or in the quantitative appraisal of current science. It seemed to us, however, that along with the technical analysis of specific procedures others might provide, there was value in examining the very concept of science indicators from the standpoints of the history, sociology, political science, and economics of science. Even so, it was with some reluctance and much trepidation that we agreed to organize a small conference to examine the field of science indicators. Our reluctance was mitigated by the thought that this would be a first venture into the *applied* historical sociology of scientific knowledge. Our invitation to the conference stated:

> We should like to pose the question, "What must one look at in order to estimate the condition of science as an intellectual activity or as a social institution?" We think of this question within a broad historical and sociological frame rather than from a delimited point of view dealing with the present inputs to and outputs of science measured in terms of men, money, and materials. We think that our discussions of Science Indicators should be problem-oriented. . . .

> At best, we will be starting an ongoing activity, designed to enlarge the scope and conceptual framework of thinking about science.

Despite our reservations and despite the obviously fledgling state of "science indicator studies," the conference was an intellectual success. Discussion was vigorous both inside and outside the formal sessions. Problems were freely aired. Research strategies were proposed. The intellectual, technical, and political problems inherent in the field of science indicators assumed a more coherent shape. Necessarily, the conference was only a beginning. But it did identify and display the need for a clearinghouse of information and action in the field (now provided through a Subcommittee on Science Indicators within the Social Science Research Council Advisory Committee on Social Indicators). The conference has also led to the present volume. As with the conference, so with the subcommittee and the book: The aim of each is to help alert interested parties, to initiate debate, to focus attention, and to define issues. In view of the novelty of science indicators and the conditions under which their systematic reporting began, it would be extravagant to expect that such early responses would be either theoretically definitive or practically exhaustive.

The papers in this volume resulted from the 1974 conference. Most of them were presented and subjected to critical response at that time. Authors then had the usual opportunities for revision. To ensure the highest possible standards for the volume, the editors also invited further commentary from outside referees. That an unevenness of tone, level, and coverage is still apparent speaks of our human weakness as editors. It also indicates difficulties inherent in a first analysis of the complex issues raised by the development of science indicators. That the volume exists at all testifies to the patience, industry, and good humor of our contributors and to their common recognition that measuring the condition of science is a matter of immense intellectual and practical importance. As such it demands the widest interdisciplinary cooperation along with vigorous discussion and exacting scrutiny.

Individually and collectively, these essays do more to raise questions than to answer them. The editors unite in the hope that others will find in this volume an invitation to serious thought on the metric of science and a stimulus to provide more developed understandings than the field yet affords. These understandings will, we trust, be sensitive to the reinstated perception of Protagoras, even as they realize our Horatian hope.

YEHUDA ELKANA
JOSHUA LEDERBERG
ROBERT K. MERTON
ARNOLD THACKRAY
HARRIET ZUCKERMAN

September 1977

Acknowledgments

We are greatly indebted to:

- O. Meredith Wilson, Director of the Center for Advanced Study in the Behavioral Sciences, and the Center Staff
- Eleanor Sheldon and Robert Parke of the Social Science Research Council
- Murray Aborn, George Brosseau, Donald Ploch, and Ronald Overmann, of the National Science Foundation
- The Van Leer Jerusalem Foundation and its Staff for their imaginative support of our venture and for never failing help in bringing it to completion.

We are also grateful to Leo Goodman and James Coleman of the University of Chicago who, along with our contributors, refereed papers in this volume. Maggie Nunley of Beersheba, Israel, edited the early drafts and made the academic prose less rebarbative. Thomas F. Gieryn and William Koerber of Columbia University helped greatly with the reading of proofs, and Mr. Gieryn and Mary Wilson Miles prepared the indexes. We are indebted to the National Bureau of Standards, for a copy of the cut, taken from R. C. Cochrane, *Measures for Progress: A History of the National Bureau of Standards* (Washington, D.C.: U.S. Department of Commerce, 1966), which appears as the frontispiece of this book to symbolize the state of the art of science indicators. Finally, we thank Barbara Thackray for helping to arrange the Conference on Science Indicators, for serving as communications hub, and for assisting in other administrative tasks.

Contents

Introduction

The Editors

Measure is a quality much admired in the abstract. However, our civilization values the ineffable as well as the quantifiable, finding utility in the tensions between such polar opposites. Specific attempts to measure particular things are, therefore, liable to encounter an ambivalent response. It has been over two millennia since Horace decreed, "There is measure in all things." Scholars uncomfortable with his perception have not displayed undue alarm. After all, they may effortlessly reach back a further four centuries and cite Protagoras' antithetical judgment, "Man is the measure of all things."

CONTEXTS OF THE SCIENCE INDICATOR REPORTS

The Horatian dictum knows its greatest successes in the field of natural science. Even there, the adoption of quantitative modes has not been especially rapid, complete, or devoid of controversy. Nonetheless, measurement has come to be perceived as vital to the character of

See p. iii of *Science Indicators 1972* (U.S. Government Printing Office, Washington, D.C., 1973). This volume will be referred to as *SI-72* throughout the present essays. However, our focus will not be on specific problems in the volume, but rather on those generic to the enterprise exemplified in *SI-72* and its successors (e.g., *Science Indicators 1974,* U.S. Government Printing Office, Washington, D.C., 1975).

1

the scientific enterprise and critical for its success. Because science and society are of a piece, it is not surprising that attempts to extend a metric from the natural to the social sphere and even to measure science itself have a rich, complex, and variegated history. In announcing its intention that the publication of "science indicators" become a regular part of its activity, the National Science Board was—whether consciously or not—placing itself within that history.

On a more immediate level, the National Science Board was also taking the critical step that linked two important intellectual movements of the past several years. The two movements in question—previously quite separate with respect to participants, ideas, and organization—are those of social indicators and of "unease with science." An example of the latter is in Theodore Roszak's *The Making of A Counter Culture* (1969), and its most central manifestation is in the Summer 1974 issue of *Daedalus*. The former has given rise to the impressively presented document entitled *Social Indicators 1972*. Much about the present state of knowledge of science indicators, about its strengths and weaknesses (both actual and potential), and not least, about the particular format of this book of essays, can be best understood in the light of this "disjuncture between," then "union of" two disparate intellectual currents.

The reality of social indicators, if not the neologism, has long been familiar in the Western world. William Petty's seventeenth-century exercises in *Political Arithmetick* come quickly to mind. Yet as a sustained intellectual movement, systematic concern with social indicators may be located primarily within the United States in the past several years. A variety of functions can be discerned from the burgeoning literature of that movement. Among these are:

1. Emulating the success achieved by economists in fashioning quantitative measures of significance to policy (e.g., unemployment, inventory accumulations, GNP, and allied "economic indicators")
2. Finding less ambitious, more empirical approaches to social science "problem solving" after the disappointed hopes of the Johnson years
3. Providing a means of discrimination within, and intellectual control of, the burgeoning information flows of "applied social science" (while creating cognitive forms appropriate to the social discourse of an expanded policy-forming apparatus)

In the nature of the case, science indicators are themselves social indicators and as such must be at least partially assimilable to the language, procedures, and assumptions around which the social indicator movement has taken shape.

A DEFINITION AND CLASSIFICATION OF SCIENCE INDICATORS

A definition is appropriate here. *Science indicators are measures of changes in aspects of sciences.* The purpose of this definition is to be heuristic, not final—a means of opening rather than closing discussion and debate. The definition suits the mood of this volume and the present state of "science indicator studies." That mood is one of disciplined eclecticism.

Science indicators will be produced, compared, and consumed by groups and individuals having varied priorities, programs, and preoccupations and dealing with a plurality of sciences. A rigid definition or an unswerving goal would have no great value (as has been slowly learned by those working in the broader field of "social indicators"). Eclecticism is as necessary as it is useful to the measurer of scientific change. Without it, there not only would be tedious wars between zealous factions but also a failure to take advantage of known, promising avenues available for the generation of measures of science. Those avenues are so various that we cannot hope for their being encompassed within any systematic, general theory of scientific change, at least in the foreseeable future.

To be useful eclecticism must be disciplined—that is, because a catholic, flexible, empirical approach is needed at this particular stage of understanding, it does *not* follow that "all measures are equal" and "anything goes" in our efforts to develop a better quantitative understanding of those processes by which science and society mutually condition each other's growth and transformation. Discipline is needed at every stage if we are to select for attention the most rewarding research sites and enable "science indicator studies" to fulfill their potential as a first example of possibilities in the *applied* historical sociology of scientific knowledge.

By way of illustration of the need for discipline in approaching science indicators, it is fruitful to reflect upon some of the distinctive categories into which such indicators can be grouped. The most important distinction is between *explicit* and *tacit* indicators. As the name suggests, *explicit science indicators* are measures of change in science, developed in detail appropriate to their context. We may further distinguish between the *discovery* and *invention* of such explicit science indicators. That the great bulk of work on science indicators in *Science Indicators 1972 (SI-72)* belongs in the "explicit-discovered" category of indicators then becomes apparent. The reasons for this are not far to seek. In the comparatively recent past many agencies, principally but not exclusively government agencies, have for their own purposes compiled annual and short-run statistical series on, for example, research expenditure, patent production, the number of Ph.D.'s awarded. Such measures are today routinely and unobtrusively produced by the system. Their use as indicators awaits only their discovery.

To depend entirely on such "explicit-discovered" indicators would be to commit the field of indicator studies to an interim empiricism of a kind apparent in *SI-72*. Thus our approaches should extend at least to "explicit-invented" indicators. Such indicators—measures that we deliberately set out to construct—will usually be "theory-laden" measures of normative interest. Examples of such explicit-invented indicators might be the citation/publication ratios of scientific literature for different fields and countries or (an as-yet-uninvented explicit indicator) the percentages of university presidents possessing Ph.D.'s in a given field of science. Finally, we can only mention the two other possible categories—implicit-invented indicators and implicit-discovered indicators—leaving their fuller discussion to some other occasion.

Enough has been said to suggest that only within a rich framework of historical and sociological understanding can an effective stance toward science indicators be developed. That stance must cope with the varieties in type and use of possible indicators and must also steer between a spurious objectivism ("the facts dictate . . .") and the sort of despairing subjectivism fashionable in the recent past. To recognize the social embeddedness of a social construct such as "science indicators" is at least to open the way toward a more distanced, dispassionate analysis. Necessary perspective may be achieved by philosophical, psychological, sociological, or historical means. All are discussed, and the last two are more fully developed in the essays in this volume. Here we can only hint at some implications of a perspective from the sociology of knowledge.

A PERSPECTIVE FROM THE SOCIOLOGY OF KNOWLEDGE

Whether pursued with scientific rigor or deliberately cast in the modes of humanistic understanding, any indicator of the state, character, or direction of change in science will necessarily reflect not only the *Ding an sich* it seeks to capture, but also the historical experience, fundamental assumptions, and present visions of the group or groups that gave it birth. Neither liberal optimism nor dismal agnosticism is permissible as the organizing framework of discourse at the administrative centers of Western nations. Instead, responsible leaders appreciate the cultural significance of science within the modern tradition and the real if intangible linkages between scientific knowledge, industrial innovation, economic prosperity, and military power. Such leaders also recognize the labyrinthine complexity of the political process, the widening range of interests demanding accommodation within that process, and the corresponding difficulty in achieving either consensus or decision on appropriate forms, levels, and characteristics for the support of science. "Indicators" may thus serve in this generation in ways not wholly dissimilar from the less quantitatively tuned optimism and pessimism of early days. That is, indicators in general and science indicators in particular may serve as modes in which to shape knowledge, to mediate perceptions, to order values, and to handle ambition.

Powerful traditions within the scientific community foster a view of science in which it is seen as primarily a matter of "results"—whether those results reside in theories, hypotheses, laws, or established facts. According to this view, science possesses great internal autonomy. Interaction with the larger society is primarily in terms (a) of decisions whether and on what scale to fund the necessarily esoteric, specialized practitioners of research, and (b) of intellectual and societal impacts of the "results" of that research. This view of science underlies much of the analysis in *SI-72*.

However, to view science as a mode of culture and hence of cognition, education, socialization, and control may be analytically more fruitful. The work of many anthropologists reminds us that different social systems yield characteristically different styles of culture, cognition, and "cosmology" (beliefs about nature and its relationships to man). Each of these characteristic modes carries with it appropriate patterns of education and socialization. These patterns maintain and reinforce the basic culture as well as its underlying social patterns. Now *science,* in the sense that we use the term (belief in natural law, empirical investigation, consensible results, and progressive understanding), is itself a belief-system characteristic of a social order that can be and has been described. According to Ernest Gellner's brilliant aphorism, "Science is the mode of cognition of industrial society," while "industry is the ecology of science."

The work of Mary Douglas suggests the possibility of constructing a typology that systematically relates social structure to varieties of cosmology. Her work also suggests ways of understanding how cosmology changes as social structure changes. For example, preferred modes of science in an industrial society may be found to be physics and chemistry. In an agrarian society the favored modes may be geology, natural history, and meteorology; in an increasingly service economy, the social, psychological, and biological (medical) sciences may be preferred. The perceived or argued "utility" of each of these modes *is part of* the cultural constellation in question. Again, there are social systems in which the prevailing cosmology and culture are not positively oriented to science at all. Equivalently, there are sectors of our own society for which scientific modes of cognition either have no meaning or have only negative implications.

Thus, if we wish to develop indicators of the state of science, we shall have to attend at least in part to the sociology of knowledge. What basic changes are taking place in our social system? Which of these changes carry implications for science as a mode of culture? To answer these questions, greater emphasis must be placed on understanding public attitudes toward science, on seeing how "images of science" in different social and professional groups relate to other aspects of their cultural experience, and on the manner of socialization in the ways of science through formal education and informal popularization.

Analysis of this kind also comes upon the difficulties inherent in a focus on "science indicators" rather than some comprehensive category such as "knowledge indicators." For instance, *SI-72* reports the growth in the numbers of natural science Ph.D.'s. Yet, as O. D. Duncan points out in his paper in this volume, such information takes on quite different aspects in a larger frame. Natural science Ph.D.'s awarded show a steady increase, suggesting a "healthy" state. However, such Ph.D.'s *decrease* relative to social science Ph.D.'s—information that indicates quite different and possibly more significant aspects of the change. Again, information on the absolute number of undergraduate science degrees holds little significance without measures of both the size and the actual alternative choices of the age cohort in question. The widening ripple of repercussions from the simple perception that "student shortage" will be the pattern of the next two decades is only the latest indication that the financing of university science must be understood within the context of the place of universities in the larger society: Finally, the funding of the academic mission of the NSF alone is a less informative indicator of the value placed on scientific knowledge than one that also includes (in both collected and disaggregated forms) the statistics for the several varieties of knowledge supported by NIH, NEH, and so on.

In short, *SI-72* rests upon an assumption of autonomy for the natural sciences that may better reflect the statutory jurisdiction of the NSF than the social reality in which the sciences actually function. The problematic nature of that assumption points toward the urgent need for better theoretical understandings of science from the perspectives of the sociology of knowledge. Were more of those understandings available, we would be able to state with greater confidence what sorts of *social and cognitive* data provide reliable indicators of coming shifts in the place of particular sciences in society, as of the whole scientific enterprise. Such understandings might also clarify the difficult questions of when a particular discipline could be examined apart from the rest of learning and when science indicators should properly yield place to knowledge indicators.

The decision to create a series of science-indicator reports came about in answer to somewhat different (but no less real or immediate) concerns than those discussed above. The National Science Board is charged by Congress to oversee the work of the National Science Foundation. Its activities lie at the interface between the ambitions of the community of academic natural scientists and the changing realities of national life, as expressed by Congress and by the Office of Management and Budget. By the early 1970s the National Science Board was understandably concerned with the relative decline in funding of the natural sciences. This decline coincided with an apparent turn away from major universities, graduate training, and pure research as foci for such support as was available. Also important was a much-reported public disenchantment with the social dislocations and possible environmental damage perceived to flow from an uncritical nurturing of the "science-technology" complex within American society. Against this background the National Science Board undertook to present as its annual report for the year, *Science Indicators 1972.*

The laudable goal was a systematic objective report on the overall state of American science. In view of the lack of previous work toward such an end and the little attention paid to the natural sciences by the social-indicators movement, the first of the biennial science-indicator reports succeeded to a surprising extent. However, the success was far from unqualified. As will become apparent from the essays that follow, *SI-72* was not only an imaginative, ambitious, and innovative venture, it was also a hurried, uneven performance. It pointed forcefully to a significant new way of conceptualizing and appraising the scientific enterprise for selected public purposes. But in places it also mixed advocacy with social reporting; conflated science with technology in confusing fashion; moved uncertainly between the presentation of available time series, the polling of opinion, and Delphic utterance; and on occasion it made insufficient use of economic and statistical techniques of analysis necessary to its stated ends. In sum, although a commendable first effort, this report on science indicators is variously flawed, the flaws making abundantly plain the need for basic improvement in the ongoing series of *science indicator* reports.

The aim of *Toward a Metric of Science* is to begin laying part of the groundwork, not the specific techniques, for such improvement by providing critical discussion of science indicators, as concept and as practice—a discussion involving historians, sociologists, political scientists, and economists of science; physical, life, and social scientists themselves; and experts drawn from the antecedent social-indicators movement.

PART **I**

ORIENTATIONS

Measurement in the Historiography of Science

1

Arnold Thackray

"E*st modus in rebus.*" So decreed the poet 2000 years ago. If Horace was right, despite one's initial skepticism, the present paper is possible, at least in the sense that there exists a subject matter awaiting investigation. The difficulty lies in demarcating that subject matter and in deciding how best to approach and analyze it. On the one hand the investigator is faced with a cornucopia of current concerns. Economists, information analysts, sociologists, and policymakers are among those actively, on occasion almost frenetically, seeking measures of science (witness the recent rash of international symposia, special subcommittees, Congressional hearings, and foundation reports) (1). On the other hand, the investigator oriented toward history soon discovers a shortage of scholarly substance. The disjunction between available tradition and present opportunity is thus the most sa-

I am indebted to colleagues and students at the University of Pennsylvania—most especially to Robert Bud, P. Thomas Carroll, Lisa Frierman, and Jeffrey Sturchio—for advice, criticism, discussion, encouragement, and unfailing support. The work reported here could not have been done without the generous aid of my coeditors and of Edward Shils. I also acknowledge partial financial support from the National Science Foundation.

lient fact confronting the tyro, eager to wield quantitative methods in the history of science. That salience, in its turn, provides the proper point of departure for this essay. In it, I endeavor to inventory the perspectives, utilities, and potentialities of previous work. I then explore the new context for scholarly inquiry and suggest some desiderata for future researches.

To apply quantitative methods in the history of science, three things are required. First, there must be a methodological conviction—a belief in the worth of those tools and procedures implied in the phrase *quantitative methods*. Second, there must be a desire to discern the patterns that occur in time and a wish to comprehend the changes, continuities, regularities, and singularities implied by the word history. Third, and most obvious—but not trivial—the subject of the methodological conviction and the orientation must be science. No existing tradition combines all three elements, but as might be anticipated, three academic traditions now being cultivated do encompass two of them. Historians of science have been indifferent to mensuration; those who have studied scientific work quantitatively have not been concerned with patterns of change in time; cliometricians and other quantitative historians have regarded science as outside or at best peripheral to their domains of discourse. Exceptions can and should be made to all these statements. As general rules, they underpin the present past of science. They also make a fulcrum by which to gain purchase on measurement in the historiography of science.

THE DEVELOPING HISTORY OF SCIENCE

The "origins" of the history of science may be traced back at least to Francis Bacon (2). However, to talk of historiography is meaningful only where there is a tradition of historical analysis, continuity of problematics, and communication among practitioners. Not until the nineteenth century did such a situation come into being within the history of science. The historiography that arose in that era was shaped by the social values, philosophical positions, and pedagogical aims of the historical practitioners in question. It would be vain to impose a simple unity of purpose on figures as varied as William Whewell, the omniscient master of Trinity College, Cambridge; Ferdinand Hoefer, a Parisian physician; and Hermann Kopp, the professor of physics and chemistry at Glessen. Together with the several groupings of lesser figures surrounding them, succeeding them, and giving resonance to their work, they may be loosely characterized as pioneers of professional science who turned to the intellectual history of their calling for rational recreation and social legitimation, as philosophers in pursuit of support for their (often Comtean) epistemological commitments, and as teachers who found a useful foil to abstract theory in the literary humanistic utilities offered by heroic biography and historical approaches. They chose their subjects and their methods accordingly. Remote periods and major theorists loomed large in the works they produced.

The complex of interests and values informing this emergent tradition of historical discussion continued among practitioners of the second and third generation. Paul Tannery—the greatest intellectual figure of the "middle period"—wrote not on government patronage or the growth of laboratory research, nor even on the emergence of scientific societies, but rather on *La Géométrie Grecque* and kindred subjects (3). Among the third generation of historians of science, George Sarton in America and Charles Singer in England were the recognized leaders. They strove mightily, but with only limited success, to create full-time positions for practitioners, primarily by establishing their field as a recognized academic discipline. Setting standards, awakening the interest of administrators, instructing scientific audiences, and creating a corpus of reliable scholarship were necessarily matters of central concern to them. Certain subtle shifts of emphasis were also apparent. There was a new stress on the internationalism of science and on the scientific method as guarantor of mankind's progress in the future. However, the intellectual focus remained toward remote periods (seventeenth-century science was seen as "modern"), toward individual biographies of major scientists (Sarton chose Leonardo), toward the history of scientific ideas, and toward the hoped-for philosophical synthesis (4).

All these emphases received renewed attention in the quarter century following World War II. During the period from 1945 to 1970 history of science was rapidly established as a university subject throughout the Western world, especially in the British Isles and North America. That institutionalization proceeded on the assumption that "historical science"—the recreation of past theories, experiments, and hypotheses in all their complexity—was the proper core of the new academic discipline (5). A corollary of great importance was the belief that competence in philosophy and natural science, rather than sociology or general history, was an appropriate adjunct to training in the history of science. The result has been intellectual work of an austere, even lofty character, often of considerable elegance and sometimes of great philosophical significance. The result has also been work increasingly distant from history as historians understand it and, by the same token, remote from concern with quantification.

No matter how imperfect the reality, the history of science may usefully be understood in terms of a central historiographic tradition. There has been reasonable agreement on important questions and continuing communication among practitioners. This is not true when those who have measured science are the object of scrutiny. The difficulties hinted at initially come sharply into focus. It is no simple task to discover and categorize occasional individuals and groups whose spasmodic attention to the measurement of science may be traced back at least a century and a half to the time when the irascible Charles Babbage set the world of European learning agog with his *Reflections on the Decline of Science in England* (6). I hope that future investigators will improve on the tentative analysis offered here. For now, it must suffice to point to work in five different genres. The genres in question are civilizational history, genius studies, statistical bibliography, the sociology of progress and, finally, policy-oriented planning and lobbying for science. This last genre deserves to be treated first, for it is the only one with science as its necessary focus and it is the oldest. For both reasons it offers the largest literature germane to this discussion.

QUANTITATIVE GENRES IN THE HISTORY OF SCIENCE

Measurement Oriented to Science Policy

Charles Babbage stands, frustrated, as the first in a succession of publicists of science who—dissatisfied with the resources of money, talent, and public esteem available to their particular interests, to their special sciences, or to science in general (within their own countries)—have invented, discovered, or assembled quantitative information to bolster their arguments. Much research is needed, particularly on the various European countries, before we can trace the history of measures of science, as displayed in the work of early lobbyists and later planners. For instance, there seems to have been a French "decline of science" debate in some ways comparable to the British one, but a generation later in time (7). Again, American, British, and French publicists agreed on the far superior status and resources of German science in the second half of the nineteenth century. What little is known of German lobbyists (e.g., those in the *Verein Deutscher Naturforscher*) suggests they dissented strongly from this view of German science (8).

Here it is sufficient to note how, in the British case, a history of this genre would run from Babbage to the efforts of the "X Club" and to the "Devonshire" or *Royal Commission on Scientific Instruction and the Advancement of Science*. It would include such economics-focused studies as Frederick Rose's widely quoted 1901 *Report on Chemical Instruction in Germany and the Growth and Present Conditions of the German Chemical Industry* and R. A. Gregory's comparison of new support for research in Britain, France, Germany, and the United States in 1910 (9). It would also note the stimulus to analysis that came with the Depression and with the influence of Marxist thought on a remarkable group of young British scientists.

In 1934 Julian Huxley published *Scientific Research and Social Needs* in which he attempted, inter alia, to calculate the total cost of research carried out in Britain. Huxley's work was followed in 1939 by J. D. Bernal's pioneering study of *The Social Function of Science*. Bernal tried to establish the number and distribution of scientists in the United Kingdom, the funding of research by government and industry, and the comparative contributions to the literature of academic, government, and industrial scientists. *The Social Function* represents both the culmination of more than a century of spasmodic endeavor by the "early lobbyists" and the bridging-point to those systematic efforts to measure science that characterize government action through the last generation. Indeed, the government-appointed Barlow Committee, whose 1946 report was to set the tone of British science policy for the next quarter-century, was staffed by left-wing scientists and confrères of Bernal such as P. M. S. Blackett, J. G. Crowther and C. P. Snow (10).

For the United States, a similar account would have to run from Alexander Dallas Bache (greatgrandson to Benjamin Franklin) and Henry Rowland through George Ellery Hale and R. A. Millikan to that gem of the lobbyist's art, Vannevar Bush's *Science: The Endless Frontier* (11). At the same time, it would be difficult and perhaps unfruitful to try to distinguish these polemical works from the more insulated policymaking and planning concerns that lay behind the work of Callie Hull and Clarence West in the 1920s, the 1933 publication of *Recent Social Trends*, the writing of *Research—A National Resource*, and finally, the creation of John R. Steelman's 1947 "Report to the President" on *Science and Public Policy* (12). However, the complex plurality of interests to be accommodated and reconciled in science, as in every aspect of American public policy, fostered the early growth of those bureaucratic "accounting" modes through which the concerns of disparate lobbies could best be handled. The contrast between the Barlow and Steelman reports offers one illustration.

By the early 1950s there was an efflorescence of official measurement of aspects of science in all Western countries. The number of people trained or employed and the money spent were the most commonly adopted measures. For instance, in the United States, the Bureau of Labor Statistics produced a 1953 report on *Scientific Research and Development in American Industry*. A corresponding growth occurred in statistical research on science by international agencies such as OEEC (later OECD) and UNESCO, among others. The increasing number of studies led to the so-called "Frascati manual" of 1963 on *Proposed Standard Practice for Surveys of Research and Development*. In 1966 a general work summarized the available knowledge of *Government and Allocation of Resources to Science*. A year later the first of a series of reports based on the International Statistical Year appeared: *International Statistical Year for Research and Development* .... The increasing interest in, and sophistication of, these quantitative appraisals is apparent from Christopher Freeman's 1969 methodological investigation of science indicators. The culmination of this phase and the beginning of a new era may be seen in the 1973 publication of the most ambitious statistical study up to that date, the U.S. National Science Board's *Science Indicators, 1972,* which has become the opening number in a biennial series. Finally, in this, as in other fields of literature, growth was followed by the appearance of abstracting services—notably the *Annotated Accessions List of Studies and Reports in the Field of Science Statistics* published by UNESCO since 1966 (13).

To a first order of approximation, this whole policy-oriented genre of literature, from Charles Babbage to *SI-72,* may be characterized by its emphasis on aggregates and on "input-output" models, focused on consumption of financial resources and production of people, papers, and patents. It may also be characterized as being more concerned with describing the demographic and economic *state* of science (or particular sciences) than with measuring and analyzing either the intellectual content of science or short- and long-run transformations in science as a part of culture. It is, in a word, ahistoric—in intention and in actuality. It is, however, significant to the quantitative historian of science for two reasons. It contains rich mines of—by now historical—data on certain social dimensions of science including financial priorities, production of scientific papers, employment in different sectors of the economy, and "outputs of credential holders." Also, the concerns of people who worked in the genre have often been canonized in the qualitative history familiar to us all (e.g., the influence of Babbage and the X-Club on our vision of nineteenth-century British science or of Rowland and Bush in creating and nourishing belief in American indifference to basic science). Because of both its present influence on our vision of the past and its possible future utilities as a source of data, this genre belongs in any discussion of measurement in the historiography of science.

The situation is somewhat different with respect to work within the canons of civilizational history, genius studies, statistical bibliography, and the sociology of progress. All stand as components within particular disciplinary traditions of intellectual discussion. All focused on science occasionally, episodically, and by way of illuming themes taking their rise in sundry contexts for divers reasons. At least in their early formulations, all were linked by having some practititioners in common and by a shared espousal of evolutionary models, metaphors, and perspectives on human behavior and social change (models that after 40 years of banishment, show signs of returning to favor). More recently, with the continued growth of the community of academics, several have spawned subdisciplines focused directly on science. The latter offer both sources of data and methods of procedure that may be significant for the future role of measurement in the historiography of science.

Measurement in Civilizational History

Civilizational history is the easiest genre to discuss, despite being the most wide-ranging and elusive. Logically, the line should begin with Condorcet and dwell on Auguste Comte, but it is Henry Buckle who is most often cited in later works. His 1857–1861 *History of Civilization in England* offers a classic account of the scientific life of Scotland. That account was not quantitative in form, but the spirit pervading Buckle's work—"to accomplish for the history of man something equivalent . . . to what has been effected by other enquirers for the natural sciences" (14)—pointed toward a comparative, classifying, and measuring approach. The desire for a historically based science of man is seen in works as varied as John Theodore Merz's *A History of European Thought in the Nineteenth Century* and Pitirim A. Sorokin's *Social and Cultural Dynamics* (15). For our purposes, the chief importance of civilizational history lies in this indirect function, informing genius studies, statistical bibliography, and the sociology of progress.

Measurement in Studies of Genius

The Swiss botanist, Alphonse de Candolle, was among the first to carry out a numerical investigation of scientific genius. His 1873 *Histoire des sciences et des savants depuis deux siècles* was stimulated by the social statistical school of Quetelet, of whom he wrote admiringly. Candolle attempted to characterize the changing scientific strength of countries by their representation among the foreign members of the Royal Society, the French Institut, and the Prussian Academy of Sciences. He sought to establish the importance of environmental factors in the growth of science (16). The following year Francis Galton, the founder of biostatistics, published his *English Men of Science: Their Nature and Nurture*. Galton wrote in direct response to de Candolle. He tried to develop a "natural history" of the English scientists of his day (17). About 30 years later Havelock Ellis extended Galton's work in his 1904 *A Study of British Genius* (18). Galton also directly influenced his student, James McKeen Cattell. Cattell proceeded to have a major impact on psychology in particular and American science in general. The latter influence was accomplished principally through his journal *Science* and through successive editions of his biographical compilations on *American Men of Science* (19).

A first list of American contributors to the quantitative study of genius (not simply scientific genius) would include Cattell, D. R. Brimhall, Edwin Clarke, Ellsworth Huntington, Scott Nearing, and Frederick Woods* before World War I, and Roy Holmes and Stephen Visher in the 1920s. The tradition they established was still giving rise to contributions such as Sanford Winston's work on American inventors and Walter Bowerman's *Studies in Genius* (which is itself a response to the 1926 second edition of Havelock Ellis's work) as late as 1947 (20). Statistical study of regularities in social background, education, occupation, and geographic location of major contributors to given aspects of high culture was facilitated and stimulated by the appearance of—inter alia—the British *Dictionary of National Biography* (1885–1901) and *Who's Who in America* (1898, etc.). The late appearance of the American analogue to the *DNB* meant that research workers in the 1920s and 1930s found *American Men of Science* (1906, 1910, 1921, etc.) a particularly tempting source. Cattell's system of "starring" the major contributors lent itself quite naturally to use in genius studies. Within this genre men of science thus became one well recognized focus of investigation (21).

In the United States in the interwar period, interest in work of this kind spilled over to students in other and differing fields. For instance, in 1923 George Sarton produced a brief statistical and graphical account of the distribution of scientific genius in different civilizations and eras (22). Civilizational analysis and genius studies were also of interest to another Harvard-based emigré, the sociologist, Pitirim Sorokin. Together with his student, Robert Merton, Sorokin used data compiled by Sarton to make a more ambitious, graphical, and statistical study of "The Course of Arabian Intellectual Development, 700–1300 A.D." (23). The study was published in 1935, the same year that Merton completed his doctoral dissertation, a fact of somewhat greater significance to this essay. Among the unusual features of the dissertation was the way he synthesized approaches and assumptions from diverse traditions, including civilizational history; genius studies; and the work of Bernal, Blackett, and Huxley (24).

The conclusions of the routine genius studies have a curiously dated character, derived from their dependence on social Darwinist views of progress and their blindness to how factors of class, race, sex, and region may operate to maintain the advantages flowing from previous privilege. Analysis of biographical guides and dictionaries has also fallen into disfavor as a psychological and sociological technique, being replaced by sample surveys, questionnaires, and intensive interviews aimed at capturing the experience of living subjects, without the ambiguities introduced by an inter-

---

*Woods, for one, was also influenced by the parallel work in France of Alfred Odin (*Genèse des grands hommes gens de lettres Français modernes*, 2 vols., Paris, 1895) and Paul Jacoby (*Études sur la selection chez l'homme*, 2nd ed., Paris, 1906).

mediary source. For obvious reasons, quantitative historians of science do not have an equal liberty of action. Available genius studies cannot be dismissed. Their data and methods provide both a source of information and a mode or procedure that must be surpassed, not simply ignored. An understanding of genius studies will also provide a partial context for appraising the only direct intervention of measurement into the historiography of science, through "the Merton thesis" and the apparently deathless controversy to which it has given rise (25).

Statistical Bibliography: Bibliometrics of Science

Statistical bibliography represents a quite different mode of investigation. Its application to science may be traced to F. J. Cole and Nellie B. Eames. In 1917 they published "The History of Comparative Anatomy: A Statistical Analysis of the Literature." Six years later E. W. Hulme further developed the possibilities inherent in quantitative approaches to the literature of science. In his *Statistical Bibliography in Relation to the Growth of Modern Civilization* he undertook an analysis of the *International Catalogue of Scientific Literature* and related the growth in periodical articles from 1901–1910 and the subsequent decline (1911–1913) to a slowing population growth and an end to trade and credit expansion in Europe. Hulme built directly on the work of Cole and Eames (26). However, independent, isolated work remained the norm, even though the 1920s and early 1930s were a kind of "heroic age" for the statistical study of scientific literature.

In the United States Alfred J. Lotka enunciated "Lotka's Law" in his important 1926 study "The Frequency Distribution of Scientific Productivity." Further contributions came from Gross and Gross, T. J. Rainoff, Hiroshi Tamiya, S. C. Bradford, and Wilson and Fred. Rainoff, for instance, used sophisticated statistical techniques to examine fluctuations in the British, French, and German contributions to the literature of physics. He then sought to establish correlations between these fluctuations and "Kondratieff waves" in the economic realm (27). After World War II and thanks in no small measure to the inventive genius of Eugene Garfield and the persuasive advocacy of Derek de Solla Price, the systematic quantitative analysis of scientific literature became a secure specialty in its own right. Within that specialty, citation analysis has become the most articulated subfield. The recent appearance of bibliographic guides to the literature on bibliometrics of science provides an indicator of these developments (28).

Measurement in the Sociology of Progress

In various ways the sociology of progress is an obverse complementary field to civilizational history. The lineage of each can be traced at least to Comte.

The strand of civilizational history, represented so persuasively by Buckle, had powerful but only indirect effects on the measurement of science. In contrast, the sociology of progress may be seen as the primary stock from which the present burgeoning subdiscipline of the sociology of science has developed.

The line of that development runs from Comte and Spencer in Europe to Lester Ward and William Ogburn in the United States. Ogburn worked with the twin convictions that "any investigation of motives is likely to be superficial without the recognition of cultural factors, which can be discerned only through historical investigation" and that "we cannot have a science without measurement. And science will grow in the social studies in direct ratio to the use of measurement." Together, these convictions provided a natural framework in which his theory of *Social Change* could be expressed and in which its emphasis on the importance of technology led to such numerical and conceptual inquiries as those found in his paper (with Dorothy Thomas), "Are Inventions Inevitable?" (29). The importance of Ogburn's historical measuring approach to the study of science and technology was recognized by his being selected in 1959 as first President of the Society for the History of Technology. He also encouraged the work of S. Colum Gilfillan (30) and provided one more ingredient for input to the creative synthesis achieved by Merton in his "Science, Technology and Society in Seventeenth-Century England."

That work may be seen as announcing the beginnings of the sociology of science as a recognizable specialty. However, it may also be seen as marking the end of an era. Its concern with science *in* society, with cultural dimensions, and with broad historical patterns were not the motifs around which the sociology of science—or American sociology, more broadly defined—developed in the 1950s and 1960s. The trend in sociology was rather to the microanalysis of the internal dynamics of small communities —the firm or the bureaucratic organization, the ghetto or the gang—and away from historical and cultural analysis (31). This trend fit well with certain developments in American science at this time. The production of Ph.D.'s in the natural sciences increased by over 375% between 1950 and 1970 to almost 9000 in the latter year alone. The budget of the National Science Foundation went from nothing to more than $400,000,000 in the same period. In 1952 NSF had approximately 100 grantees, in 1969 it had 9000 (32). Indicators of the growth in scale and complexity of the scientific enterprise could be multiplied without effort.

Questions of organization, reward, morale, and procedure *within* the burgeoning ranks of American academic science enjoyed a new saliency. The common sociological concern with institutions also fitted this situation. The intellectual emphasis on the internal dynamics of the newly discovered scientific community meshed well with the broader drift of sociological theory and with the *perceived* experiences of scientists themselves. The structural-functional paradigm of the workings of that community, which Merton articulated in a series of masterly essays, also served the latent function of providing an intellectual program, a set of problematics, and a cognitive identity to the emergent North American group of practitioners in the sociology of science (33).

The appearance of bibliographical guides serves as an indicator of the maturity this genre has now achieved and makes surveying recent work within the confines of this paper as unnecesssary as it would be unrealistic (34). The distance traveled may be seen by comparing Bernard Barber's pioneering 1952 study of *Science and the Social Order* with Merton's 1973 *Sociology of Science.* The growing sophistication of analytical tools is apparent in the contrast between Edward Shils's seminal discussion of "The Scientific Community" in 1954 and the subsequent studies of Warren Hagstrom and Diana Crane (35). Prevailing standards of devotion to, and performance in, the quantitative study of science may be seen from two recent books on stratification and competition in the physics community. One offers 44 tables in its 174 pages of text, the other 42 tables in 261 pages. Examples could be multiplied (36).

The sociology of science now boasts a lively North American tradition of measurement and analysis of the internal dynamics of (usually academic) scientific communities. No clear demarcation can be made between this tradition and that of the bibliometrics of science in regard to the study of subjects such as the "productivity" of and communication channels among scientists. More obviously sociological is work on the norms of science; on gatekeeping, refereeing, and priority disputes; on reward systems and social stratification; and on the organization and differentiation of disciplines and specialties. Not altogether surprisingly, this sustained attention to the "internal sociology" of science has not passed without comment. The critics have usually been residents in Europe—most often in Britain—and hence remote from the advantages and the limitations shared by a group of practitioners enjoying a common paradigm (37). Their unease over the ahistorical character of recent sociology of science is the item of greatest interest in the present context.

CONVERGENCE OF HISTORICAL AND SOCIOLOGICAL INQUIRY

At this juncture several general observations are in order. As should be manifest, the five genres—civilizational history, genius studies, statistical bibliography, sociology of progress, and policy-oriented work—together offer a rich array of data and techniques to the historian of science concerned with quantitative method and systematic knowledge. The past generation may be characterized as one in which the various disciplinary traditions potentially associated with the quantitative study of science maintained or increased their mutual distance. This is especially true for the history of science and the sociology of science. Each was preoccupied with establishing its own cognitive and professional identity, and found distance advantageous. The third observation is that a profound change is now in the making. Its results will be an altogether new level of attention to measurement as a technique for understanding past science. Its causes may be found on various levels and in varied contexts.

The most powerful (because most general) cause lies in the rapprochement now taking place between historical and sociological studies. As historians become more concerned with regularities and less with singularities, sociologists become more amenable to the search for models of change as well as ones of order (38). A second cause is attributable to the shift now under way in the cultural valuation of scientific knowledge and technological change. Science is becoming less an example of objective truth and a guarantor of spiritual progress and more a source of problems and of uncertain troubling knowledge. The shift toward a more ambivalent view of science implies renewed attention to science as a cultural phenomenon (39). That attention will come from both the publicists and planners and from research workers sensitive to the social milieu in which their isolated worlds hang suspended. A more proximate cause for attention to the dimensions of the past lies in the consolidation of fields made manifest in the formation of a new society, the Society for Social Studies of Science (4-S). The rapid early growth in its membership signals the emerging scholarly interest in the social aspects of science (40).

In conclusion, it is appropriate to consider the third of those groups with which this analysis began—the cliometricians and other less militant quantitative historians. Given a concern with historiography—with past and future ways in which history has been and may be written—and given a focus on the potentialities of quantitative methods, to seek illumination from the historiography of history itself is only natural.

What the inquirer discovers may be both sobering and encouraging. The narrow sectarian wars waged by men of fanatic temperament and limited vision should give him pause before he engages in *Methodenstreit* for or against quantitative methods (41). The failure of quantitative methods to bring in some golden age of historical understanding should tell him at least as much as the impressive roster of achievements now associated with the spread of quantitative methods in demographic, economic, family, political, urban, and social history. But the progress made in the prosopography of the Classical world; the routine production of historical series of demographic and economic statistics; the recreation of the world we have lost in our family life; the uncovering of regional, sectional, and local voting patterns and their political implications; the articulation of patterns of mobility in the early industrial city—these are quantitative achievements to make the historian of science envious as he contemplates the rudimentary state of his knowledge and understanding (42).

In fact, a trickle of quantitative work in the history of science has begun to find its way into print. In the past five years quantitative studies have been published by Robert V. Bruce; Clark A. Elliott; Paul Forman, John L. Heilbron, and Spencer Weart; Karl Hufbauer; Sally Gregory Kohlstedt; Roy M. MacLeod; Margaret W. Rossiter; Steven Shapin; and Arnold Thackray, among others (43). However, a simple measure shows that such studies could as yet scarcely be said to carry all before them. Of the 200 articles published by *Isis* in the past 10 years, only five have been quantitative in nature, even in the most generous interpretation. One might say that the first research desideratum is simply "more"—except, as Kingsley Amis warned, perhaps "more means worse."

The foregoing analysis suggests that the time is ripe not simply for "more"—that is, more studies that extend into history an examination of the forms, growth, and funding of isolated scientific specialties—but also for quantitative historical studies that

1. Establish the "enduring trends" in education, deployment, and productivity of scientists and in the character and diffusion of scientific knowledge
2. Investigate the "enduring trends" in the socialization, work and status of the technical cadres which both consume and support science.
3. Disaggregate those trends by country and discipline
4. Examine both short-run variations in and conjunctures of those trends

There is also need for studies that treat science as an integral part of culture by

1. Relating those trends to other *historical* changes in the societies in question
2. Systematically examining the ideological functions of scientific knowledge and belief within particular groups
3. Examining the social supports for, and political organization of, the communities of professional science

4. Sustained prosopographical analysis of "preprofessional" groupings and of the nonscientific characteristics of professional associations

5. Exploring the "low life" of science as expressed in the magazines and other lingering manifestations of popular culture

Finally, studies are needed that

1. Incorporate a comparative element
2. Make explicit use of models of the growth of aspects of science
3. Bring quantitative evidence to bear on definite hypotheses

In all this the quantitatively inclined historian of science will be sensitive to the call for "the development of a New History of Science, of a Social History of Science." More than a quarter of a century has passed since Henry Guerlac sounded that call. It is even longer since Walter Pagel noted, "It has become a fashion to discuss the sociological implications of science and its history." Perhaps the appearance of the successive *Science Indicator* reports and the informed interest with which they have been greeted may be taken to indicate that the fashion, now once more returned, will become a decisive transformation in the nature of historical investigations of scientific knowledge. If this is to be, it will require many hands working together to remedy a situation that is as true now as it was 40 years ago: "There are only a few attempts to prove the thesis by exact figures and methods" (44).

NOTES

1. A partial inventory of American activity would have to include the June 1974 conference (the proceedings of which are reported in the present volume); the August 1976 Berkeley Symposium on Quantitative Methods in the History of Science; the Subcommittee on Science Indicators first convened by the SSRC Committee on Social Indicators in the autumn of 1974 (chairman, Harriet Zuckerman); the work of the Science Indicators Unit of the National Science Board, culminating in *Science Indicators 1972* (Government Printing Office, Washington, D.C., 1973); and *Science Indicators 1974* (Government Printing Office, Washington, D.C., 1975); and finally, the round of hearings centered on the National Science Board reports—U.S. Congress, House, Committee on Science and Technology, Subcommittee on Domestic and International Scientific Planning and Analysis, *Hearings on "Science Indicators,"* 94th Cong., 2d sess., (No. 75, Government Printing Office, Washington, D.C., 1976).

2. See, for example, Thomas S. Kuhn, "The History of Science," in David L. Sills, Ed., *International Encyclopedia of the Social Sciences,* Vol. 14, Macmillan and Free Press, New York, 1968, p. 75.

3. Paul Tannery, *La Géométrie Greque,* Gauthier-Villars, Paris, 1887.

4. For Sarton's attempts to promote the history of science in the United States, see Arnold Thackray and Robert K. Merton, "On Discipline Building: The Paradoxes of George

Sarton," *Isis*, **63**, 473–495 (1972). On Charles Singer, see A. Rupert Hall, "Charles Joseph Singer, 1876–1960," *Isis*, **51**, 558–560 (1960).

5. See the inaugural lecture by Margaret Gowing, "What's Science to History or History to Science?" Clarendon, Oxford, 1975.

6. Charles Babbage, *Reflections on the Decline of Science in England*, B. Fellowes, London, 1830. For a brief discussion on the "decline of science" debate in England, see A. D. Orange, "The Origins of the British Association for the Advancement of Science," *British Journal for the History of Science*, **6**, 152–157 (1972). J. B. Biot, "[Review of] Reflections on the decline of science in England . . . ," *Journal des Savans*, 41–49 (January 1831); and [Gerard Moll], *On the Alleged Decline of Science in England . . .*, T. & T. Boosey, London, 1831 provide some indication of the range of European reaction to Babbage's philippic.

7. Harry W. Paul, "The Issue of Decline in Nineteenth-Century French Science," *French Historical Studies*, **7**, 416–450 (1972); Robert Fox, "Scientific Enterprise and the Patronage of Research in France, 1800–1870," *Minerva*, **11**, 442–473 (1973). For an illustration of the degree to which French declinist arguments have become part of received knowledge, see Robert Gilpin, "The Heritage of the Napoleonic System," in *France in the Age of the Scientific State*, Princeton University Press, Princeton, 1968, Ch. 4, pp. 77–123.

8. Frank Pfetsch, *Zur Entwicklung der Wissenschaftspolitik in Deutschland, 1750–1914*, Duncker und Humblot, Berlin, 1974.

9. See Ruth Barton, *The X Club: Science Religion and Social Change in Victorian England*, Ph.D. dissertation, University of Pennsylvania, 1976. In 1868 Leone Levi sought to demonstrate the improvement since Babbage's time. Levi calculated the membership and income, by field, of major scientific societies in the United Kingdom in his paper "On the Progress of Learned Societies," *Report of the Thirty-Eighth Meeting of the British Association for the Advancement of Science*, John Murray, London, 1869, Notes and Abstracts of Miscellaneous Communications, pp. 169–173. See also his "A Scientific Census," *Nature*, **1**, 99–100 (1869), and his "The Scientific Societies," in *Report of the Forty-Ninth Meeting of the B.A.A.S.*, John Murray, London, 1879, pp. 458–468. See also, *Great Britain, Parliamentary Papers*, Royal Commission on Scientific Instruction and the Advancement of Science ["Devonshire Commission"], *Reports* [Cds. 318 (1871), 536 (1872), 868 (1873), 884 (1874), 1087 (1874), 1279 (1875), 1297 (1875), 1298 (1875)], and *Minutes of Evidence, Appendices and Analyses of Evidence*, 3 Vols., [Cds. 536 (1872), 958 (1874), 1363 (1875)]; Frederick Rose, *Report on Chemical Instruction in Germany and the Growth and Present Condition of the German Chemical Industry*, Great Britain, Parliamentary Papers, Vol. 80 (Accounts and Papers, Vol. 44, 1901), Diplomatic and Consular Reports, Miscellaneous Series, No. 561, Cd. 430–16, 1901; R. A. Gregory in *Fifth Annual Report*. British Science Guild, London, 1911, pp. 25–96.

10. Julian Huxley, *Scientific Research and Social Needs*, Watts, London, 1934; J. D. Bernal, *The Social Function of Science*, Routledge, London, 1939; and *Great Britain, Parliamentary Papers*, Vol. 14 (1945–1946); *Reports from Commissioners, Inspectors and Others*, Vol. 5, May 1946; *Scientific Man-Power: Report of a Committee Appointed by the Lord President of the Council* [Barlow Committee Report], Cd. 6824 (1946).

11. See, for instance, Bache's "Remarks Upon the Meeting of the American Association at Charleston, S.C., March 1850," *Proceedings of the A.A.A.S.*, **4**, 159–176 (1850); Henry Rowland, "A Plea for Pure Science," *Science*, **2**, 224–250 (1883). The shifting relationships between scientific lobbyists and their potential patrons may be traced in Howard S. Miller's *Dollars for Research: Science and Its Patrons in Nineteenth-Century America*, University of Washington Press, Seattle, 1970. For the twentieth century, see Daniel J. Kevles, "George Ellery Hale, the First World War, and the Advancement of Science in America," *Isis*, **59**, 427–437 (1968); Lance E. Davis and Daniel J. Kevles, "The National

Research Fund: A Case Study in the Industrial Support of Academic Science," *Minerva*, **12**, 208–220 (1974); U.S. Office of Scientific Research and Development, *Science: the Endless Frontier*. Report to the President on a Program for Postwar Scientific Research, by Vannevar Bush, Government Printing Office, Washington, 1945. The context of Bush's report and much else of relevance to the patronage of science in the twentieth century is detailed in A. Hunter Dupree, *Science in the Federal Government*, Harvard University Press, Cambridge, 1957. See also Daniel J. Kevles, "The National Science Foundation and the Debate over Postwar Research Policy, 1942–1945," *Isis*, **68**, 5–26 (1977).

12.   Hull and West provided annual compilations of "Doctorates Conferred in the Sciences by American Universities" for the National Research Council from 1920–1933; see, for example, West and Hull, "Doctorates Conferred in the Sciences by American Universtities in 1921," *Science*, **55**, 271–279 (1922). See also, President's Research Committee on Social Trends, *Recent Social Trends in the United States*, Foreword by Herbert Hoover; William F. Ogburn, Director of Research; 2 Vols. McGraw-Hill, New York, 1933; U.S. National Resources Planning Board, *Research—A National Resource*, Report of the National Research Council to the National Resources Planning Board, Government Printing Office, Washington, D.C., 1941; and U.S. President's Scientific Research Board, *Science and Public Policy . . .*, A Report to the President, by John R. Steelman, Government Printing Office, Washington, D.C., 1947.

13.   U.S. Bureau of Labor Statistics, *Scientific Research and Development in American Industry: A Study of Manpower and Costs*, Government Printing Office, Washington, D.C., 1953. [This report was the forerunner of the annual series *Research and Development in Industry* (1957–) compiled by the National Science Foundation]; Organization for Economic Cooperation and Development, Directorate for Scientific Affairs, *The Measurement of Scientific and Technical Activities: Proposed Standard Practice for Surveys of Research and Development* ["Frascati manual"], OECD, Paris, 1963; idem, *Government and Allocation of Resources to Science*. OECD, Paris, 1966); idem, *International Statistical Year for Research and Development: A Study of Resources Devoted to R & D in OECD. Member Countries in 1963–64*. 2 Vols., OECD, Paris, 1967; Christopher Freeman, *The Measurement of Scientific and Technological Activities: Proposals for the Collection of Statistics on Science and Technology on an Internationally Uniform Basis*, UNESCO Statistical Reports and Studies, No. 15, UNESCO, Paris, 1969; and idem, *Measurement of Output of Research and Experimental Development: A Review Paper*, UNESCO Statistical Reports and Studies, No. 16, UNESCO, Paris, 1969.

14.   Henry Thomas Buckle, *History of Civilization in England*, 2 Vols., J. W. Parker, London, 1857–1861; quotation from Vol. 1, p. 6.

15.   John T. Merz, *A History of European Thought in the Nineteenth Century*, 4 vols., W. Blackwood, Edinburgh and London, 1903–1914; Pitirim A. Sorokin, *Social and Cultural Dynamics*, 4 vols., American, New York, 1937. Two American examples of this genre are John William Draper, *History of the Intellectual Development of Europe*, Harper, New York, 1863; and Edward Eggleston, *The Transit of Civilization from England to America in the Seventeenth Century*, Appleton, New York, 1901.

16.   Alphonse de Candolle, *Histoire des sciences et des savants depuis deux siècles*, H. Georg, Geneva, [1872] 1873. See also, S. Mikulinski, "Alphonse de Candolle's *Histoire* and its historic significance," *Organon*, **10**, 223–243 (1974); and Robert K. Merton, "The Sociology of Science: An Episodic Memoir," in R. K. Merton and Jerry Gaston, Eds., *The Sociology of Science in Europe*, Southern Illinois University Press, Carbondale, 1977. Candolle's book was also undertaken in part to answer Francis Galton's 1869 study, *Hereditary Genius*.

17. Francis Galton, *English Men of Science: Their Nature and Nurture*, Macmillan, London, 1874. See also, Ruth Schwartz Cowan's "Introduction" to the reprint of this work, Cass, London, 1970; and Victor L. Hilts, "A Guide to Francis Galton's *English Men of Science*," *Transactions of the American Philosophical Society*, **65**, Part 5, 1–85 (1975).

18. Havelock Ellis, *A Study of British Genius*, Hurst and Blackett, London, 1904.

19. J. McKeen Cattell, Ed., *American Men of Science: A Biographical Directory*, Science, New York, 1906. Subsequent editions edited by Cattell appeared in 1910, 1921, 1927, 1933, and 1938. Albert Theodor Poffenberger, Ed., *James McKeen Cattell, 1860–1944, Man of Science*, 2 vols., Science, Lancaster, Pa., 1947, provides a guide to Cattell's interests and influence.

20. An exhaustive listing is not possible, but see, for instance, J. McKeen Cattell, "A Statistical Study of Eminent Men," *Popular Science Monthly*, **53**, 359–378 (1903); and "Homo Scientificus Americanus," *Science*, **17**, 561–570 (1903). Further discussions by Cattell are in *Science*, **24**, 658–665 (1906);**24**, 699–707 and 732–742 (1906); **32**, 633–648 and 672–688 (1910), and so on. See also the articles of Frederick A. Woods, including his 1909 call for "A New Name for a New Science," *Science*, **30**, 703–704 (1909) in which "I proposed the term historiometry for that class of researches in which the facts of history have been subjected to statistical treatment according to some method of measurement more or less objective or impersonal in its nature." Woods' work is available in collected form in his *Mental and Moral Heredity in Royalty*, Henry Holt, New York, 1906; and his *The Influence of Monarchs. Steps in a New Science of History*, Macmillan, New York, 1913; quotation, p. 407. See also Scott Nearing, "The Geographical Distribution of American Genius," *Popular Science Monthly*, **85**, 189–199 (1914); and "Younger Generation of Scientific Genius," *Scientific Monthly*, **2**, 48–61 (1916). Ellsworth Huntington, *Climate and Civilization*, Yale University Press, New Haven, 1915; Edwin L. Clarke, *American Men of Letters: Their Nature and Nurture*, Columbia University Press, New York, 1916; Stephen S. Visher, *Geography of American Notables: A Statistical Study . . . to Evaluate Various Environmental Factors*, Indiana University Studies, **15**, Indiana University, Bloomington, 1928; Roy Holmes, "A Study in the Origins of Distinguished Living Americans," *American Journal of Sociology*, **34**, 670–685 (1928–1929); Sanford Winston, "Biosocial characteristics of American inventors," *American Sociological Review*, **3**, 837–849 (1937); and Walter G. Bowerman, *Studies in Genius*, Philosophical Library, New York, 1947.

21. Stephen S. Visher, *Scientists Starred 1903–1943 in "American Men of Science." A Study of Collegiate and Doctoral Training, Birthplace, Distribution, Backgrounds, and Developmental Influences*, The Johns Hopkins Press, Baltimore, 1947, represents the goal of this mode of inquiry.

22. See George Sarton, *Carnegie Institution of Washington Yearbook*, **22**, 335–337 (1923).

23. Pitirim Sorokin and Robert K. Merton, "The Course of Arabian Intellectual Development, 700–1300 A.D.," *Isis*, **22**, 516–524 (1935). Sorokin's own earlier interest in genius studies is apparent in his "American Millionaires," *Journal of Social Forces*, **3**, 627–640 (1925) and "American Labor Leaders," *American Journal of Sociology*, **33**, 382–411 (1927).

24. Robert K. Merton, "Science, Technology and Society in Seventeenth-Century England," *Osiris*, **4**, 360–632 (1938) [reprinted with a new introduction, Howard Fertig, New York, 1970].

25. For the most recent contribution to the debate, see Charles Webster, *The Great Instauration: Science, Medicine and Reform, 1626–1660*, Duckworth, London, 1975.

26. F. J. Cole and Nellie B. Eames, "The History of Comparative Anatomy: A Statistical

Analysis of the Literature," *Science Progress*, **11**, 578–596 (1917); and E. W. Hulme, *Statistical Bibliography in Relation to the Growth of Modern Civilization*, Butler & Tanner, Grafton, London, 1923.

27. Alfred J. Lotka, "The Frequency Distribution of Scientific Productivity," *Journal of the Washington Academy of Sciences*, **16**, 317–323 (1926); R. L. K. Gross and E. M. Gross, "College Libraries and Chemical Education," *Science* , **66**, 385–389 (1927); T. J. Rainoff, "Wave-like Fluctuations of Creative Productivity in the Development of West European Physics in the Eighteenth and Nineteenth Centuries," *Isis*, **12**, 287–319 (1929); Hiroshi Tamiya, "*Eine mathematische Betrachtung über die Zahlen Verhältnisse der in der 'Bibliographie von Aspergillus' zusammen gestellten Publikationen*," *Botanical Magazine*, **45**, 62–69 (1931); S. C. Bradford, "Sources of Information on Specific Subjects," *Engineering*, **137**, 85–86 (1934); and P. W. Wilson and E. B. Fred, "The Growth Curve of a Scientific Literature," *Scientific Monthly*, **41**, 240–250 (1935).

28. See Derek J. de Solla Price, *Little Science, Big Science*, Columbia University Press, New York, 1963; and *Science Since Babylon*, enlarged ed., Yale University Press, New Haven, 1975; Eugene Garfield, I. H. Sher, and R. J. Torpie, *The Use of Citation Data in Writing the History of Science*, Institute for Scientific Information, Philadelphia, 1964; Henry G. Small and Belver C. Griffith, "The Structure of Scientific Literatures," *Science Studies*, **4**, 17–40 and 339–365 (1974). See also Ming Ivory, Janice LaPorte, Henry Small, and Janet Stanley, *Citation Analysis: An Annotated Bibliography*, Institute for Scientific Information, Philadelphia, 1976; and Francis Narin, *Evaluative Bibliometrics: The Use of Publication and Citation Analysis in the Evaluation of Scientific Activity*, Computer Horizons, Cherry Hill, N.J., 1976.

29. William F. Ogburn and Dorothy Thomas, "Are Inventions Inevitable? A Note on Social Evolution," *Political Science Quarterly*, **37**, 83–98 (1922). See also his *Social Change with Respect to Culture and Original Nature*, B. W. Huebsch, New York, 1923. Reprinted with an Introduction by O. D. Duncan, University of Chicago Press, Chicago, 1964. Quotation at p. xi of the reprint edition.

30. S. Collum Gilfillan, *The Sociology of Invention*, Follett, Chicago, 1935. Reprinted 1963 and 1970.

31. See W. Richard Scott, "Organizational Structure," *Annual Review of Sociology*, **1**, 1–20 (1975); Joseph Ben-David, "The State of Sociological Theory and the Sociological Community: A Review Article," *Comparative Studies in Society and History*, **15**, 448–472 (1973).

32. Letter to "Dear Colleague" from H. Guyford Stever, Director, the National Science Foundation. June 30, 1976.

33. See Norman W. Storer "Introduction," in Robert K. Merton, *The Sociology of Science*, University of Chicago Press, Chicago, 1973; and Jonathan R. Cole and Harriet Zuckerman, "The Emergence of a Scientific Specialty: The Self-Exemplifying Case of the Sociology of Science," in Lewis A. Coser, Ed., *The Idea of Social Structure: Papers in Honor of Robert K. Merton*, Harcourt Brace Jovanovich, New York, 1975, pp. 139–174.

34. Joseph Ben-David and Teresa A. Sullivan, "Sociology of Science," *Annual Review of Sociology*, **1**, 203–222 (1975); I. Spiegel-Rosing, *Wissenschaftsentwicklung und Wissenschaftssteuerung: Einführung und Material zur Wissenschaftsforschung*. Athenäum Verlag, Frankfurt/M, 1973, pp. 133–309.

35. Bernard Barber, *Science and the Social Order*, Free Press, Glencoe, Ill., 1952; Edward Shils, "The Scientific Community: Thoughts after Hamburg," *Bulletin of the Atomic Scientists*, **10**, 151–155 (1954), [amended reprinting in Edward Shils, *The Intellectuals and the Powers*, The University of Chicago Press, Chicago, 1972, pp. 204–212]; Warren O.

Hagstrom, *The Scientific Community*, Basic Books, New York, 1965); and Diana Crane, *Invisible Colleges: Diffusion of Knowledge in the Scientific Community*, The University of Chicago Press, Chicago, 1972.

36. Jonathan R. Cole and Stephen Cole, *Social Stratification in Science*, The University of Chicago Press, Chicago, 1973; and Jerry Gaston, *Originality and Competition in Science: A Study of the British High Energy Physics Community*, The University of Chicago Press, Chicago, 1973. Most recently Harriet Zuckerman has extended this sociological, quantitative approach to cover phenomena distributed over a seventy-five year period. There are 7 figures and 33 tables in the 302 pages of text in her *Scientific Elite, Nobel Laureates in the United States*, Free Press. New York, 1977.

37. See, for example, Michael Mulkay, "Some Aspects of Cultural Growth in the Natural Sciences," *Social Research* **36**, 22–53 (1969); S. B. Barnes and R. G. A. Dolby, "The Scientific Ethos: A Deviant Viewpoint," *European Journal of Sociology*, **11**, 3–25 (1970); and M. D. King, "Reason, Tradition and the Progressiveness of Science," *History and Theory*, **10**, 3–32 (1971). See also the September 1976 issue of *Social Studies of Science*.

38. Among recent indicators of this rapprochement are the establishment in 1974 of the Social Science History Association and the appearance of such hortatory monographs as Richard Hofstadter and Seymour Martin Lipset, Eds., *Sociology and History: Methods*, Basic Books, New York, 1968; David S. Landes and Charles Tilly, Eds., *History as Social Science*, Prentice-Hall, Englewood Cliffs, N.J., 1971; and Lee Benson, *Toward the Scientific Study of History: Selected Essays*, Lippincott, Philadelphia, 1972. Earlier European interest underlay the foundation of two journals: *Annales: Economies, sociétés, civilisations* (1929; refounded 1946) and *Past and Present* (1952). More recently, the *Journal of Social History* (1967), the *Historical Methods Newsletter* (1967), the *Journal of Interdisciplinary History* (1970), and *Social Science History* (1976) have been founded in the United States.

39. Jerome R. Ravetz, *Scientific Knowledge and its Social Problems*, Clarendon, Oxford, 1971; Barry Barnes, *Scientific Knowledge and Sociological Theory*, Routledge and Kegan Paul, London, 1974.

40. The formal organization of the Society took place in San Francisco in August 1975. In little more than a year 4-S, or the Society for Social Studies of Science, had attracted about 500 individual members. This record compares favorably with the initial experience of the History of Science Society. See "Fiftieth Anniversary Celebrations of the History of Science Society," *Isis*, **66**, 443–482, especially 452–457 (1975).

41. An obvious example here is the brouhaha surrounding Robert W. Fogel and Stanley Engermann's *Time on the Cross*, 2 vols., Little, Brown, Boston, 1974, which dealt with the antebellum American slave economy, using cliometric methods. For one recent overview of the debate, see Charles Crowe, "Slavery, Ideology and 'Cliometrics,' " *Technology and Culture*, **17**, 271–285 (1976).

42. For an introduction to this voluminous (and rapidly growing) literature, see the following: Lawrence Stone, "Prosopography," *Daedalus*, **100**, 46–79 (1971); William O. Aydelotte, *Quantification in History*, Addison-Wesley, Reading, Mass., 1971; William O. Aydelotte, et al., Eds., *The Dimensions of Quantitative Research in History*, Princeton University Press, Princeton, 1972; Robert W. Fogel, "The Limits of Quantitative Methods in History," *American Historical Review*, **80**, 329–365 (1975); Peter Laslett and Richard Wall, Eds., *Household and Family in Past Time*, Cambridge University Press, Cambridge, 1972; V. R. Lorwin and J. M. Price, Eds., *The Dimensions of the Past: Materials, Problems, and Opportunities for Quantitative Work in History*, Yale University Press, New Haven, 1972; Edward Shorter and Charles Tilly, *Strikes in France, 1830–1968*, Cambridge University Press, Cambridge, 1974; Ronald Syme, *The Roman Revolution*, Clarendon, Oxford, 1939;

Stephan Thernstrom and Richard Sennett, Eds., *Nineteenth-Century Cities: Essays in the New Urban History,* Yale University Press, New Haven, 1969; Charles Tilly, Louise Tilly, and Richard Tilly, *The Rebellious Century,* Harvard University Press, Cambridge, 1975; and E. A. Wrigley, Ed., *Nineteenth-Century Society: Essays in the Uses of Quantitative Methods for the Study of Social Data,* Cambridge University Press, Cambridge, 1972.

43. See, for example, Robert V. Bruce, "A Statistical Profile of American Scientists, 1846–1876," in George H. Daniels, Ed., *Nineteenth Century American Science: A Reappraisal,* Northwestern University Press, Evanston, Ill., 1972, 63–94; Clark A. Elliott, "The American Scientist in Antebellum Society: A Quantitative View," *Social Studies of Science,* **5,** 93–108 (1975); Paul Forman, John L. Heilbron, and Spencer Weart, "Physics *circa* 1900," *Historical Studies in the Physical Sciences,* 5 (1975); Karl Hufbauer, "Social Support for Chemistry in Germany during the 18th Century," *Historical Studies in the Physical Sciences,* **3,** 205–232 (1971); Sally Kohlstedt, *The Formation of the American Scientific Community,* University of Illinois Press, Urbana, 1975; Roy M. MacLeod, "The Support of Victorian Science," *Minerva,* **9,** 197–230 (1971); Margaret W. Rossiter, "Women Scientists in America before 1920," *American Scientist,* **62,** 312–323 (1974); Steven Shapin, 'Property, Patronage and the Politics of Science," *British Journal for the History of Science,* **7,** 1–41 (1974); Arnold Thackray,* "Natural Knowledge in Cultural Context," *American Historical Review,* **79,** 672–709 (1974).

44. "Rapport de M. H. Guerlac," *IX Congres International des Sciences Historiques. Rapports,* Paris, 1950, pp. 182–205, quotations from p. 202; and Walter Pagel, "Review of Merton's *Science, Technology and Society,*" *The Cambridge Review* (1938).

Science Indicators and Social Indicators

2

Otis Dudley Duncan

The approximate coincidence in time and the ostensible lack of formal connection between the two publications *Science Indicators 1972* (1) and *Social Indicators 1973* (2) afford a pretext for these remarks, which are more hortatory than analytical. What are, have been, and should be the terms of a relationship between these two enterprises?

It quickly becomes evident that there is very little information about science in *Social Indicators 1973*. Table 3/19 reports on numbers of degrees earned, by level and subject area, in 1952, 1962, and 1972. From these data one can learn (among other things) that the growth in number of degrees (1952-1972) was at a lower rate in the natural sciences than in the social sciences, the humanities, or the "education and other" category at the bachelor's and master's level and at the same rate as the social sciences at the doctor's level, though below the rates of the humanities and education. (*SI-72* likewise examines degree production. Its figures pertain to a shorter time span, 1959-1960 to 1970-1971, but provide annual data within this period. No comparisons with non-science fields are given.)

Two small tables (3/9 and 3/11) in *Social Indicators 1973* portray group differences (by sex, color, parental education, and community size and type) in science achievement test scores for 17-year-olds in the 1970 National Assessment of Educational Progress. From these tables, we learn that boys do better than girls on science achievement tests; whites do better than blacks; students whose parents went to college do better than those whose parents did not; and the most favorable situation for learning science is a high school in a large metropolitan area attended in disproportionate numbers by off-spring of parents with professional or managerial jobs. (There are no data on public knowledge of science in *SI-72,* although it includes an interesting study of public attitudes toward science.)

The paucity of information on science in *Social Indicators 1973* is not attributable to accident or oversight. It follows logically from the decision to focus that volume on measures of "individual and family (rather than institutional or governmental) well-being." We have here the *reductio ad absurdum* of the definition of a social indicator offered in 1969 in *Toward a Social Report* (3), one of the official federal publications ancestral to *Social Indicators 1973:* "A social indicator . . . may be defined to be a statistic of direct normative interest which facilitates concise, comprehensive and balanced judgments about the condition of major aspects of a society." But there is no precedent or sanction in *Toward a Social Report* for limiting the purview of social indicators to measures of *individual* welfare or for eschewing information oriented to the welfare of the collectivity. (That could be a misreading of one or another of the terms, *direct, normative,* or *condition.*) On the contrary, one of the seven chapters of *Toward a Social Report* is "Learning, Science, and Art." The main quantitative indicators of "scientific capability" alluded to in this chapter are numbers and rate of increase in numbers of scientists and engineers, dollar amounts and percentage of GNP expended on research and development, counts of the number of scientific specialties recognized by NSF (695 in 1958 and 1235 in 1968), and the balance of payments by and to the United States for use of patents and technical expertise. (Some version of three of these four series can be found in *SI-72*).

Even more attention was given to science in the immediate predecessor of *Social Indicators 1973,* the first and only report of the short-lived National Goals Research Staff, *Toward Balanced Growth: Quantity with Quality,* issued in 1970 (4). A full chapter is devoted to "Basic Natural Science." The textual discussion emphasizes the need for basic science as a requisite for "balanced growth," the faltering tempo of government financial support for research, and the emergence of some public hostility toward the aims and methods of scientific inquiry. The statistical appendix to the volume, a somewhat haphazard selection of "trends and projections," includes time series on number of scientists and engineers, number of patents, total R & D expenditures, cumulative number of abstracts published in four scientific fields, R & D expenditures by sector (federal government, industry, and educational and nonprofit institutions), projections of basic science expenditures, and balance of trade in technology-intensive products. There is also a set of "selective illustrations of the speed for introducing technical development into social use" ranging from 112 years for "photography" to two years for the "solar battery." (Except for the last item, at least rough counterparts to most of these series also appear in *SI-72*).

A more remote but still authoritative precedent can be found in the 1960 report of President Eisenhower's Commission on National Goals, *Goals for Americans* (5). Although both the content and the title of an early chapter, "A Great Age for Science," emphasize the humanistic value of science as "an adventure of the human spirit," the application of science for physical welfare is not neglected. Formal presentation of statistical indicators is not featured, but we do find summary data on R & D expenditures, a breakdown of which suggests to the author, Warren Weaver, a lag in the support of basic science, in contrast to applied science and development. He also discusses the distribution of NSF fellowships by field and the proportion of the Federal support of basic research dispensed through military establishments—an undesirably high proportion, according to Weaver. (These kinds of information are also found in amplified form in *SI-72*).

Both the official social reports listed and the prolegomena and prototypes thereof have typically featured assessments of the state and progress of science (often bracketed with technology) or its role in society.

What we can now recognize as the first production model of the "annual social report" called for in so many discussions of social indicators since the late 1960s appeared as a series of special issues on recent social changes, edited by W. F. Ogburn and published by the *American Journal of Sociology* annually for eight years, beginning with a report describing changes through 1927. Most of these collaborative works include a chapter on "Inventions and Discoveries," written variously by Ogburn himself, Clark Tibbetts, or S. C. Gilfillan. In the initial volume (6) Ogburn suggests the justification "that the origins of most of the innumerable social changes occurring today lie in new inventions of a mechanical nature and in the scientific discoveries of natural science." The main content of the chapter is a list of that year's outstanding developments, roughly classified by field. According to this record, for example, the year 1927 is credited with the discovery that liver extract can be used to cure pernicious anemia and with H. J. Muller's report of how he speeded up the evolutionary process in fruit flies by the application of X-rays. Several dozen of these items are abstracted from five main sources. Ogburn freely acknowledges that judgments of potential significance are unreliable and that one is not likely to have a complete record of a year's events within the first few months of the following year. Still he urges that because the general importance of scientific developments for social change is so great, even an incomplete chronicle should have value.

I don't know what kind of evaluation historians and sociologists of science today would put on Ogburn's paradigm for reporting the progress of science. No doubt, it had the defects he acknowledged and others as well. Still, in perusing a list of actual (or purported) discoveries, one does not have the impression—which occasionally becomes overpowering in leafing through *SI-72*—that one is witnessing Hamlet without the Prince. Could it be that the quest for "indicators" of the "state of the entire scientific endeavor" is taking precedence over the attempt to find out what that endeavor itself actually produces? Ogburn paid attention to the content of science because he believed that it is the actual findings of scientific investigations—not the institutional framework, the personnel, the machinery, nor the social support of science—that are most consequential for society.

The chapter on invention and discovery in the monumental *Recent Social Trends* (7)—the 1933 report of a committee appointed by President Hoover—expands the approach of Ogburn's earlier reports. The focus is on the social consequences of inventions. Again, many specific, mostly recent inventions are mentioned, although no fixed time period is covered. Brief comments are offered concerning uses and applications, actual or prospective. To illustrate the influence of inventions, the chapter lists 150 presumed or probable effects of the radio. The chapter concludes with a statement of principles thought to apply to the interaction of invention and society and an enumeration of some problems of public policy concerning invention. The main statistical indicators in the chapter are a time series on the number of patents and some computations of growth rates describing change in the use of various products and processes.

In another chapter of *Recent Social Trends* we find a different kind of treatment. In his assessment of "changing social attitudes and interests" Hornell Hart relies mainly on indices of the subject matter and content of printed publications. An elegant chart shows that between 1900 and 1930 the circulation of religious publications declined from about 4½ to less than 1% of the total circulation of "representative periodical groups," while popular science periodicals rose from about 1 to nearly 4%. He attempts to infer a relatively declining interest in pure science and an increasing interest in applied science from classifications of the topics of magazine articles listed in *Reader's Guide to Periodical Literature*. A more detailed content analysis of the actual articles themselves reveals a more rapidly increasing "index of attention" to applied research than to pure research over the period 1900–1930 in seven leading general periodicals.

Hart's work has lain fallow for a long time. Shortly after the appearance of *Recent Social Trends,* the modern public opinion poll emerged. Although expensive, it seemingly offers a more reliable method for ascertaining trends in public attitudes than does the classification of communications content. Still, analysis of media messages might well supplement poll data in the construction of science indicators as well as social indicators in general (cf. Stone, 8).

Three volumes are widely recognized as basic documents in the contemporary social indicators movement (which appears to me to be a reincarnation of the spirit that produced *Recent Social Trends*). These are the symposia edited by Bauer (9), Sheldon and Moore (10), and Gross (11). It may be instructive to note the treatment of science indicators in these works.

Bauer's *Social Indicators* is, of course, the proximate source of our very nomenclature. It is suggestive that the occasion for his project was quite specifically linked to science. Bauer and associates were commissioned by NASA to assess the impact on society of the program of space exploration. Finding themselves unable to carry out this assignment, they redefined the problem on the grounds that "the problem of measuring the impact of a single program could not be dealt with except in the context of the entire set of social indicators used in our society." In the event, although Bauer and his principal coauthors, Biderman and Gross, do not get very far into the specifics of "the entire set of social indicators," they do imply that science indicators belong in the set. Bertram Gross's table of "grand" and "intermediate" abstractions—one of several "frameworks" for social indicators he considers—represents "scientific or technological progress" as the intermediate counterpart to the grand abstraction, "reason" (9, p. 264). Gross does not, however, submit proposals for the measurement of such progress. In his preliminary assay of the adequacy of selected sets of statistics, Albert Biderman adopts as his framework the Eisenhower commission's statement of national purposes and values (5). He reports:

> For only 59 percent of the goal statements are any indicator data in these sources judged pertinent. Table 2.1 lists the numbers of indicators available in the *Statistical Abstract* and *Historical Statistics* for the goals in each of the eleven *major* goal areas treated in the commission's report. Interestingly, the two domestic goal areas probably most directly related to space activity—"arts and sciences" and "technological change"—are among those for which these sources provide the least indicator coverage (9, p. 87).

A better performance, at least on the score of "relevance," is posted by Gendell and Zetterberg's *Sociological Almanac* (12). Biderman finds that 14 tables in this publication are relevant to the "arts and sciences" goal area and that at least one relevant table can be found in it for each of seven out of the eight specific goal statements in this area. Biderman's criterion of relevance is, by design, exceedingly loose. All we can conclude is that some kind of beginning had been made in the provision of a set of science indicators—that is, statistics relevant to national goals in the area of the arts and sciences—as of a decade or so ago.

Each of the other two symposia on social indicators features a whole chapter bearing directly on our topic. It is fair to say that neither Daniel Bell's "The Measurement of Knowledge and Technology" (10, Chapter 5) nor John McHale's "Science, Technology, and Change" (11, Chapter 9) was written by a specialist in social measurement. One would not consult these chapters for technical guidance in planning a compilation of indicators or a quantitative social report. Each author, on the contrary, is a gifted generalist and interpreter. Each does present a rather lavish display of examples of statistical series, but these are, precisely, *illustrations* of science indicators rather than the vehicles for structured analyses of changes in science or its relations with society. The illustrations are drawn from the same kinds of sources used by the compilers of *SI-72*, although it would be tedious to document the substantial overlap. Both authors interweave the themes that the advancement of scientific knowledge is dependent upon societal support, ideological as well as financial, while the progress of science is registered pervasively, if often indirectly, in consequent social changes. Regrettably, McHale's proposal for delineating "first, second, and third-order consequences of technological change" (11, p. 242) hardly represents an improvement on the similar formulation by Ogburn in *Recent Social Trends* (7, pp. 158ff). But to reinvent the wheel is no doubt better than to get along without wheels altogether.

What, then, have we learned from our "Cook's Tour" of the social indicators literature? I hope it is clear that no case has been made for the separation of science indicators from social indicators, either in previous practice or in principle. Perusal of the two official statistical publications suggests that the interrelations of science and society fall between the stools when an artificial disjunction is imposed by bureaucratic mandate. In *SI-72* society does, of course, play a sort of role—principally as an apparently exogenous source of manpower and funds for the conduct of the science enterprise. Then too, but almost as an afterthought, it includes a chapter that reports on a survey of public attitudes toward science and technology. Here the consequence of proceeding in isolation from the broader social indicators endeavor is apparent, for if that endeavor does command any one highly developed skill, it is the conduct and analysis of public attitude surveys (cf. 13). What is one to make of this statement, "The intercorrelations among indicators of socioeconomic status (i.e., ethnicity, income, education) also make it difficult to relate variations to specific subgroups" (1, p. 100), except to conclude that the analyst is not acquainted with recent developments in multivariate statistical techniques specifically designed to cope with this kind of question as it typically arises in survey data?

But if *SI-72* falls short in depicting the social context of science, *Social Indicators 1973* fails completely to recognize the role of science in society—not to mention the pursuit of science as a cultural value in itself.

What is at issue, of course, is not the matter of scope and limitations of two publications, which is of interest only as a symptom. No doubt there is room for a specialized publication dealing with science manpower and government support of science and another dealing with statistics of the health, education, and welfare of the American population. But those of us who want to see the idea of social indicators developed to its potential and the idea of social reporting realized to the benefit of the society (14, 15) cannot be satisfied with arbitrary criteria that issue in the hopeless fractionation of social reality. Science and society are of a piece, and one cannot altogether trust sets of indicators that would, if only implicitly, have it otherwise.

REFERENCES

1. *Science Indicators 1972: Report of the National Science Board, 1973,* National Science Board, National Science Foundation, Government Printing Office, Washington, 1973.

2. *Social Indicators 1973,* Office of Management and Budget, Government Printing Office, Washington, 1973.

3. *Toward a Social Report,* Department of Health, Education, and Welfare, Government Printing Office, Washington, 1969.

4. *Toward Balanced Growth: Quantity with Quality,* National Goals Research Staff, Government Printing Office, Washington, 1970.

5. President's Commission on National Goals, *Goals for Americans,* Prentice-Hall, Englewood Cliffs, N.J., 1960.

6. W. F. Ogburn, *Recent Social Changes,* The University of Chicago Press, Chicago, 1929.

7. President's Research Committee on Social Trends, *Recent Social Trends in the United States,* McGraw-Hill, New York, 1933.

8. P. J. Stone, "Social Indicators Based on Communication Content," *Fall Joint Computer Conference,* 1972, pp. 811–817.

9. R. A. Bauer, Ed., *Social Indicators,* M.I.T. Press, Cambridge, 1966.

10. E. B. Sheldon and W. E. Moore, Eds., *Indicators of Social Change,* Russell Sage, New York, 1968.

11. B. M. Gross, Ed., *Social Intelligence for America's Future,* Allyn & Bacon, Boston, 1969.

12. M. Gendell and H. L. Zetterberg, *A Sociological Almanac for the United States,* Bedminster Press, Totowa, N.J., 1961.

13. A. Campbell and P. E. Converse, Eds., *The Human Meaning of Social Change,* Russell Sage, New York, 1972.

14. O. D. Duncan, "Developing Social Indicators," *Proceedings of the National Academy of Sciences*, **71**, 5096–5102, December, 1974.

15. Roxann A. Van Dusen, Ed., *Social Indicators 1973: A Review Symposium*, Social Science Research Council, Center for Coordination of Research on Social Indicators, Washington, 1974.

Can Science Be Measured?

3

Gerald Holton

The mind comprehends a thing the more correctly the closer the thing approaches toward pure quantity as its origin.

Johannes Kepler,
Letter to M. Maestlin, April 19, 1597

THE COMING OF SCIENCE INDICATORS

When the chairman of the National Science Board (NSB) sent *Science Indicators 1972 (SI-72)* (1) to President Richard M. Nixon in January 1973 for transmittal to Congress, his covering letter stressed that the Report, prepared by the NSB staff in NSF,

> ... presents the first results from a newly initiated effort to develop indicators of the state of the science enterprise in the United States .... If such indicators can be developed over the coming years, they should assist in improving the allocation and management of resources for science and technology, and in guiding the Nation's research and development along paths most rewarding for our society (2).

Two years later these hopes had become more concrete. In December 1975 the NSB sent to President Gerald R. Ford the second such volume, *Science Indicators 1974 (SI-74)* (3). Longer by about

39

100 pages than *SI-72* and containing more sections and analyses, it was fundamentally built on the earlier model. But in Mr. Ford's own letter of transmittal to the Congress he said that from his point of view this series of reports should not be viewed as merely an academic effort to develop and perfect indicators for their own sake. Instead, they would serve to show that:

> The nation's research and development efforts are important to the growth of our economy, and future welfare of our citizens, and the maintenance of a strong defense. The nation must also have a strong effort in basic research to provide the new knowledge which is essential for scientific and technological progress (4).

It was a unique launching of a young discipline. Science indicators have suddenly become a tool for high-level scrutiny of the quantity and quality of the research enterprise. This occurred partly because of the national expectations about research and partly because the scale on which it is supported has become large enough for it to be highly visible, as some of the "indicators" in these volumes show. By 1974 the expense for basic research was nearly $4 billion (two-thirds supplied by the federal treasury), and a total of $32 billion was spent on national research and development (R & D) (including defense and space) of which over one half came from federal sources. Clearly, any enterprise that large—one employing over a half million scientists and engineers and commanding 15% of the relatively controllable portion of the federal outlay—is subject to calls for accountability of its performance and to justification of the national investment in terms of returns that the taxpayer can appreciate.

The developing science indicators are therefore placed at the intersection of a variety of pressures and hopes. From the academic's point of view, they are merely a new subset in the long-established field of social indicators (measures of population, health, education, income, crime, etc.). Hence, they are a specific form of statistical series that measure changes in significant aspects of society. But even while they are being shaped, they are unlikely to remain unused in science-policy decisions. As stated in *SI-72*, these indicators can "assist also in setting priorities for the enterprise, in allocating resources for its functions, and in guiding it toward change and new opportunities . . ." (5).

In fact, these two science indicator publications received a considerable amount of public attention in the press, with the most recent one lending itself to such headlines as "U.S. Science Lead Is Found Eroding . . ." (6). The chairman of the NSB noted, "The indicators appear to be of increasing usefulness to the executive and legislative branches of the government in dealing with science policy issues, and in providing an objective basis for discussions of public policy affecting R & D activity" (7). He indicated that they had played a role in the budget decisions made by the President concerning basic research funding for fiscal year 1977 (8).

These two publications, therefore, are of extraordinary interest both to scientists and to other scholars, not only for the wealth of data (usually in the form of time series) and what they reveal on subjects ranging from the relative share of research support, publications, or students in different research fields to the measures of public attitudes toward science and technology, but also because the planned use of science indicators may come to influence their professional lives deeply. It thus behooves them, during this formative period in the development of the science-indicator movement, to examine the concepts and methodologies in the measurement of the quantity and quality of science and technology, with a view to their critique and improvement (9). It is to this end that the following remarks are addressed. On the one hand, the need for sound indicators cannot be doubted, and their usefulness will depend on the credibility they have within the scientific community. On the other hand, this is the time to discern and correct shortcomings—before they are so firmly set that they are difficult to dislodge. One recalls here Einstein's remark:

> Concepts which have proved useful for ordering things easily assume so great an authority over us that we forget their terrestrial origin and accept them as unalterable facts. They then become labeled as "conceptual necessities," "a priori situations," etc. The road of scientific progress is frequently blocked for long periods by such errors. It is therefore not just an idle game to exercise our ability to analyze familiar concepts, and to demonstrate the conditions under which their justification and usefulness depend (10).

Thus, without foregoing entirely a detailed analysis of portions of SI-72 and SI-74, I shall concentrate chiefly on some fundamental epistemological problems adhering to any attempt to make "indicators," no matter for what purpose, and I shall refer to the two publications, particularly to the stage-setting SI-72, for actual examples.

QUANTITATIVE AND QUALITATIVE INDICATORS

The idea of making quantitative indicators of anything at all fascinates some people and repels others as being dangerous or absurd. This difference is caused largely by thematically incompatible—and, therefore, often unresolvable—personal views concerning the ability of quantifiables to lead to or attest to the deepest reality. The history of science is full of cases where some purists claimed that only when you measure something do you really know what you are talking about, while others held that quantification (other than for classification or similar purposes) distorts the full, natural sense of things by confining phenomena in a straightjacket. Like all thematic tensions, this one is healthful for the eventual progress of a science because it tends to juxtapose deeply felt, articulated models and methods (11).

Most scientists today, if forced to choose, would lean toward the first of these two poles. But thoughtful spokesmen would also perceive and verbalize the consensus on a hierarchical order: The ultimate use of quantitative measures within science is that they can guide one eventually to an understanding of the basic features that are associated with, but *not* exhaustively motivated, described, or explained by conceptions expressible in numeric terms—basic features described in such terms as simplicity, symmetry, harmony, order, and coherence.

Similar polarizations concerning quantification and a similar attempt at resolution may be noted in the new field of technology assessment. In his article "Technology Assessment from the Stance of a Medieval Historian," Lynn White has put the issue memorably:

> Some of the most perceptive systems analysts are pondering today how to incorporate into their procedures for decision the so-called fragile or nonquantifiable values to supplement and rectify their traditional quantifications. Unhappy clashes with aroused groups of ecologists have proved that when a dam is being proposed, kingfishers may have as much political clout as kilowatts. How do you apply cost-benefit analysis to kingfishers? Systems analysts are caught in Descartes's dualism between the measurable *res extensa* and the incommensurable *res cogitans,* but they lack his pineal gland to connect what he thought were two sorts of reality. In the long run the entire Cartesian assumption must be abandoned for recognition that quantity is only one of the qualities and that all decisions, including the quantitative, are inherently qualitative. That such a statement to some ears has an ominously Aristotelian ring does not automatically refute it (12).

Since the most respectable parts of the "hard" sciences, at least on the surface, seem to be chiefly preoccupied with quantifiable measures, nothing is more natural than to develop indicators about science that themselves consist of quantifiable measures. This tendency would also fit well with both the ongoing "scientification" of the modern social studies through the introduction of more mathematical methods and operationally demonstrable concepts and the ever-growing desire of administrators to rationalize the allocation and use of resources (see Ezrahi, this volume). In any case, the desire to formulate some indicator-measures, on any basis, is almost irresistible. Who in this field would not want to find out how many scientists there are, how much money they spend, and how many papers they publish?

Who in the Congress or in science administration would not like to see how the science enterprise in this country compares with that of other nations or to what extent the expenditures for basic research bear fruit in terms of industrial developments that can claim to be "outputs" of basic science? And above all, should one not have some measure of how "good" the efforts of our scientists and engineers are?

Fortunately, *SI-72* took note of the need for developing indicators of *quality* on the very first page of its text, stipulating that intrinsic measures should include those of the "quantity and quality of associated human resources," and that extrinsic indices would center on "the achievement of national goals . . . and the consequent impacts on that elusive entity, the 'quality of life.' " The ambition is laudable—the more so as the quantification of qualities, ever since Nicole Oresme's attempt in the fourteenth century, has been a precarious task.

In measuring qualities, three types of opportunities exist that are not available when the usual quantitative indicator is made by folding a large number of individual contributions into one grand total. The first is the identification of *one or a few crucial events,* perhaps in a mass of noisy data. The quality and health of a science, as we know from personal experience, and the abundant evidence of historical study, not infrequently depend on nothing so much as the appearance of one paper of high imagination or the founding of a professional society, the accession of a new journal editor, or the entrance of a new patron. (For example, nothing was more significant for the quality of science in Italy in the 1920s than the appointment of young Enrico Fermi.) Insofar as singular events that have the potential of transforming a field can be identified quickly, they should find some place in any publication that claims to indicate the health of science at a given time—even if these events are not quantifiable and this potential sometimes may not be realized.

Second, even in a consensual activity such as science, contemporaneous estimations of quality are difficult. The full significance of an important event does not necessarily appear immediately, and identification of quality often requires historical perspective. Very few physicists in 1905 were aware that it was one of the greatest years in the history of their field. Sadi Carnot's fundamental paper of 1824 on thermodynamics was not recognized as important until 1834—and then only by one scientist, Émile Clapeyron. Another decade passed before the scientific world began to appreciate (through William Thomson's work) the merits of the 1824 publication. Moreover, the intellectual structure of Carnot's work has been adequately understood by historians of science only in the past 15 years.

Similarly, the profound implications of Max Planck's work of 1900 did not become immediately clear even to Planck himself. It began to be generally understood—at different rates in different countries—only after the 1911 Solvay Congress. And suggestions of the existence of what are now called "Black Holes" lay about, essentially neglected, in the literature for decades before current interest focused on them. (Conversely, some sensations of the moment have a habit of disappearing in time.) Instances can be adduced *ad infinitum*.

Although on the whole we may be more alert to "significance" today than in the nineteenth and early twentieth centuries, it is not realistic to hope that the quality of scientific work and of scientific life will be estimated safely "in real time." Therefore, the assessment of the state of science for a given year, while useful and interesting, has to be reexamined and updated in the light of changing results of informed historical scholarship.

Third, the methodological problem of determining the quality of a subject field is still far from solved. Some attempts exist (see Kochen, this volume), such as the recent re-review by a panel of peers of the quality of previously granted research proposals (13). But the methodology for developing measures of quality seems to require much further experimentation. One possibility (14) is to evaluate the quality of publications in a field or the quality of a journal by appropriate random sampling—a technique not far removed from that for obtaining measures needed as a part of any scientific research itself. A properly chosen, rather small sample of, say, 100 articles in a field or even a few hundred pages of a journal volume might be evaluated in depth by a high-grade panel of assessors, somewhat the way a set of research-funding proposals is ranked both relatively and absolutely.

Looking to *SI-72* and *SI-74* for indices of quality, we find very few. One is the listing of the number of Nobel prizes received by U.S. scientists from 1901 to 1974. There are, of course, difficulties with this simple quantification of quality. The assumption that the award is given on the basis of an international search for work of the highest quality, uninfluenced by political or other extraneous considerations, is certainly sounder in science than in some other fields. But the number of awards per year is so small that fluctuations over a short period are not likely to be meaningful. Other obvious problems exist—how to count prizes if they are shared, to which country to credit a prize given by a migrating scientist, or how to treat the time delay between the publication and the honoring of it. Should one pay more attention to the fact that the United States has the largest number of prizes since 1901; or that the rate of receiving prizes in the United States has become smaller since 1951–1960; or that in relation to the population as a whole, U.S. scientists rank far below those of the Netherlands, Switzerland, and the United Kingdom? In fact, what precisely is the link between the quality of research and the Nobel prize (15) ? All these questions hint at the larger problem, to be developed below, that numbers by themselves may not indicate anything or may mislead.

Another attempt to measure quality, with due acknowledgment of the difficulties and dangers, was the use in the *Science Indicator* volumes of literature indicators, comparing "the relative standing of a given Nation's literature in the references of a large sample of the world's scientific literature. The belief was that the most significant literature will be most frequently cited by subsequent investigators. . . . By this measure the United States ranked or tied for first place in each of eight major scientific fields" (16). But it was also acknowledged that this method of assessing the quality of scientific literature is subject to factors that may severely slant the results—for example, differences in publishing habits in different countries, editorial policies, and representatives of journals in the sample being analyzed (17).

There are additional attempts at quality measures—for example, in the interesting distinction between the degrees of innovations calculated for several hundred specific innovations, from improvements of existing technology to "radical breakthroughs" (and the resulting finding that the "radicalness" of major U.S. innovations has been slipping markedly over the past two decades) (18). The innovations themselves "were selected by an international panel of experts who rated the innovations based on their technological, economic, and social importance" (19). But obviously quality measures are still in an early stage indeed, in this as in all other fields; hence the opportunity for improving them is that much larger.

BULK MEASURES VERSUS FINE-STRUCTURE SPECTROSCOPY

Enterprises such as the making of economic indicators or science indicators—almost by definition and surely by virtue of the vastness of the task—tend to gravitate to the use of aggregative, composite indicators, to bulk or gross measures rather than to attempts to look for fine or detailed structure. To use an analogy, the preference is to measure the intensity of a source by cumulating the energy emitted at all wavelengths, rather than passing the beam through gratings or prisms to lay out the spectrum and measuring the detailed energies radiated at different wavelengths. Of course, important questions exist for which no fine details are needed, in which, therefore, a "pre-Newtonian" treatment of the beam (i.e., without spectral resolution) is adequate; in any case, that is usually the far simpler measurement to make. However, such bulk measurements could not have led to the optical discovery of Fraunhofer lines, Balmer lines, blackbody radiation, hyperfine structure, and so on—all important to the advance of physics.

Here again, contrary thematic preferences enter the researcher's design. Some are "lumpers" and will be naturally predisposed to seek bulk measures, others are "splitters" and will seek fine structure. To some, the decay of a society may be indicated by, say, the rate of fall of the GNP or the net trade balance; to others the same information is contained in significant details, such as the observation that the mastiff is starving at the master's gate. To some, the rather smooth exponential rise of the curve of the *cumulated* number of science abstracts as a function of time is the raw material for speculation; to others, the essence of the work lies in filtering out the few, sporadically appearing significant articles from the steady stream of the banal, or even lies in the strange sometimes large fluctuations of the annual output of articles (20).

By training and preference, research scientists tend to be more interested in identifying significant detail than in aggregating data. Their interest in a new idea is often aroused by what might seem a specific minor anomaly, an unexpected mismatch along the edges when expectations and reported findings are superimposed, and they usually test their theories by referring to carefully selected phenomena in a well-defined corner of phase space. Thus, a peculiarity noted in the "fine structure" of phenomena (e.g., the observed differences in the intensity of artificial radioactivity found on two different laboratory tables in Fermi's group in 1934) is more likely to signal the existence of a fruitful new area of work. However, gross properties are more likely to be useful for exhibiting, confirming, or summarizing the state of an existing perhaps well-established theory. They also succeed in a field such as economics because of the possibility of converting very different products to the common denominator of price (21).

When bulk measures are used in scientific research, it is generally assumed that they do not mix together incommensurable populations. In fact, this can be known only *post hoc*—that is, when an adequate theory has determined what the pure and the mixed populations, respectively, consist of. I greatly doubt that we are near enough to such an understanding in the case of science indicators, and I am struck by the tendency throughout most of *SI-72* and *SI-74* to opt for bulk measures (time-line series for "mathematics" or "engineering" or "social sciences," taken as a whole), even when more detailed spectroscopy of data was available in the literature (22).

Science Indicators seem to be looking to thermodynamics for their models. That is a dangerous exemplar to follow. Thermodynamic variables such as pressure, volume, temperature, and compressibility are indeed wonderful indicators of the state of a sample: They have clear operational meanings, and they are connected to a body of knowledge that allows these measures to achieve interest by virtue of demonstrating that matter behaves according to laws. Above all, they are independent of any detailed understanding (or ignorance) of the microstructure of matter. But even after the resolution of those enormous conceptual struggles, spanning more than a century, that were needed to develop notions by which the laws of thermodynamics could be formulated and tested (23), thermodynamics does not remain a clean and simple model when it turns out that mixed rather than pure systems are being dealt with.

To illustrate the tendency toward bulk measures and the predominant influence of "lumpers," consider the very title of the reports *Science Indicators*. According to a doubtlessly apocryphal story, when Jerome Wiesner was in charge of the Office of Science and Technology, he had his staff run a contest to discover a less clumsy, shorter term for identifying the activities referred to as "sciences and technologies." The largest number of proposals suggested that the single word "science" be used. To be sure, calling this activity "science and technology" instead of "science" is still a gross oversimplification, splitting into an arbitrary two sections a large set of different activities that exist in a more complex space and do not fit along one line. The very fact that there is no good name for the set of activities for which indicators are being sought is itself an indicator of the complexity of the activities.

If nothing else can be done to solve this problem quickly, at least a note of caution is needed in the present use of bulk measures and global statements—for example, qualifying the claim that the indicators can be used for describing "*the* state of *the* science enterprise" (italics added for emphasis) (24); or remedying the absence of a listing of all the separate sciences and technologies that have been included in the gross measures.

EMPLOYMENT STATISTICS

A significant example of insufficient fine structure in *SI-72* is found in the subsection "Supply and Utilization" under "Science and Engineering Personnel." This example offers a remarkably brief treatment of one of the chief preoccupations of many scientists in the very year for which these indicators were designed (as well as the previous two or three years)—namely, the substantial decrease of financial support for science and technology, with consequent unemployment, underemployment, and holding actions. A report issued by this nation's chief research-support agency, and addressed to the President of the United States for transmittal to the Congress, bears an extra measure of obligation to present in the clearest light just those problems that may be most difficult or embarrassing for the life of science. Obviously, few official "indicators" will be scrutinized more skeptically by scientists, scholars, and their professional societies than employment data. A report that is thought to underplay the highly visible and tragic problem of unemployment and underemployment risks the label of cover-up, the more so as the claim was made that the indicators will reflect the impacts on the "quality of life."

One notices, for example, that space was available in *SI-72* (p. 59) to make such a statement as "the overall science and engineering unemployment rate was still only about half of that for all workers"—a euphemism produced by mixing into one gross indicator two entirely differently prepared populations (not to speak of the fact that the high unemployment rate in the general work force—officially between 5% and 6% throughout 1972—was neither humanly nor politically satisfactory). But to allow a full picture of the national situation to emerge and to permit the identification of specific disaster areas that merit special help—surely an important function of indicators—it would have been much more to the point to analyze the details of unemployment and underemployment, including those ("dropouts," "overqualified") who have been excluded from the employment statistics, those who are subprofessionally employed, and so on.

Dr. Alan C. Nixon, recently president of the American Chemical Society, has pointed out (25) the discrepancies between simplified, aggregative figures and figures based on detailed analysis. Thus the American Chemical Society's 1973 report on chemists' salaries and employment statistics indicates in a survey response of chemists that 1.7% overall, and 1.5% of the Ph.D.'s, were unemployed. Nixon notes, however, that if the definition of people with employment problems is allowed to include transient or part-time employees, etc., the figure is 8.3% of the ACS membership and 9.5% for the Ph.D. members. He also points out the effect this situation has on the life of the scientist in terms of the rapid loss of a scientist's or engineer's capacity to stay in the profession, once the ability to remain updated is impaired.

Nixon's analysis is only one example highlighting the artificiality of some of the concepts in general use for official employment statistics. Thus the fine Report, *Work in America,* prepared by a special task force to the Secretary of Health, Education and Welfare (26), warns:

> The statistical artifact of a "labor force" conceals the fluidity of the employment market and shifts attention from those who are not "workers"—the millions of people who are not in the "labor force" because they cannot find work. . . . Although this narrower concept of a "labor force" is useful for many economic indices, it is inadequate as a tool for creating employment policy.

For example, it excludes from consideration "people who answer 'no' to the question 'Are you seeking work?' but who would in fact desire a job if one were available and under reasonably satisfactory conditions." The Task Force Report indicates that such people range from those in school or training programs because they have been unable to find suitable jobs to many women and older people who no longer even look for jobs.

Other evidence points to a need for more detail in order to arrive at an accurate picture of the employment situation for scientists, especially in the 1970s (27). Thus, an American Institute of Physics (AIP) survey of new physicists in the summer of 1972 showed that of the Ph.D.'s who had graduated the year before, 7 to 8% were still seeking jobs, and an additional small percentage was changing professions (28). Among more recent graduates, 16% of the Ph.D.'s and 19% of the terminal M.A.'s were still seeking employment. Of the 1970–1971 B.A. recipients in physics, surveyed by AIP in December 1971, who did not go on to graduate work or into military service, "only 14 percent found jobs that make extensive use of the physics training received in college" (29).

Similarly, a survey of the American Astronomical Society, made in 1973-74, showed that unemployment was both serious and still rising (30). Of the holders of Ph.D.'s granted between September 1969 and September 1972, 7.5% were known to have serious difficulties in finding full-time employment in astronomy (up from 5% about a year earlier). The same problem was faced by 13.5% of the postdoctoral appointees and by about 18% of the Ph.D. recipients in the most recent year. The findings for the years following 1972 were not bright, either. In the 1974–1975 academic year about 200 newly trained astronomers were vying for about 20 faculty openings.

Although such dismaying details, well known in general to everyone within the profession, were not considered sufficiently important to raise a warning flag in *SI-72*, they were significant in shaping the decisions of young people choosing a career. To many of them, the portents were quite clear. Here is an example that cannot be found in the global indicators of *SI-72* (31): From 1961 to 1970 the number of doctorates granted in physics rose, roughly parallel to the rise in the total student population, to about 1550. There the curve turned down, and the projection of new Ph.D.'s, based on students in the pipeline, is 950 in 1978, a steep descent to 60% of the output of 1970 (32).

These trends lead to contrary but equally arresting predictions. Some observers now warn of the coming of a severe shortage of trained physical scientists and engineers in the 1980s. Others see in the recent downturn an early response to the general exacerbation of the employment problem in the next decade: "If matters continue as they are . . . there will be 2½ college graduates competing for every 'choice' job [in the next decade], not to mention the additional 350,000 Ph.D.'s who will be looking for work" (33).

One of the reasons announced for the existence of *SI-72* was to "provide an early warning of events and trends which might reduce the capacity of science . . . to meet the needs of the Nation." The lack of careful attention to the employment picture is therefore a serious matter if one believes that high on the list of needs of the nation one must place reasonably healthy employment opportunities, including retraining programs for scientists and engineers and other help for the vulnerable parts of the profession—even at a time when the political clout of the science community is not high, as was the case in the early 1970s. The concept of a policy of reasonably full employment at or near a level of trained competence is neither simple to implement nor generally agreed upon. Yet, by 1973 it was no longer merely a matter for politicians of different ideologies to disagree about in the United States. The "full employment" movement was becoming stronger, as measured by the bills introduced in Congress, and conclusions, such as the one in the HEW report *Work in America,* that the provision of the opportunity for each family's "central provider to work full-time at a living wage . . . should be the *first goal* of public policy" (italics added for emphasis) (34). But as measured by *SI-72* and *SI-74,* the time for the idea had not yet come for scientists and engineers.

The "quality of life" of scientists and engineers, including that of the younger ones who did find new jobs or held on to makeshift arrangements in a distressed employment market, would also have been worth some attention. Fine-structure indicators can be well developed—for example, the AIP Placement Register of job-seeking physicists at the (Spring) American Physical Society meetings shows that the ratio of employers to registrants present had dropped to 0.07 for 1972, roughly one-seventh of what it had been six or seven years earlier (35).

Another detail of the employment situation that concerned many scientists in the early 1970s was, of course, the employment and training for employment of women and minorities. It is remarkable that in this Age of Affirmative Action there was hardly a word on the subject anywhere in *SI-72,* even though NSF surveys elsewhere have gathered useful data (36). (*SI-74* does have four pages on the employment of women and minorities in science and engineering.) Evidence exists that although women scientists constituted only 9% of the total number of physical scientists (chemistry, physics, atmospheric and space science) in 1970, their rate of unemployment was double that of men (37). The rate of inflow of women and minority scientists would have been useful indicators in *SI-72* (in physics in 1971–1972, 6% of the graduate students were women, and about 1% were black, the latter being the same as that for the annual output of all doctorates in science and engineering). Even the modest starts made toward improvement, such as the National Institutes of Health (NIH) Minority Biomedical Support program begun in 1972, would merit notice as an example of the detail in the fine structure of the life of science, where policy changes begin to be translated into reality.

To summarize: An unhealthy clash may develop between some global and optimistically slanted indicators put out by a government agency and the much soberer details found in surveys made by professional societies, as well as in the articles and letters in scientific journals. A real danger also exists that "science" and its preoccupying problems, as they emerge from "official" science indicators, constitute an entity qualitatively different from science and its problems, as perceived by individual working scientists, members of the governing boards of professional societies, and the like.

POLITICAL REALITIES

To avoid the charge of naiveté, anyone reviewing a U.S. government publication such as SI-72 should keep two political realities in mind. Everything that a government appointee wishes to publish involving the mission of his agency must be cleared by the Office of Management and Budget (OMB). Moreover, in the specific case of SI-72, the report had to undergo scrutiny by OMB and other White House staff in the days when OMB was not loath to impound funds appropriated for agencies such as the NSF and when OMB's Deputy Director was Frederick V. Malek, a member of H. R. Haldeman's White House "management team" with wide-ranging mission and power. (Malek had earlier been Special Assistant to President Nixon, and he has been identified as the developer of the "Departmental Responsiveness" program that aimed at obtaining political leverage through grants and contract programs of federal agencies (38). Many of the weaknesses and omissions in the published version of SI-72 may well stem from these two historic facts. SI-74, published after the disappearance of Nixon, Haldeman, and Malek from the government scene, while still cautiously worded, is a substantially more useful and explicit document; critiques received after the publication of SI-72 may have helped toward this end.

An inherent generic limitation in such government reports in general and a special ambiguity in the status of the NSB deserve comment here. The scientific community, and to some extent the Congress, see the NSB as an independent board of scientific statesmen, in a position to speak out forthrightly to the President and the Congress on the state of science and the interests of the scientific community and whose scholarly and scientific training and credentials should certify the independence and quality of any report they issue. In contrast, the Administration sees the NSB as a part of the Administration, subject to its political discipline, and therefore required to conform in any of its public pronouncements to "the President's Program," as is the case for full-time government administrators.

The result is evidently an uneasy accommodation in a constant tug-of-war between the NSB and the OMB, the agency in practice responsible for enforcing conformity with the President's Program on the part of all presidential appointees in the Executive Branch. The final text of a report is the result of long and complex negotiations between the NSB, the NSF Director, and the OMB. To put the matter bluntly, in preparing the report, the NSB may have tried to seek a compromise between what it would have liked to say and what it thought could get passed, feeling that even a necessarily watered-down or negotiated version of the report would have a considerable positive public impact with respect to the support of science.

Another issue may have inhibited the emphasis on questions of unemployment and the like. Such discussions might open the Board to the charge of being "self-serving" in its report—that is, with respect to any statement that could be interpreted as favoring the economic interests of the scientific community or groups of individual scientists.

To be sure, there is a dilemma here. My criticisms of the report imply that society has a measure of obligation to help provide jobs for scientists that make good use of their training, or at least that the existence of a body of scientists whose talents are seriously underutilized represents a notable symptom of a lack of health in the scientific enterprise. But how far can this dilemma be generalized? Is the case of science different from that of, say, scholars and artists (39) ? In either case, should a group of scientists speak up only for scientists or, conversely, should they arrogate to themselves the task of speaking up also for nonscientists? Questions such as these, and the more serious problems associated with the ambiguous positions of the NSB that place it on a political tightrope when issuing a report, suggest that the public record cannot by itself be an adequate basis for assessing the health of science. Once again, we see merit in the argument on behalf of supporting independent assessments that originate in the academic sector.

NUMERICAL DATA VERSUS INDICATORS

Gathering censuslike data on the state and performance of the professions is useful and interesting, regardless of the various ultimate uses to which these data may be put. This operation is, in fact, performed regularly, for example, by NSF, National Academy of Sciences-National Research Council, and the professional societies. Indeed, most, if not all, the data used in *SI-72* (with the exception of the Delphi experiment and the commissioned study on public attitudes) seem to have been taken from previously available sources. What, then, is the difference between data and indicators?

To put it in terms of an analogy, the measurements of a rapidly declining blood pressure of a patient remain data until they are seen by someone with enough understanding of physiology to recognize them as indicators of a change in the state of health. I propose that the term *indicator* is properly reserved for a measure that explicitly tests some assumption, hypothesis, or theory; for mere data, these underlying assumptions, hypotheses, or theories usually remain implicit. Indicators are the more sophisticated result of a complex interaction between theory and measurement.

Even "taking data," as is well known, is not possible without at least an implicit theory. But an implicit theory is often fallacious, as the discovery of inadequacies in "simple" census data have shown (see Kruskal, this volume): Significantly more people in certain categories were discovered to have been invisible to the takers of the U.S. Census than had been allowed for in the hypotheses underlying the method of measurement.

If data are believed to be necessarily objectively factual, this merely indicates that the existence of an implicit and possibly erroneous theory has been overlooked. For example, I once worked near a superb experimental physicist, one of whose publications typically was on the tensile strength and other properties of 50 alloys. He ground out these data at a phenomenal rate. It turned out, however, that what he was doing was testing an implicit theory that the small amount of impurities in these particular samples of the alloys has a negligible effect on the properties of the metal. His measurements were very sound—that is, they were repeatable on these particular samples. But the level and type of impurities were found later to have a large effect on the properties, severely limiting the applicability of the measurements to the general case.

At an early stage, the plausibility of some data for use as indicators may be a good guide. Eventually, however, there must be some explicit theoretical base for choosing some data, discarding others, and noting the absence of needed data or of needed fine structure. No such *explicit* theoretical base appears in *SI-72* or *SI-74*. Instead, one can occasionally glimpse an *implicit* theory of science (to be discussed in the next section).

The attempt to put data ahead of an explicit theory reminds me of a great discussion reported by Werner Heisenberg. In 1926, just after his first major contribution to quantum physics, Heisenberg met Einstein. Heisenberg explained that he had abandoned such hypothetical conceptions as the unobservable electron orbits; "since a good theory must be based on directly observable magnitudes," he thought it more fitting to base his theory on observable frequencies of the emitted light. Moreover, Heisenberg thought he was faithfully following Einstein's own model: "Isn't that precisely what you have done with relativity?." Einstein's reply took Heisenberg completely aback:

Possibly I did use this kind of reasoning . . . but it is nonsense all the same. Perhaps I could put it more diplomatically by saying that it may be heuristically useful to keep in mind what one has actually observed. But on principle, it is quite wrong to try founding a theory on observable magnitudes alone. In reality the very opposite happens. It is the theory which decides what we can observe. You must appreciate that observation is a very complicated process. The phenomenon under observation produces certain events in our measuring apparatus. As a result, further processes take place in the apparatus, which eventually and by complicated paths produce sense impressions and help us to fix the effects in our consciousness. Along this whole path—from the phenomenon to its fixation in our consciousness—we must be able to tell how nature functions, must know the natural laws at least in practical terms, before we can claim to have observed anything at all (40).

To be sure, those who make science indicators need not regard themselves bound by the credo formulated by theoretical physicists. Yet, the story has some application to these volumes, which are full of observations but studiously avoid discussing an underlying theory.

It is significant that the term *science indicators* was, to my knowledge, first published in a paper by Harvey Brooks in which he proposed "four models of the research system"—the Polanyi-Price model, the Weinberg model, the social overhead investment model, and the Toulmin model—and linked these models with "indicators for more quantitative planning and for providing a means by which the output of the system might be measured or assessed" (41).

In a related paper two years later Brooks noted a number of "unresolved issues of science policy"—overspecialization; centralized versus pluralistic management; the place of engineering, mathematics, and physical sciences in new national priorities; critical size versus dispersion in research; the integrity of the self-regulatory systems of science; models for the support of graduate education; and the relation between U.S. and world research efforts (42). Each of these issues is of course susceptible to discussion in terms of indicators developed for the purpose and shaped by the model of the research system.

For example, even in the carefully more "upbeat" *SI-72,* and certainly in *SI-74,* one notices evidence of a sense of the ending of the "endless frontier" in the United States, the decrease in momentum and advantage with respect to other countries, and the glum future of contracting prospects for political and financial support for science and technology. (Even in the 1930s, the average annual growth of academic science in the United States was 6%, and for R & D funds as a whole, 9%.) If one also believes that industry depends on, and therefore has extra responsibility for, basic research, one will notice with dismay that in real dollars the expenditure by American industry in support of its own basic research declined so markedly that in 1974 it was down to 68% of the level for 1966, having returned to the level for 1960 or 1961 (43). An obvious next step would be to obtain indicators to test the economic forces, among others, that have progressively discouraged long-term research in industry—for example, how support for research in industry has been related to profitability and tax structure. None of this was discussed.

Moreover, with the burden shifting correspondingly to the universities, it is the more ominous to see that expenditures for basic research per scientist and engineer in doctorate-granting universities were almost 30% lower in constant dollars in 1974 than in 1968. (The total R & D expenditure from all sources as a percent of GNP in the United States had dropped correspondingly, from 3% in 1964 to 2.3% in 1974 and, at this writing, is estimated to be 2.2% in 1976. Of that total, the share provided by federal funds for the "Advancement of Science"—fundamental research and science instruction—in the early 1970s was 3% in the United States, versus 16% for the United Kingdom, 25% for France, 41% for West Germany.) Since these NSB volumes frankly put the main emphasis on the normative aspects of indicators, one hopes that in the future there will be detailed work to analyze such trends—for example, by presenting the details of basic research allocations by fields in the budget of the various sponsoring federal agencies. In this manner, one might test to what degree the original expectations and mandates of the various agencies are continuing to be fulfilled and to what degree their missions have been changing.

Indicators, then, are to be thought of as the more meaningful the more they lend themselves to being tools for confirmation or refutation in handling questions, hypotheses, and theories that interest scholars and practitioners concerned with the state of the sciences or engineering, from their various platforms in science-policy studies, the history or sociology of science, and the like. The indicators cannot be thought of as given from "above," or detached from the theoretical framework, or as unable to undergo changes in actual use. They should preferably be developed in response to and as aids in the solution of interesting questions and problems. (To be sure, the possibility cannot be dismissed that the theoretical basis eventually may have to be amended—useful concepts can survive theory shifts, as we know from many cases in the history of science—or that a "natural history" sort of data accumulation, with weak grounding in theory, may occasionally provide measures useful for later, serious work.)

It follows that different models or perceptions (of science, of the science-society feedback loop, and even of the prognosis concerning the good or bad uses to which science indicators might be put) will produce spectra of different indicators, as well as different views of the limitations inherent in some indicators or of the way in which indicators are functionally dependent on the input and output expectations.

This prospect of diversity is not dismaying in the least. The major attraction of science indicators may reside in the possibility that eventually they may help test aspects of different theories and models of the scientific process. If *SI-72* or *SI-74* had started from some of the current theories, significantly different indicators would have been developed. For example, starting from Robert K. Merton's categories, indicators would be required to help measure the degree of universalism, organized skepticism, and so on, and the effort would be to determine whether the system in fact attends to these norms and to its own claims or pretensions regarding them—measured by the degree of openness to (including funding of) young talent, the degree of international collaboration in science, the health of the peer-review process, and the like. It is chiefly these comparative tests of different models that will permit development of better models concerning the life of science.

I do not believe that the NSB would strongly resist such ideas. In his testimony, its chairman recognized that "The objectives of R&D . . . are intimately bound to the social system that convert and incorporate their results" (44). In fact, *SI-74* already contains material for testing this perception. Thus, it was found in *SI-74* that in the United States, the percent distribution of scientific literature for selected fields shows that clinical medicine has the largest share and chemistry and physics have nearly the smallest share, whereas in the USSR it is exactly reversed.

There are two other reasons for welcoming the admission of a diversity of models and of corresponding indicators. One is that in the absence of conscious pluralism, *one* theory is likely to establish itself or, at least, discourage the others. The possibility that the evaluation of science may be captured by one ideological group is by no means an idle fancy. In fact, the philosopher of science Imre Lakatos, the leader of a sizable group of adherents, repeatedly and frankly announced such a goal. In a posthumously published discussion, Lakatos identified his own school of thought and, especially, his "methodology of research programmes," as a version of "Demarcationism" (45). A chief role of the "demarcationist philosophy of science" is to "reconstruct *universal* criteria" that help to distinguish "progressive" research programmes from "degenerating" ones. Thus, Lakatos warned, not only "medieval 'science' " but also "contemporary elementary particle physics and environmentalist theories of intelligence might turn out not to meet these criteria. In such cases, philosophy of science attempts to overrule the apologetic efforts of degenerating programmes."

Lakatos gave the label *Elitism* to the contrary conception—that is, the notion that in distinguishing between worthy and unworthy work ("progress and degeneration, science and pseudoscience"), the "only judges are the scientists themselves." That idea he scorned as stemming from an "undemocratic, authoritarian" school of philosophy, whereas his own demarcationists "share a democratic respect for the layman." The layman has to be helped, of course; for this very purpose "a statute book, written by the demarcationist philosopher of science, is there to guide the outsider's judgment." But *guidance* leaves too much to chance, and so *direction* is perhaps safer after all: The demarcationists, Lakatos announced, will "lay down *statute law* of rational appraisal which can direct a lay jury in passing judgment."

It is a breathtaking ambition. Far from being dismissed out of hand, it was reported recently that one major private research funding agency in the United States for a time considered using the idea in selecting among its applicants' proposals. Needless to say, there will not be universal acclaim if this (or any other) particular philosophy of science takes over the indicators by which the NSF measures the quality of science. The absence of any explicit theory to guide the making and use of indicators may not be good; but the adoption of a single one is likely to be worse.

Another argument for plurality is that, precisely as in the sciences themselves, indices may be developed eventually that are *invariant* with respect to theoretical models. These will, of course, be the most useful ones: Among other advantages, they and only they allow rival theories to be put to meaningful quantitative tests.

Even now, quite opposite theories of science can lead to demands for the same indicators. For example, in the early 1970s the OMB's view was evidently to regard the professional person as a statistic in the free market; from that viewpoint, a rapid decrease in the expected number of new physics Ph.D.'s is seen as a salutary adjustment of an "oversupply" situation (46). On the opposite side, to someone concerned that the intellectual and industrial bases provided by science are being eroded, the same curve has a different and much more ominous message. But both parties will agree on the *need* for such manpower indicators.

Advocates of different theories of science will also agree in their concern for the institutional strengths of science. There was an "Institutional Capabilities" section of *SI-72* (47). One would look there to find measures of the diversity of centers of research and of the strength of the professional societies and their scientific journals. However, this first effort at compiling an "official" array of science indicators did not address such questions; for example, there is not even a hint that a number of professional societies are in severe financial trouble (48). Yet development of indicators on professional health is badly needed, and it may well be possible to phrase such indicators in a form invariant with respect to theoretical framework.

As a practical matter, an NSB publication by itself cannot be expected to go very far in using and comparing a large number of tenable conceptual frameworks (though one would hope that their existence would be noticed).

Whatever the NSB elects to do in this connection, scholars not on the staffs of Washington agencies should be encouraged and helped to work on such problems on a proper scale. Science indicators will be far healthier when provision is made for institutional programmatic steps to support academically centered research.

THE MODEL OF SCIENCE IMPLICIT IN SI-72

What, then, is the theoretical framework implied in the *Science Indicators* volumes? Part of the answer comes from what is said there, part from what is not said; part comes from the sequence, and the rest from the style of the contents. Since on this score *SI-72* and *SI-74* are closely similar, the examples are drawn from the model-setting first volume.

By way of background, it may be useful to recall that the year 1972 was full of turbulent worldwide scientific activity. To give a few examples from physics and astronomy only, there was the identification by a Columbia University team of the strong cosmic ray source in the Crab Nebula as being identical with the pulsar there. Canadians recorded remarkable bursts of radio emission from Cygnus X-3. A Bell Telephone Laboratory group showed how to make the catalytic removal of pollutants from automobile exhaust much cheaper, and at Brookhaven, progress was made toward superconducting electric-power transmission. Fusion reactions became a more probable source of power owing to successful experiments in laser-produced implosion and adiabatic compression of the plasma. The world's largest particle accelerator became operational and gave its first results; so did two new colliding-beam storage rings.

A beautiful puzzle was found in neutrino astronomy, where the difference in observed and expected neutrino flux from the sun is so large that the model either of the sun or of the neutrino's stability has to be revised. The periodic table was extended by the discovery of traces of naturally occurring plutonium-244. By laser-ranging, the distance to the moon was determined to about 6 inches, and the method could now be applied to measuring continental drift. Accurate radio-ranging of Venus, Mercury, and Mars gave preliminary data for another test of gravitational effects in Einstein's theory of general relativity. More Black Hole theories attracted fascinated attention. And three American physicists shared the 1972 Nobel Prize for the development of a theory to explain superconductivity, one of them (Bardeen) becoming the first person to win two Nobel Prizes in the same field.

There is no indication of any of this in *SI-72*, thus illustrating, perhaps, the difference between "knowledge indicators" and Science Indicators (49). Another "big story" in science in 1972—the continuing decrease of support in funds, facilities, educational prospects, and the like—does appear there in aggregate indicators, presented in a deadpan style. Thus, the curve for federal obligations for academic research-and-development plants is shown to drop precipitously, from about $105 million in 1965–1966 to about $17 million (in constant dollars) in 1971 (50). It is a drop that has had enormous impact on the life of scientists in the United States—but there we find only a paragraph acknowledging the existence of the slaughter. (Nor does *SI-72* mention the dismissal of the President's Science Adviser and the dismantling of the science advisory apparatus set up by President Eisenhower some 15 years earlier.) As in other pages, it is as if the authors had been on the verge of telling what some of these indicators in fact *indicate,* but then drew back.

No such hesitation appears in regard to two other aspects of the image of science implicit in *SI-72.* One is the primarily Baconian, rather than Newtonian (and primarily "hard sciences" rather than natural *cum* social sciences), view of "science." This view is signaled from the outset: The second paragraph speaks of "the capacity of science—and subsequently technology [thereby revealing its trickle-down theory of science-technology relations]—to meet the needs of the Nation." Interestingly enough, the experts polled in the Delphi experiment (in Chapter 6) had their priorities the other way. Their desired "criteria for use in determining total funding levels for basic research" were found to be first, the "potential for fundamental new insights" and second, "science needed to generate technological solutions to major societal needs" (51).

In the same Delphi experiment, the tenth and last of the "criteria" for funding considered important among scientists was found to be "Competitive pressure: activity of greater emphasis existing in other nations." But whereas international competition is the least and smallest part of the image of science reflected by the panel of scientists, *SI-72* gives the impression that it is the first and most significant part. This comes about through the placement of a long section entitled "International Position of U.S. Science and Technology" at the very beginning of the book. (In *SI-74* it is renamed "International Indicators of Science and Technology.") The accent is on "international comparisons"—for example, in expenditures, manpower, national origins of literature, literature citations, the "Patent Balance" as a measure of "inventive output" by U.S. nationals versus patents awarded by the United States to nationals of certain foreign countries, relative productivity in manufacturing industries, unit labor cost in manufacturing industries, balance of trade in technology-intensive products, and similar indicators for transnational comparison. One catches a motivating theme for this whole section in such findings as " . . the deficit balance in the high-technology area developed with Japan in the mid-1960s and persisted in the following years, with the largest increase (almost 120%) occurring in 1971" (52).

What this view of science and technology emphasizes is, of course, the competitive, not the cooperative, element. To be sure, scientists have been known to champion this view too, particularly when trying to get their governments to underwrite large expenditures for scientific facilities. However, we find here no counterweight of the kind so familiar to working scientists—namely, the evidence that actual scientific research is not a national but a transnational enterprise.

Consider a concrete example. *The Physical Review Abstracts,* a semimonthly publication of the American Physical Society, prints abstracts of articles that have been forwarded from the editorial office of *The Physical Review* to its printer for early publication. A quick check of a typical recent issue shows that—leaving aside minor items entitled "Comments," "Addenda," etc.—140 research papers are listed (53). Of these, 51 papers (about 36%) are collaborations with persons at foreign addresses or are written entirely by foreign authors. If one also allows for the fact that many an author listed at a U.S. address is perhaps not a "U.S. national"—as a glance at names such as Shyamalendu M. Bose and Subodh R. Shanoy tends to indicate—one would find an even more gratifying degree of internationalism emerging from the publication of a U.S. professional society.

However, when the country of origin of the *journal* is chosen as a measure of the "National Origins of Literature" (as in the citation data given on pages 7–11 in *SI-72*), not only does a rather different view of the U.S. share of the literature result, but one is almost automatically prevented from seeing or asking about measures of transnational *sharing* in actual scientific work. Future issues of *Science Indicators* could well try to define indices of such cooperation among both individuals and institutions and the relation of such cooperation to the viability of science itself. Already *SI-74* shows much improvement over *SI-72*—for example, in the interesting tabulation that in the science journals of the six major R&D performing nations taken together, almost 60% of all citations to the science literature, in eight selected scientific fields, are to articles published *outside* the country of the particular journal (54).

Eventually one may even hope that analogous efforts in other countries, of which the OECD surveys were precursors, would produce a body of work that allows the sciences and technologies to be studied as specific examples of a "world system." We know well that from the vantage point of many sciences, each location on earth is only a specialized sample of the universe as a whole. There are common physical laws and uniform physical properties of matter, stretching from one end of our globe to the other, as indeed they stretch from one end of the galactic cluster to the other. This fact imposes some necessary features on the organization of science as done in different parts of our own world, and makes science an interesting candidate for the study of "world systems." The sciences and technologies may already be the best operational exemplification of the concept itself.

Moreover, in terms of training, research models, communication, prestige-sharing, and the like, the larger scientific community in the twentieth century is organized on transnational lines and has aspects of universality analogous to those characterizing the phenomena it studies. The recruitment and colleagueship system at the best universities is a case in point. Scientific data collections are also increasingly produced by international effort. One can well imagine science indicators specially designed to measure these and other properties of the world system.

While the "official" view or model of science that emerges from *SI-72* understandably does not stress these aspects, it also omits some indicators that one would expect to find on any model. The relatively low place assigned to education is striking. Thus we are shown that federal R&D expenditures for education (in constant 1958 dollars) have risen from an almost negligible amount in 1963 to about $100 million in 1972 (.8% of the total federal R&D) (55). But there is no indication that this amount is only about .1% of the annual expenditure of the whole education "industry" (now around $100 billion, of which the federal contribution is somewhat less than 10%), with very little educational research being done in addition to that supported by federal funds. Symbolic perhaps of the ever-ambivalent attitude of NSF to supporting science-education research, the NSB report even neglected to mention the substantial and useful role the NSF itself has played in fostering such research, for the NSF was not on the list of sponsors of education R&D given in *SI-72*. (56).

However, the place of education and the dissatisfaction with its state loom large in the conception of science that emerged in the poll of scientists and other experts in the Delphi experiment, recounted in Chapter 6 of *SI-72*. When asked for a list of "National Problems Warranting Greater R&D Effort," the "inappropriateness and expense of education" came up in the eighth category (out of 21 categories, ranked in terms of "areas which could benefit from science and technology"); 62% of the panelists—most of them distinguished scientists, educators, or administrators—thought that it was an area warranting major increases in R&D (57). Similarly, 92% of the Delphi panel thought that one of the primary "Factors impeding Technological Innovation" is that the "education of scientists and engineers [is] inappropriate for innovation" (58). Indicators that would respond to such concerns would have to be far more extensive and more detailed than the scanty ones now given.

A section of *SI-72* is given to a survey of public attitudes toward science and technology. Among the interesting preliminary results is the early-warning indication of public concern with the ethical and human values impacts of science and the basic ambivalence or multivalency about science. Thus, while a large and rising majority agrees it feels "that science and technology have changed life for the better," 39% of the sample of the U.S. population expressing an opinion also did not think that "overall, science and technology do *more* good than harm," and 34% of the group expressing an opinion felt "the degree of control that society has over science and technology should be increased" (59).

It would seem that the public, the Delphi panel, and the NSF authors of *SI-72* are synchronized, at least on the need for a better public understanding of science. Thus, the Delphi panel listed "negative public attitudes toward technology" high among the factors impeding technological innovation and expressed high concern with "lack of understanding of process of discovery" and with "lack of public understanding of role of basic research" (60). Among the criteria for allocating basic research funds among scientific fields, the panel of experts gave considerable weight to whether the work "fosters public understanding of basic research" (61).

Considering all these concerns, it seems reasonable to look forward to the development and inclusion of more detailed indicators of public attitudes toward science, as well as indicators of an assessment of the actual level of public understanding of specific scientific materials and processes. An educational assessment of scientific understanding by various age groups in the United States already exists in the literature (62). Such findings might well serve as one element of such a survey. Ultimately, any model of science and technology must allow a visible place for the intellectual benefits reaped by the population at large and not merely because that population pays for these developments.

The Baconian view of science implicit in the *Science Indicators* and their preoccupation with the "harder" sciences and their technological fruits prevented any examination as to whether the balance of funding between the natural and social sciences is appropriate. There are many indications elsewhere that it is not. For example, the "success ratio" (ratio of total funding applications to the successful ones) is generally far lower in the NSF programs in the social and behavioral sciences than in the "hard" sciences programs. Only 5% of the NSF's total research obligations (of $591 million) in FY/75 was made available by the NSF for basic research in all of its 13 social and behavioral sciences programs from anthropology and psychology to history of science (63). Since the NSF provides one-third or more of all available basic research funds from federal sources, there are not many alternatives for the applicants, and the health of the sciences as a *balanced whole* therefore depends greatly on the willingness and success of NSF to obtain the needed funds for the large and growing group of America's social and behavioral science researchers. This is obviously a highly charged topic, one that soon would raise the question as to whether the Congress was wise to be persuaded some years ago not to fund a separate Social Sciences Foundation.

It is tempting to note other individual aspects of the two publications. For example, one could go on to praise some others of the ingenious ideas or suggestive findings. Conversely, one could go on to identify other needed improvements (64), and questionable treatment of data (65). But in sum, what do the first issues of *Science Indicators* indicate? There is no doubt that the volumes already mark the significant beginning of an important enterprise that will have repercussions on science policy, on scholarly work in the history and sociology of science, and above all on the life of science itself. But in addition to giving a wealth of data, *do they show that the sciences and technologies, their quantities and qualities, can be "measured" in a meaningful way?*

The volumes leave one with the overwhelming feeling that although a positive answer to the question is not yet guaranteed, it is highly probable. The problem of making and testing indicators is immensely attractive. Good researchers are entering the field. And the volumes themselves have allowed for the identification and the further development of the bases and scope of the enterprise—by strengthening the theoretical foundations and making them more explicit; by grounding indicators in stated problems; by perceiving the fine structure and the qualitative changes more clearly; by infusing an enlarged spirit of independence (e.g., in dealing with negative developments); and by bringing together independent research scholars in the sociology of science, the history of science, and related areas, in order to cross-check, extend, and further professionalize the making of science indicators.

Notes

1. *Science Indicators 1972: Report of the National Science Board 1973,* National Science Board, National Science Foundation, Government Printing Office, Washington, D.C., 1973. The report constitutes the fifth in the series of annual reports to the President and to Congress, as mandated by Congress in its amendment of the National Science Foundation Statute in 1968 (Public Law 90-407).

The Congressional charge to the board with respect to its annual report, was that it focus on "the status and health of science and its various disciplines [including] an assessment of such matters as national scientific resources . . . progress in selected areas of basic scientific research, and an indication of those aspects of such progress which might be applied to the needs of American society." Quoted by Norman Hackerman, Chairman, NSB, in a statement before the Subcommittee on Domestic and International Scientific Planning and Analysis, Committee on Science and Technology, U.S. House of Representatives, May 19, 1976, page 2.

2. *SI-72*, p. iii.

3. *Science Indicators 1974: Report of the National Science Board 1975*, National Science Board, National Science Foundation, Washington D.C., 1975.

4. Release by the Office of the White House Press Secretary, February 23, 1976.

5. *SI-72*, p. vii.

6. *New York Times*, March 14, 1976, p. 1.

7. Hackerman, *op. cit.*, p. 8.

8. *Ibid.*, p. 19. See also *Science*, **191**, 1033 (March 12, 1973).

9. The NSB has invited such efforts and the NSF has helped support various studies in the field (e.g., on "limitations to scientific literature citations, investment in innovation, technology transfer, economics of publishing scientific literature, social indicator utilization by executives, international scientific activities, and many others"). Roger W. Heyns in the May 19, 1976 Hearing cited in note 1. Heyns also noted that from the start of the Science Indicator series, a major purpose was that of "stimulating social scientists' interest" in the field. *SI-74*, p. viii, expresses appreciation for the seminar that has resulted in this volume, noting that "the present efforts to assess U.S. science are still only in the early stages of maturity . . ." and that the "reports to follow in this series will aim to sharpen concepts, refine their treatment, and seek new measures of the state of science. It is hoped that all those interested in science indicators will participate in the search." This paper is based on an earlier version given at the seminar; I wish to acknowledge helpful comments by the editors of this volume, by C. S. Gillmor, S. D. Ellis, and B. A. Stein, as well as partial support by a grant from the NSF.

10. Albert Einstein, "Ernst Mach," *Physikalische Zeitschrift,* **17**, 2 (1916).

11. A recent example is the discussion in population biology concerning the relative merits of special "tactical" models with scaled graphs and abstract "strategic" models with scaleless diagrams. Cf. R. M. May, "Review of 'The Mathematical Theory of the Dynamics of Biological Populations,' " *Science*, **183**, 1188–1189 (March 22, 1974); and R. Mitchell and R. M. May, "Letters to the Editor," *Science*, **184**, 1131 (June 14, 1974).

12. L. White, "Technology Assessment from the Standpoint of a Medieval Historian," *American Historical Review,* **79**, 1–13 (1974).

13. H. Brooks, "The Physical Sciences: Bellwether of Science Policy," in J. A. Shannon, Ed., *Science and the Evolution of Public Policy,* Rockefeller University Press, New York, 1973, p. 125.

14. Based on an idea suggested in private conversation by E. M. Purcell.

15. See, for example, H. Inhaber and K. Przednowek, "Quality of Research and the Nobel Prizes," *Social Studies of Science,* **6**, 33–50 (1976); and Harriet Zuckerman, *Scientific Elite*, Free Press, New York, 1977.

16. Hackerman, *op. cit.,* p. 15. See also *SI-72*, p. 7, which reports that "an effort was made to estimate the relative 'quality' or 'significance' of the literature" by calculating the average number of citations per published article in selected fields for different countries. The result was the "upbeat" conclusion that the United States ranked first by a large margin in physics and geophysics, molecular biology, systematic biology, psychology, ,engineering, and economics; higher than all others except the USSR, by a relatively narrow margin, in mathematics; and second only to the USSR, by about the same narrow margin, in chemistry and metallurgy. The results in *SI-74* are not very different, except that economics was dropped, and the USSR was now first both in mathematics and chemistry.

17. In "Supplementary Comments," on the last page of *SI-74*, one member of the NSB spelled out his dissatisfaction on these points. Thus, he notes that scientists in the USSR have in most fields by far the highest self-citation index; that a small-scale, high-quality effort such as that in certain fields in France would be swamped in a citation index; and that in some countries the editorial and space limitations on publications allow little room for citations.

18. *SI-74*, p. 106.

19. Hackerman, *op. cit.*, p. 13.

20. L. J. Anthony et al., *Reports on Progress in Physics*, **32**, Part II, 764–765 (1969).

21. Cf. Statement of Robert Parke at the May 26, 1976 Hearing before the Congressional Subcommittee identified in Note 1.

22. In fact, the baseline for a great deal of "spectroscopy" (data on publication, funds, manpower, etc.) for the various subsections of fields was available in publications such as *Physics in Perspective*, National Academy of Sciences, Washington, D.C., 1972; and *Physics Manpower 1973*, American Institute of Physics, New York, 1973; but they were not used or even referred to bibliographically in *SI-72*, and only slightly in *SI-74*.

23. Y. Elkana, *The Discovery of the Conservation of Energy*, Hutchinson, London, 1974.

24. *SI-72*, p. iii.

25. A. C. Nixon, "Letters to the Editor," *Science*, **184**, 1028–1030 (June 7, 1974).

26. *Work in America*, Report of Special Task Force to the Secretary, Department of Health, Education and Welfare [Elliott Richardson] MIT Press, Cambridge, 1973, pp. 155–156.

27. The support for an annual science manpower survey, long provided by NSF, was discontinued some years ago just when the job crunch became notable, and professional societies had to assume the support to continue the enterprise just when many of them were least in a position to do so.

28. American Institute of Physics (A.I.P.), Publication No. R-207.5, American Institute of Physics, New York, April 1973.

29. A.I.P. Publication No. R-211.3, American Institute of Physics, New York, 1972.

30. Editor, "Current Trends in Ph.D. Production and Employment Among Astronomers," *Bulletin of the American Astronomical Society*, **6**, No. 2, 233–234 (1974).

31. *SI-74* had more details on enrollments, although not on employment.

32. A.I.P., *Physics Manpower 1973*, American Institute of Physics, New York, 1973, p. 47. See also *Newsletter of the Forum on Physics and Society*, **4**, No. 1, published by A.I.P. (February 1975).

33. A.I.P. Publication No. R-211.3, p. 136.

34. *Ibid.*, p. 184. The "full employment" movement is evidently becoming stronger in the United States. See, for example, E. Ginzberg, Ed., *Manpower Goals for the American Democracy*, Report of the American Assembly, Prentice Hall, Englewood Cliffs, N.J., 1976.

35. It is significant that Chapter 6 of the National Academy of Sciences (NAS) report, *Physics in Perspective*, Vol. 1, NAS-NRC, Washington, D.C., 1973, was entitled "The Consequences of Deteriorating Support"—with discussion of the decline of federal support for research in Chapters 4, 5, and 10, and the effects "on the livelihood of highly trained people" the subject also of detailed data in Chapter 12, devoted to manpower.

   I have used many examples from physics, although I am aware that this field in some ways has been the one hardest hit by the deterioration of science support since the late 1960s. My choice is necessitated chiefly by the fact that I am most familiar with this field. But I agree with the remark on this point in the NAS report: "To a considerable degree, of course, these effects [of deteriorating support] apply to any field of scientific research and are not unique to physics. Although our discussion is rooted in physics, the problems we address have much broader relevance" (p. 452).

36. National Science Foundation, "Undergraduate Enrollments in Science and Engineering," Government Printing Office, Washington, D.C., January 5, 1972, NSF No. 71-42.

37. B. Henderson, "Unemployment Among Scientists," *Geotimes*, **16**, 16 (September 1971).

38. For the text of Malek's memoranda, see "New Federalism," Record of Hearings of Committee on Government Operations, House of Representatives, January 29, 1974. Also see *New York Times*, 13 August 1974.

39. A recent report by the National Board on Graduate Education pointed out that while the proportion of new Ph.D.'s having no specific job prospects rose from 4.5% in 1968 to 17.2% in 1973, the figures for new Ph.D.'s in English who had no job increased from 3.9 to 21.5% during that period. Cf. *New York Times*, 21 January 1976.

40. W. Heisenberg, *Physics and Beyond*, Harper & Row, New York, 1971, pp. 62–63. In a letter to this author, elaborating further on this account, Heisenberg referred to his Principle of Indeterminacy, formulated the year after this exchange with Einstein. Heisenberg noted wryly that the Principle added a refinement to Einstein's dictum, for it tells what one *cannot* observe.

41. H. Brooks, "Models for Science Planning," *Public Administration Review*, **31**, 367–368 (May/June 1971). According to R. Heyns' testimony (in the May 19, 1976 hearing cited in Note 1), the NSB had established an internal Committee on Science Indicators by 1971.

42. Brooks, *op. cit.* (note 13).

43. *SI-74*, p. 71. The level might be even lower if one used the correct deflator for basic research, one that properly includes the increasing complexity of research tools and management.

44. Hackerman, *op. cit.*, p. 5.

45. Imre Lakatos, Review of S. Toulmin's *Human Understanding, Minerva*, **14**, 128–129 (Spring 1976).

46. See Brooks, *op. cit.*, (note 13), p. 127.

47. This section was dropped in *SI-74*; some of the subsections were treated in the new section "Industrial R&D and Innovation."

48. Here, too, fine structure is of interest. Thus, the American Physical Society and the American Association of Physics Teachers are both essential societies with somewhat overlapping memberships and mandates. But the latter has for some years been on the verge of financial bankruptcy, while the former has enjoyed a handsome income (total net income for 1972 was $400,000; for 1973, $244,000; the accumulated Income and Reserve Fund of the APS stood at $2.38 million at the end of 1973 (see "Summary of American Physical Society Income and Expense," *Bulletin of the American Physical Society*, **19**, Series II, 1069–1071 (November 1974).

49. The distinction is suggested in the Introduction to this volume.

50. *SI-72*, p. 73.

51. *Ibid.*, p. 84.

52. *Ibid.*, p. 17.

53. The issue of *Physical Review Abstracts* being analyzed here is Vol. 5, No. 15 (August 1, 1974). As an aside to citation-index scholars, I note that under the masthead is the injunction: "This journal is not to be cited."

54. *SI-74*, p. 8.

55. *SI-72*, pp. 25 and 111.

56. *Ibid.,* p. 26.

57. *Ibid.,* p. 85.

58. *Ibid.,* p. 88; see also the list of 15 suggested "currently or potentially important graduate educational changes," p. 93.

59. *Ibid.,* pp. 97–98. In *SI-74* the results are slightly more positive toward science. On the subject of simultaneously held but contrary public opinions about science, see the essays by A. Etzioni and Clyde Nunn and by Dorothy Nelkin in G. Holton and W. Blanpied, Eds., *Science and Its Public: The Changing Relationship,* D. Reidel, Dordrecht and Boston, 1976.

60. *SI-72,* pp. 88, 90, 91.

61. *Ibid.,* p. 92.

62. For example, *National Assessment of Educational Progress, Science Results,* Education Commission of the States, Denver, 1970 and after; and articles such as A. Ahlgren and H. Walberg, "Changing Attitudes towards Science among Adolescents," *Nature,* **245,** 187–190 (September 28, 1973).

63. These and similarly disturbing data are in the report, *Social and Behavioral Science Programs in the National Science Foundation,* NAS-NRC, Washington, D.C., 1976.

64. A listing of additional concerns might include these: Are not the existing divisions of science and of current candidates for financing taken too much for granted, so that, for example, research in agriculture and nutrition, or fundamental learning theory, or certain aspects of energy research, continue to get less attention than they need? Were the expectations for more social science research support actually fulfilled on an adequate scale? How adequate do scientists think the current review process is in publication and in research-support application? Should the science indicators not also be extended to cover more thoroughly fields that NSF does not traditionally fund, for example, clinical medicine? Would it be useful to include such indications of benefits of scientific research as the rapidly decreasing death rates in the United States for specific diseases, or the rise in the yields of specific crops? See Nestor E. Terlickyj, *State of Science and Research: Some New Indicators: A Chart Book Survey,* National Planning Association, Washington, D.C., 1976; and N. E. Terleckyj, *State of Science and Research: Some New Indicators,* National Planning Association, Washington, D.C., 1971.

65. An example of the dubious use of data is the finding in Figure 4-14 (*SI-74,* p. 101) that the share of the "major innovations" in the United States has been by far the lowest and constantly dropping, since 1953, for the group of companies that have 5000–9999 employees (now down to a mere 3% of the major innovations), whereas it has been by far the largest, and constantly rising, for the group of companies having 10,000 employees or more (now about 45%). However, if one calculated the ratio of Innovations per Employee (I/E), the picture would look very different, as indicated by these figures (kindly made available to me by Barry Stein).

Employment Size of Firm	I/E All Industry	I/E Manufacturing Industry Only
1–99	4.5 ($\times 10^{-6}$)	21.4 ($\times 10^{-6}$)
100–999	10.4 ($\times 10^{-6}$)	18.5 ($\times 10^{-6}$)
1000–9999	9.2 ($\times 10^{-6}$)	12.9 ($\times 10^{-6}$)
> 10,000	9.5 ($\times 10^{-6}$)	11.8 ($\times 10^{-6}$)

Data for Employees (E) from *General Report on Industrial Organizations, 1967, Part 1,* Census Bureau; omits CAOs and support (nonproductive activities). Data for Innovations (I) from Report by Gellman Research Associates, Inc., *Indicators of International Trends in Technological Innovation,* Final Report, Gellman Research, Jenkintown, Pa., April 1976, cited as the source of the data on Innovations in *SI-74* itself. Thus, far from supporting the simple but erroneous conclusion one might draw from *SI-74* that major innovations are correlated with large company size, the data when properly presented tend to a very different conclusion—as has been previously shown by the work of J. Jewkes, J. Smookler, and others. See Barry A. Stein, *Size, Efficiency, and Community Enterprise,* Center of Community Economic Development, Cambridge, Mass., 1974, Chapter 2, particularly pp. 31–37.

Toward a Model for Science Indicators

4

Derek de Solla Price

Some twenty-five years ago I published a first little paper on the exponential growth of scientific literature. It had been inspired by my baptism as a historian of science during which I had read through all the volumes of the *Philosophical Transactions of the Royal Society of London* since its beginning in 1665. As the reading progressed, I had begun to wonder at the increasing weight of each annual installment and, in the spirit of my recent prehistoric past as an experimental and theoretical physicist, I posed myself the question that James Clerk Maxwell asked so often as a child, "What's the go of it? What's the *particular* go of it?"

A *geistesblitz* of this sort leaves an indelible impression on the memory. I can easily recall my fascination that the tally of papers published each year grew so inexorably, so regularly, for the quarter of a millennium from Boyle and Newton to the current research front. Immediately the analogy jumped to mind that it must be rather like examining the path of a falling projectile or the motion of a planet in a Keplerian orbit. The masterstrokes of Galileo and Newton had been in perceiving that behind the mere empirical description of the motion there lay a simple law, a law of nature from which all else

This work has been supported in part by a National Science Foundation Grant No. SOC 73-05428.

followed and from which one might predict phenomena not yet observed and "explain" all that had been.

In those days as now, my interest in the philosophy of sciences was perhaps shallow and naive, my appetite for major discovery brash as that of any physicist (1-4). Had I known that others had several times considered head-counts in science and had recorded the regularity of growth, it would not much have diminished my sense of deep discovery. What appealed to me was the existence of a simple underlying principle that would produce such exponential growth, that in some sense knowledge had bred new knowledge at a constant rate. Small statistical fluctuations there certainly were from year to year. One could see immediately that some fluctuations were to be expected, like random errors in determining the exact place of the planet or projectile, and others might be due to systematic second-order effects, like the resistance of the air or the effects of minor gravitating bodies. One would have to work with these later, after the big picture was quite clear.

It took only a few days to convince myself that the phenomenon was no artifact of this one particular journal. The exponentiality and, almost precisely, the same single parameter of the growth constant emerged wherever I looked, in the great abstract journals of physics and chemistry and, in particular, in bibliographies of subjects. It was no artifact either of a process of counting papers, for the same thing was evidently true of the fragmentary information then available (in Singapore in 1950) about the numbers of scientists, the numbers of journals, and even the sort of economic data about science and technology that could be gathered from the pioneering work of Bernal (4).

I wish I could say that from this time onwards I never looked back. I did publish, but the paper fell flat (5). Although I maintained a file of statistical data and added new material from time to time, ranging far beyond the original discovery of a basic law of growth of science, I did not take up the matter again until it forced itself into my work as a historian of science, being needed to explain why science had progressed as it had in certain periods and gradually suggesting to me more and more links between historical explanation and the field of sociology of science about which I was gradually becoming more aware (6).

THE ROLE OF SCIENCE INDICATORS

The purpose of this autobiographical account is to indicate that, from the beginning, my attitude has been that of the basic (social?) scientist, attempting to discern general underlying principles that must govern this great mass of somewhat hairy empirical indicators that emerge as soon as one looks quantitatively at the historical and modern development of science. I find it strange that so many people fail to see the difference between such an attempt to find underlying principles and the methodologies, similar only in their quantitation, to fit arbitrary mathematical curves to empirical data so as to describe the data and perhaps to extrapolate (7). For me, the richest approach must be to take the empirical data that is to hand, much of it produced from technological practice, and then to find the most simple first-order pattern that exists within it; then to proceed both in the direction of looking for the more complex and higher-order patterns that modify the first and also to suggest models of underlying structure that produce such a pattern and test these models by further recourse to the data. To continue the analogy with terrestrial mechanics, one starts with the empirical data on the ballistics of projectiles that antedate Galileo and Newton; then one shows that a simple law of uniform gravitational acceleration will explain the bold outlines. Later one shows that simple laws of air resistance improve the agreement with empirical data and, still later, one shows that uniform gravitational acceleration derives from the universal inverse square law of gravitation, and air resistance laws derive from the statistical mechanics of gases.

It must be added at this point that, as is usual in the historical development, the primary strong flow of information runs from the technology to the scientists; only secondarily, and after understanding has been achieved, does this new wisdom succeed in producing change and innovation in the technological process. Probably because far too high a proportion of the historians—and, indeed, the sociologists of science—have been theoreticians rather than experimentalists, there exists a chronic underestimation of the part played by the craft of experimental science. It is precisely in the region populated by instruments, big and small machines, gimmicks, techniques and goos, that the interaction runs high, hectic, and startling, and that useful technological innovation is endemic.

Somewhat similar considerations seem to link (or should link) the search for the scientific understanding of the growth and development of science with the related technologies of science policy—that is, the attempts that are made to manipulate, administrate, control, and plan this growth and development. I believe that the main road to scientific understanding leads from the artisan's workshop to the scholar's study. In this case our best way to erect a framework of underlying theory is to take whatever empirical data have been generated in the attempts to plan and administer science policy. Almost ideal for this purpose is the publication *Science Indicators 1972*. It was published in response to frequent and repeated pleas from those concerned with the science policy of the country that there should be some sort of state-of-the-art survey for science and technology that would parallel the

President's annual State of the Union message. Presumably, a topical treatment of specific scientific and technological advances and retreats would have been too invidious to survive, so what we have is essentially the quantitative overall picture, served up as well as it could be compiled from the hundreds of reports full of NSF fiscal and administrative tabulations for past and present. To these have been added some contracted-out studies of the unobtrusive indicators of scientific publication and citation that have been generated in ever-increasing quantity as a by-product of scientific information technologies.

With all its blemishes, it seems to me a noble job of work for which we should be most grateful. The only way I know to express that gratitude is by attempting to use the data to draw and test conclusions and thereby emphasize the need for better data the next time the job can be done. It can indeed, as I hope to show, be taken as the text for a sermon on science indicators in general. However, not all of *SI-72* can be handled in this fashion. One experience of participating in a Delphi experiment has been amply sufficient to convince me that this procedure cannot be more than sophisticated opinionating. If one needs opinions rather than understanding, this procedure may be useful and, possibly just once, as a concession to the futurologists, it was appropriate to include opinions about science policy as well as the hard facts of the case. We know, alas, that there is often an almost perfect, but negative, rank correlation between expert opinions on what priorities there should be in national science policies (8) and the implicit priorities revealed by actual budget distributions. This might be a better world if nations acted on the basis of priorities perceived by such experts with and without sophistication, but evidently they do not.

Then again, in the main body of *SI-72* are quantitative statements that fail to evoke more than a "Gee Whiz" response; the reader is told a fact and duly records it but cannot react specifically to such a piece of data. For example, we are told (p. 16) that "aircraft and parts had a 9:1 export/import ratio in 1971." This seems a rather precise statement, but if the book had said 2:1 or 20:1, my response would have been similar. To be meaningful, a statistic must be somehow anticipatable from its internal structure or its relation to other data. Since I have already said that for me the general search for an overview of science indicators means the establishment of a set of relatively simple and fundamental laws that are capable both of refinement and underpinning, this new consideration insists that we must try and make the system cohere. Laws must be related to other laws, statistics to other data, in order to produce the best interlocking system with the minimum number of independent parameters.

One great virtue of *SI-72* is that it is one of the first instances in which a government bureau has been bold enough to excite more than a "Gee

Whiz" response by comparing data for the United States with that for other nations. It is not, of course, completely novel for we have had before excellent, indeed much better controlled, cross-national studies from the Science Policy office of OECD. There have also been monumental compilations of international science statistics produced from UNESCO, but the consensus is that these are useless because of the usual UNESCO reason: Each nation has autonomy, virtually unquestioned, over its own brands of interpretation of ground rules, and numbers seem to blow in political winds far too violently for orderly comparison.

Finally, *SI-72* gives all of us a *locus classicus* for good hard data (at least fairly good, and hard in places) that begin to demolish the tedious literature about what science indicators would be like, good or bad, if only they existed. It is notable that the social structure of the scientific community is such that it exhibits a reaction (9), often quite violent, in the presence of any process that seems to threaten the autonomous judgment by colleagues familiar with their world and operating within the internal system of norms that have been described classically by Merton (10). Perhaps the best examples are a scattering of angry letters to editors against evaluation of scientists and their work by citation analysis and related methods. Since no such general citation analyses of this sort have ever been generally published and even the few limited examples are not generally known to the complainers, it seems that the reaction represents an uninformed fear and various rationalizations derived therefrom. I would argue in all such cases that it would be better to make conclusions and decisions on the basis of knowledge rather than ignorance. Citation counts and other such science indicators in general may not be very useful; they may run counter to other evidence, but it is surely safer to have the option of using them or not.

BOUNDARIES FOR A SYSTEM OF SCIENCE INDICATORS

The most vexing questions in the consideration of science indicators lie not in random or systematic error in the indicators themselves but in the definition and partition of the universe that is to be considered. Practices differ fundamentally from nation to nation, and policymakers and theoreticians habitually confuse one definition of "science" with another that has totally different boundaries. Concepts run all the way from the widest possible one, in which we attempt to consider all the scientific and technological activity in the nation, to a rather narrow and much smaller system of "research."

It is not very useful to use very wide and all-embracing definitions of science and technology for our indicators. Most people in the nation have had some sort of scientific training if they have had any education at all, and a large part of the national wealth is generated and spent in connection with "technological" products ranging from food to electric light bulbs. We can, however, take the narrow line that we wish to have indicators only for the most traditional sort of scientific research, which is usually conducted by Ph.D.'s working in or near universities and a few high technology industries and federal bureaus, and publishing their work in the learned journals. As a rough gauge of this, we may note that in the United States in 1971 there were about 200,000 such people in all fields of science and engineering.* Allowing about $20,000 per year for each scientist and engineer, the total bill would come to $4 billion, which amounts to less than half a percent of the national GNP and only about one-seventh of the national R&D expenditure.

The other six-sevenths of the national R&D expenditure is due to some 325,000 more scientists and engineers who are, in fact, largely engineers working in the fields of defense and space and describing their work as "applied research" or as "development." The cost per person seems also somewhat greater—about $52,000 per person. The 525,000 R&D workers are, however, only a small part—about 30%—of the total population of "scientists and engineers," the majority being taken up with non-R&D activities such as production, management, and teaching. Even this population seems to be below the largest group that could be claimed for the professions of science and engineering since the present rate of production of 270,000 bachelor's degrees in science and engineering each year implies a pool of magnitude twice or even thrice the registered total of 1.73 million scientists and engineers. Thus, the group of people to be considered ranges from about 5 million, with at least a bachelor-level qualification, to about 0.2 million qualified in and working at the research front in science or technology. The amounts of money involved cover a similarly wide range.

To throw further light on the matter of definition, let us consider a distribution (unfortunately not included in *SI-72*) for the way in which the population of "Qualified Scientists and Engineers" are (were?—my data are for 1964, but probably better and more recent figures are available) distributed by employer and function (12). About two-thirds of all qualified scientists and engineers are engaged in activities related to the processes of production, entirely in industry. If government employment is also counted, three-fourths of the scientists and engineers are so employed. Although the rubric R&D is seen to include 40.5% of scientists and engineers in this table, "research" alone, both allegedly basic and allegedly applied, accounts for only 12.0% of the population of scientists and engineers.

*I have deliberately allowed here for a continuant population of about 100,000 together with about three years' worth of transients totaling another 100,000. This should be an outside upper limit for this population; it is almost certainly an overestimation (11).

Table 1. Distribution of Scientists and Engineers by Employer and Activity

Percentage of U.S. Total	Industry	Government	University	Total
Research	5.5	1.5	5.0	12.0 ⎱ (2)
Development	24.0 ⎫	4.0	0.5	28.5 ⎰
Production	34.0 ⎬ (1)	3.5	0	37.5
Administration	7.5 ⎭	3.0	1.0	11.5
Teaching	0	0	8.0	8.0
Total	71.0	12.0	14.5	97.5 [a]

[a] 2.5% "others" ignored; (1) 65.5% in production-related economic activity; (2) 40.5% in all R&D.

If one is to seek systematically for indicators of input and output for the fields of science and technology, one must choose to define population groups of personnel by their types of output. The time-honored procedure, hallowed by the Frascati convention (13) and now adopted by UNESCO, OECD, and most member nations including the United States, is to consider R&D as the fundamental universe and to break it down into Basic and Applied Research and Development activities, and Industrial, Governmental, and University sectors of employment and funding. Time-honoring and hallowing is, I hope, not irrevocable in a discipline so tender and young. The trouble has been that although most of the research budget and manpower is relatively easy to define and measure, it is rather small. Moreover, in the age of the atom and of space, with issues of energy, pollution, cancer, defense, and other issues with political glamor, it is difficult to relate "research" to such objectives and to outputs not more momentous than mere scientific papers and research reports. There has been an obvious aggrandizement of research policy activity (and those concerned with it) to take in as much possible of the valuable technologies with all the development and even some of the cost of delivery included. There is obviously much more political punch for an R&D policy that costs 2.6% of the GNP or 7.7% of all federal expenditures, than one which is less than one-fifth of this, far below the critical size for voter interest or politician concern.

Then again, so much of what seems conceptually vital in input and output indicators for entire sectors have become subject to reporting conventions that seem, at the least, questionable. For scientists and engineers in universities and colleges, it has become standard to ignore all the inputs of their institutional teaching salaries, public and private, and also the outputs of their teaching at the undergraduate and graduate levels. We make the assumptions, virtually unjustifiable, that one can separate their activities on some basis, for example as 2/9 researcher, 7/9 teacher. Similarly in industry, especially in the aircraft and electronic industries which account for a prodigious fraction of all national R&D expenditure, the fiscal conventions can hardly reflect with any accuracy whatever boundaries exist between the frontiers of technological research and the prototype building activities of essentially production engineers.

THE DISAGGREGATION OF R FROM D

One way out of much of this arbitrariness in convention would be to define the universe of personnel in terms of the measurable outputs that constitute the bottleneck in science indicator theory; inputs have been easy enough to define and estimate at will. A major plank in this program is that we shall have to disaggregate the R from the D since these have quite different conceptual bases and functions in different social systems. The D component is obviously closely related to the output of industrial production, especially in those high technology industries (mostly aerospace and electronics and to some extent chemical industry) where there is a high rate of change of products on the market. The D manpower is largely engineers similar to, though no doubt much superior to, those engaged in routine production in the same industry. The D money, whether paid by industry itself or by government, goes into the salaries of these engineers and the materials and services they consume.

We may therefore regard D expenditure as being a type of overhead on production costs that must occur in those highly innovative industries where the products competing for the market are very different from those of a few years ago. It is apparent from the classical study by Freeman (14) that the R&D expenditures in various industries in the United States and the United Kingdom vary with the growth rate of the industry. In fact, the R&D investment as a percentage of net output varies as about the square or cube of the percentage growth rate. Aircraft and electronics with growth rates in excess of 10% per annum require a 25% investment, whereas lumber, food, and apparel industries with a growth rate of only 1% per annum require investment at less than 0.2% of net output. In high technology areas the profitability seems to reside in a highly competitive growth produced by product innovation, and this process demands development costs that increase rapidly with the advantage sought so that delicate marginal considerations apply. It would, as a limit, take 100% of all net output to maintain such an innovation industry that it would grow (by Freeman's figures for 1935–1958) at some 15% per annum.

Ideally, I believe, the D sector of our R&D indicator study should be left to the traditional economists and, perhaps, also to political scientists as a study of the habits of the nation in purchasing more or less available technologies from the industries and their staffs of engineers. We purchase production of all the various products and we purchase more or less deliverable innovative change in these products at what must be rather predictable and explainable costs in manpower and money. Of the present 2.6% of GNP that is spent on R&D, the greater part, at least 1.6% of GNP, represents the decisions to purchase high technologies, many in the defense sector where the purchase of next year's model comes at a premium. Now much of so-called "science policy" represents in fact a "technology policy" composed exactly of such decisions to purchase available or almost available technology. These decisions are not made so much by any sort of research as by the huge power of government procurement.

I take it that what we the people want to do with these government and industrial decisions to purchase sundry expensive technologies is to have information so as to argue about the wisdom, efficiency, democracy, and priority-setting exhibited. For this purpose we need, for the relatively few high technology industries involved (almost all of them giant corporations capable of massive development investment), a list of costs of the various specific technologies involved. Minimally, we need the short list of the most expensive investments that together make up perhaps half of the 2.6% GNP we have previously been including under an R&D budget and an establishment of scientists and engineers responsible for all the other sectors of activity.

If one now considers the remainder of the communities of scientists and engineers, without those deployed almost entirely in the procurement of technologies through production and the related development efforts, it is seen to consist very largely of people whose output is a socially valuable service such as teaching, the delivery of medical care, agricultural extension, quality control, and data processing, for example. Many of those so employed in services are also engaged in research, scientific or technological, with or without special funding for this sector of their activity. The reason for this is that in most cases the value of the service depends rather critically on the quality of the scientific or technical knowledge that is transmitted or applied. With knowledge of this sort growing exponentially, any corpus of knowledge once learned will obsolesce or prove inadequate with a half-life of a decade or less, and only continuing research activity provides a clear enough indication that the person performing the service has the totality of knowledge in his field at his disposal. It is rather like the Archimedes Jar used for specific gravity experiments; the only way one knows that a mind is full is by watching it running over.

I am not quite sure, but lean toward the iconoclastic position that this need to have people who "know everything" rather than any value of research in itself should justify the greater part of all this research activity. It is particularly noteworthy that in the United States, as in most other developed capitalist economies, scientists or engineers are rarely employed for research rather than for work in connection with production of some valuable service. When the job is "research," one often finds that the product of the labors constitutes some sort of service rather than research itself. Apart from these cases we have a very few exotic institutes of advanced study and miscellaneous research professorships whose product can be compared with that of the few other creative people who society can afford to support for their music, poetry, and painting. Such activity is admirable but rare, even with great educational and cultural opportunity, and if the only problem of science policy and planning were the support of such genius, we should not need more than a middle-sized philanthropic foundation, this without political in-fighting.

Interestingly enough, the pattern just described is not followed by such countries as Canada, Australia, India, and the academy-dominated socialist countries. In all these cases a considerable part of the research budget is channeled through an official government academy or council for the support of a special, and usually large, part of the population of scientists and engineers who are then maintained and paid salaries to produce research out of any context of teaching or other service. In the case of former colonial countries, the historian might explain this position as a historical relic of a particularly wise British science policy. It seems reasonable to adopt such a tactic in a country where there is an urgent need to build up a scientific and technological expertise at a rate far in excess of that which could be used in the limited services of education and health and other such activities. Without such a direct employment, the best minds would have been swept down the brain drain to countries who had such jobs ready. This argument can, however, hardly continue to apply once the country has left its peripheral status and developed to a point where other experts begin to be imported to staff universities, colleges, hospitals, and other service institutions. In the academies of the socialist countries, there seems to have been a steady move to shift the status of the academicians from being a kept elite without specific service to being the equivalent of teachers of graduate students and performers of specific services of social and economic value, sometimes to the point where such functions, especially the useful services, minimize or preclude the adequate amount of research that is needed to keep abreast of the growing scholarly knowledge. I find it entirely plausible that such imbalances between research and service functions could be the chief contributory cause of a nation suffering inferior economic conditions and quality of life than would be expected from the ability of its people and their investment in the R component of their R&D.

If it is true, then, that only a small part of all research is supported in most countries purely for cultural and aesthetic value as a magnanimous gesture to the creativity of the human race, the greater part must be tied either to the valuable goods of industries' production sector or to the valuable services. Just as we might have a balance sheet showing the costs in manpower and money of the technologies we produce by using our scientific and engineering talent in innovative high technologies, so we might produce a balance sheet showing the national deployment of the valuable services in the delivery of, for instance, education, health care, agricultural and geological extension, meteorological advice, and data processing. Outputs in such areas are obviously difficult to quantify, but it cannot be altogether impossible since market prices seem well established and not altogether independent of quality estimates. At all events, the basic principle seems to be that the trained experts are hired by private individuals, by government, and by industry to provide these services. For manpower planning and for estimates of the quality of our lives, we need to know the deployment of our resources in all such services; and for the sector of government purchase of them, it is particularly desirable to have the information to see if priorities seem to be changing in their quality and quantity. The Rothschild reform (15) which caused such a flurry of excitement in British science policy seems to me to be a step in the right direction for a proper philosophy of the service function of scientists and technologists. In brief, what Rothschild proposed was that in such activity one should contract for the particular valuable service desired and treat the contract like any other by being not at all satisfied if there is no service or if there is one different from that expected. The Rothschild debate was not only on this point; much of it arose from its being the first red light signaling an end to free exponential growth of the support of science in that nation, as the Mansfield amendment did in the history of science policy in the United States. Then again, many observers were most concerned that ruthless or stupid application of this principle might cause a shock wave to run through the entire scientific and technological community that would play havoc with a system that had grown up under different ground rules. Though these are reasonable objections, I feel that the principle was correct and that, in general, in these cases we pay for the service itself, which can be valued so much better than the more nebulous process of "research."

A DEFINITION OF RESEARCH

This brings me to define research in more rigorous terms so that it may be distinguished from the activities of scientists and engineers engaged in giving service or producing goods. In the latter cases the personnel have the economic property of their labor to sell, but in the former case there is no such available property for selling because all the newly created knowledge is given away in the very process of creation. This position arises because of a vital anomaly within the social system of science. The act of creation in scientific research is incomplete without publication, for it is publication that provides the corrective process, the evaluation, and perhaps the assent of the relevant scientific community. Merton has ably delineated this essential point that gives the world of scientists a different flavor, a different set of internal machinery, from most of the other labors of mankind (16). Private property in science is established by open publication; the more open the publication and the more notice taken of it, the more is the title secure and valuable. The value, however, is such that the scientist cannot participate in the economic process of buying or selling that which has been effectively made universal property through open publication. There is no possible market value in Boyle's Law, or in the structure of DNA, or even in the discovery of fission or the properties of semiconductors, in spite of the obviously strongly related patents and industries. The Curies could not have patented the natural law of radioactivity, though they could have sold the radium that was the product of their labors, or they might have been able to patent their special process for purifying this substance.

There is an old children's riddle that asks: "What is it that you can only keep if you have already given it?" The answer is supposed to be "a promise," but the same anomalous property of property is true for scientific knowledge. Making use of yet another old maxim, if two people share $10 belonging to each, each comes out of the exchange with a pair of five dollar bills and no net gain; but if the two people share 10 pieces of knowledge belonging to each they come out with 20 pieces of knowledge apiece, a doubling of each holding. There exist in the world fanatic collectors who believe that the value of their art works or manuscripts are diminished by each person permitted to see them. But by and large, the attitude toward scientific creativity holds that knowledge is not thus diminished. On the contrary, since every researcher has had the experience some time in life of the false discovery, the accidental omission of a factor of π, the ignorance that some discovery had in fact been made by others long ago, the highly significant perception that turns out to be trivial, we must all doubt the value of any scientific contribution until it has been examined by our peers. For reasons such as these, strongly built into three centuries of development of norms of scientific behavior, we must regard the end product of scientific research as the openly published scientific paper or its functional equivalent. There are variants of the paper since in some fields the monograph or even the research book are more appropriate, and in others the unit of contribution might be a brief note or a new data parameter to be added to a tabulation. In all these cases I argue that the publication is not in any way a surrogate or an epiphenomenon for the job that has been done; it is in a strong sense the final product of the research.

I say, therefore, that when a person with research-front training in any branch of science or technology labors and produces something new, and the principal product is such an open publication, then we must say that research has been performed. However, the result of similar labors might well be a new process or a product, valuable advice, medical care, a weather report, a biological assay of some pharmaceutical, or something else that some industry, government, or private customer is willing to pay for. In that case we have instances of production or of service. The new knowledge or skill obeys within society the normal legalities of property. It can be bought or sold as any other goods, perhaps even licensable or with title of ownership, as with a patent or a service given by some controlled profession. There must be some sort of gray area where we cannot quite tell if the principal product is the one thing or the other. A rather good test of whether it is a principal product is provided by ascertaining if this product by itself enables the producer to dispose of this particular project or activity and proceed with something fresh. For example, quite often when an innovative service has been given, such as a series of instances of clinical practice or the design of a new product, the professional will wait until the property has been established and the customer has been satisfied with a title to his good and then publish as an auxiliary benefit to the "researcher" or as an advertisement for the new product or procedure. In instances of this sort I think it evident that the prime product is service, or even production, and that the publication that comes secondarily must at best be regarded as para-research activity.

Although the anomalous property characteristic is sufficient to define research, it seems much more difficult and not very useful to subdivide such research into such categories as "basic" versus "applied" or as "scientific" versus "humanistic." For all national fiscal records, the basic/applied distinction—though defined in fine lofty language in the Frascati Convention (13, especially p. 11, paragraph 25)—proceeds, in fact, by assigning all work in such fields as High Energy Physics, Physical Chemistry, and Pure Mathematics to the "basic" category, and all work on topics such as Nuclear Physics, Chemical Engineering, Electronics, and Metallurgy to the "applied" category. Since most of the nations work on about the same mix of all available regions of good firm scholarship, it becomes an artifact of the definition that the more numerous applied subjects give indicators that are all about twice those of the basic subjects. Nations differ greatly in how much they spend in development, for they vary in their manufacturing capacities in the high technologies, particularly as world suppliers of armament, but all nations seem to have something like the applied/basic ratio of about 2/1 (17). For this reason, I would prefer indicators that covered the whole of research, when necessary breaking this down by subject classifications as broad or as narrow as we wish.

For anyone who has ever worked in a language with terms like *Wissenschaft* or *nauka*, it is apparent that there is no universal agreement on the boundaries of science as opposed to other varieties of scholarly learning such as history, law, folklore, and the various branches of philosophy. Probably the biggest technical hang-up lies in the area known as the social sciences. Rather large numbers of social scientists become very nervous about insinuations that they might not be in some sense "scientific" and that they are, therefore, of questionable authenticity. Getting away from pejorative uses of such definitions, one might, as I have shown elsewhere (18), get some sort of illumination by seeking indicators for the special qualities of those subjects that are designated "science" in the Anglo-Saxon world. It would seem as if such subjects as chemistry, physics, astronomy, and most of biology have a kind of pattern of rather tight-knit cumulative structure, in which contributions are made in a positive, definite, and finished form (at least, temporarily!), and each advance is made on the basis of immediately prior contributions. In the humanities, by contrast, the very relation to the individual human creative scholar makes each viewpoint different so that each scholar works in the nutrient fluid of all that has gone before. The scientist economizes efforts and achieves relative certainty and rapidity of cumulation by working almost (but not quite entirely) from the research skin of recent work. Perhaps the difference is shown best by the comparison of Boyle's Law or Planck's Constant with any symphony by Beethoven or any painting by Picasso. The law and the constant are objective knowledge that could have been discovered by others in any language, and (following Merton again) that is why we assign such property eponymously. One can hardly imagine Beethoven or Picasso being involved with a priority dispute, for their creative contributions are not significantly exterior to their human personalities. Boyle or Planck certainly have individual styles and personalities of exposition, but when they engage in research, scientists act *as if* they are discovering something that is already there. There is only the one world to discover, and therefore, physics in Kansas must be identical with physics in Keflavik. Some branches and twigs of science might indeed be of very local interest, such as the botany of Brazil or the demography of Denmark, and might be publishable only in and of interest only to the domestic scientific community. But in the nonscientific, humanistic brands of scholarship, a Marxist historian might have little community of discourse with a capitalist colleague, and if they do intersect in chosen topics, two philosophers of different schools of thought might engage in what the kindergarten teacher calls "parallel play."

An activity called "scientific research" can be distinguished from other activities engaged in by scientists and engineers; moreover, there is a reasonable basis for separating this community activity from more general areas of scholarship which lack the peculiar characteristics of anomalous property laws and a creativity that identifies itself as discovery. Because the word *technology* is now left over from the schema but remains far too valuable to lose completely, I prefer to use it to refer to the labor of scientists and engineers whose end-product obeys normal property laws rather than the anomalous open publication of the world system of scholarly information. The outputs of technology are economically valuable goods and services; the outputs of science are publications into the world system of scientific information. I am a technologist when I teach or give science policy advice, and I am a scientist when I write.

When we have normal property laws *and* innovation, we must speak of technological research, examples of which might be the making of new pharmaceuticals, a new fighter jet plane, or a nuclear reactor or minicomputer, or in the field of services, a new surgical procedure of a disease diagnostic. In general such technological research is similar to the scientific variety in that we need a good mental knowledge of the existing state of the art at the research front before one can innovate, and in the fact that innovations proceed on the basis of the last lot of innovation. In a certain sense technological research necessitates the use of scientific knowledge since this, after all, is the analytical wisdom of all the technology already achieved. What does not happen, however, is a direct process of first making some sort of advance in scientific understanding and then attempting to "apply" it. More importantly it seems that in general technological research occurs when there is a likely purchaser for the new technology, not merely someone to pay for technological research.

AN INDICATOR SYSTEM FOR SCIENTIFIC RESEARCH

Having now delineated a system of scientific research whose output is the world corpus of scientific papers, we are able to set aside the technological activities of innovating and providing the products and services of our high technological age. The products and services can be accounted on a cost-benefit system. What seems needed for good policy is an analytical budget showing the goods and services that have been purchased with the nation's limited resources of expert labor and how the implied priorities compare with those of other countries in value and in shifting trends. These expenditures must determine the economic and military health of the nation and much of its quality of life, and on the whole, policy is made by decisions to purchase. Technological innovations proceed largely from market pull rather than from the push of new understanding. There is naturally a strong link between the technological system and that of scientific research in the overlapping of manpower, particularly the segment that educates and trains the scientific and engineering manpower for all future activities. Technology policy is constrained not only by what goods and services the nation is able to afford but also by the availability of scientists and engineers whose training is long and whose field of applicability limited and rather difficult to shift or extend.

Scientific research does not, therefore, constitute a closed system of expert manpower. The process of carrying out research and training others to do such work is strongly associated with the teaching of more people and with the production of people who have been brought to the research front but do not remain there. Such people are not, in any sense, wasted by a failure to continue research. Indeed, it might be said that the socially desired product is not the scientific researchers whose output is the world knowledge system, but rather those whose labor results in the valuable products and services. Be that as it may, the study of scientific research productivity shows that one is dealing with a process having a large flow-through.

The demography of scientific research seems to have a pattern similar to that of a medically backward country in which the birthrate and the death-rate are both very high and, in consequence, where most of the population consists of infants who are just passing through this existence. The vital characteristics and the gross value of the rates seem to a first order of approximation to have been constant for several centuries and to be much the same in all countries of the world, whatever their social and educational states and systems. The net birthrate of people qualified in the scientific research front seems to be carried along by the exponential growth of world scientific knowledge, which is probably the case, as mentioned at the outset, because each piece of knowledge seems to breed new problems and new knowledge at a rather constant average rate that is of order of magnitude 7% per annum.

The gross birthrate of the scientific population is, in fact, much higher because most of those born to the research front (i.e., publishing their first scientific paper) do not continue to publish. Again, to a rough approximation, the position seems to be that of every 100 scientists publishing in a given year in any field of science, 33 are newcomers to the field and, of these, 22 will not continue at the research front. The net gain of 11 people is then offset by about four "deaths" of people who were at the publishing front but who discontinued. It is this attrition process (extremely strong in beginning research careers) that is responsible for most of the observed variation in the productivity of research scientists. Some fields consist of rather small unit papers and others of large ones, so that there must be some productivity differences from field to field, even for constant work rates. Then again, some people spend more time on research than others, and there must be some small differences of individual efficiency. But, by and large the pattern is set by the attrition process and rapid exponential growth which combine to give the research population a distribution of productivity.

A noteworthy feature of any such hyperbolic distribution (19), whatever its mathematical pattern and whatever mechanism might lie behind it, is that a relatively small and stable core of the publishing population is responsible for a large fraction of the papers and that the bulk of the population contributing the rest flows through rapidly. We do not, in fact, have a complete analysis of this interesting demographic picture, but we do know empirically that the productivity distribution laws are regular and constant over large historical, geographical, and topical spans. We also know empirically that the general behavior is consistent with the process just described for both those with low productivity and high transience and those who form a stable and highly productive group. The presence of such a productivity theory, even in its present primitive condition, makes it possible to relate all indicators of manpower to those of publication in a simple and linear way.

For example, in any given year, over the whole field of science, there are about equal numbers of papers and authors. At present an average of two authors' names appear on the by-line of each paper, and each author emits about two authorships per year so that the factors of two cancel each other. The number of authors is, however, not the totality of those present in the population of publishing scientists. For every 100 authors who publish in a given year, there are about 33 more who do not publish in that year but will again in the next few years. Since, however, 33 of this year's authors are new recruits, the 67 "old hands" together with the 33 silent authors make a total of 100 established authors who are continuing, if intermittent, publishers. We find this approximate numerical equality between the numbers of papers published, the number of authors participating in such publication, and the number of continuant authors residing in the more or less stable nontransient scientific community (11).

For an order of magnitude of the population involved, one may note that the source index of the *Science Citation Index*, based on more than 2000 of the world's most cited journals, records a world population of 362,000 authors in 1970, of which about 100,000 were in the United States (20). On the preceding model, this implies about 33,000 newly recruited authors per year in the United States, which may be compared with the annual science output of about 18,000 Ph.D. and 50,000 M.A. graduations. The 100,000 total for the continuant author population may be compared with the 171,800 Ph.D.'s in "science and engineering" or with the 68,400 scientists and engineers involved in R&D in the colleges and universities. My feeling is that the 100,000 figure is more reliable and more comparable with the figures in other countries, and rests on a sounder conceptual basis than the other manpower figures, with their wide range of fiscal definition and far from unobtrusive mode of generation.

THE WORLD DISTRIBUTION OF SCIENCE

At this point it is possible to get some basic pattern and sense out of this national indicator of the crude size of the scientific research community. In conjunction with the *Current Contents* series of publications, the Institute for Scientific Information (which also publishes the *Science Citation Index*) issues an annual directory that lists the names of all source authors on a geographical basis by country, city, and by institution for the address given by the author in the by-line. Taking the countries alone, we may readily have measures for those who publish in this corpus of 2000 journals including all the best-known international organs in virtually all fields of the natural sciences, technologies, and the social sciences. When the countries of the world are ranked by their numbers of authors, it turns out that the distribution is remarkably stable from year to year, even though the actual lists of authors change greatly and include large numbers of transients each year. In spite of obvious difficulties—with many authors listed twice or more with variant addresses and other such systematic "observation errors"—it is immediately obvious that the rank order is simply that of economic size with remarkably little perturbation.

Even at the first issue of this directory, based only on a single year of data, it could be seen that with only moderate scatter, this indicator of research-scientific activity was a reasonable monotonic function of the economic indicators (21–22). The difficulty of using the standard economic indicator of GNP or some variant thereof, either directly or on a per capita basis, is that of estimating conversion rates of nations that have a controlled socialist economy. Even with rough values, however, they clearly follow the general trend, and the best fit for all data is that, the per capita economic activity. This result can now be improved upon enormously, partly through the cumulation of several years of the geographical directory and partly through better comparative data for the world's countries. Using the eight years of data, we can get good average values not only for the magnitude of the scientific research indicator but also for its rate of change. Next, from studying the corpus of cross-national political and economic indicators, I have come to realize that the single measurable entity that correlated best with all others, that was easiest to obtain without conversion factors, and that was indeed most readily available is the national use of electrical energy measured in kilowatt-hours. A very few nations have obvious anomalies —Norway, for example, with its vast hydroelectric power facilities—but otherwise this seems an excellent and well-behaved indicator of wealth and development. As it happens, it obeys empirically the same sort of distribution law as that of our scientific research indicator (23).

The per capita use of electricity seems to vary as the 3/2 power of the per capita economic activity (GNP). Armed with this result, it is obvious that the per capita science and per capita electrical power should be linearly related. The empirical data that we now have shows that the agreement is indeed excellent. In fact, the scatter is much less than for a correlation of the science with money because there is no need to use the somewhat arbitrary conversion factors for socialist economies. One can indeed do even better, for if the per capita indexes of scientific research and electricity are correlated linearly, then a multiplication of both by that country's population will lead to the result that the absolute size of the scientific research activity of any nation is proportional to the absolute size of its electrical power consumption. The result is excellent with rather little scatter. For those nations above and below the trend, there is agreement with intuitive expectation of idiosyncratic behavior in having for some special reasons "too much" or "too little" of the scientific activity as related to the industrial and economic development.

This investigation does more than tell us that science and development are related for the nations of the world. In the first place, the functional relation with small scatter validates both of the independent measures. If agreement is this good, then both indexes must be measuring something real. Next, the fact that the function is so simple suggests that we are near to the proper variables of whatever simple and natural laws exist. I am thus inclined to suppose that scientific research and use of electrical energy both depend linearly on the population size and some index of development. The economic wealth, in contrast, does not follow such a simple law but rather some nonlinear function which in its central region seems be as the 2/3 power of the development index. The main point here is, however, not the detailed result but simply the validation we have for this index of scientific activity. We know now what values to expect for a country in terms of its other social, economic, and industrial phenomena, at least in principle, and we are entitled to be surprised if a nation seems to be performing more or less scientific research than this expected value. It is rather difficult to make such an estimate for the United States, for as the world's most developed nation and the largest in absolute size of that development, it falls at the extreme upper end of any distribution. The indications are, however, that the United States is, if anything, falling slightly short of the main trend; the deviation is, however, not strong enough for one to be confident of such an estimation.

This falling short may be a recent phenomenon, one that is increasing in a way that should cause concern. The time variation of the indexes of scientific research activity for the nations of the world shows that the relative growth rate varies inversely as the index of development. The most developed nations are growing much more slowly than the least developed ones and may, indeed, be declining in size. Although the scientific activity of the United States may still be increasing each year, the increase is much less than that of the rest of the world, which, after all, has an absolute majority over even the largest country. Thus, the share of the United States in the international knowledge pool is declining, perhaps quite rapidly. I lean towards the opinion that there is, indeed, a general saturation of development that is proceeding in the more developed nations, with the United States leading the field in this trend. Over and above this, however, there seems to be some local and idiosyncratic cutting back that has the obvious potential of being extremely damaging to the whole scientific and technological complex, not only that of the scientific research activity. For this reason, it becomes urgent that this type of indicator be studied further.

Since the United States and all other major countries have almost exactly the share of the world's scientific research (publishing) activity that one would expect in terms of their populations and states of general and economic development, it follows that the same should be true of any particular field of scientific activity. The data given in *SI-72* (Figure 4, pp. 8–9; Table 4, p. 104) seem to show that this is far from true. I do not much like the methodology of this study, particularly the manner of choosing source journals, and year-to-year results seem to show merely a scatter rather than any real time variations. Nevertheless, the deviations of most of the fields of science from any apportioning by the average shares of the nations involved seems very large and a little mysterious. The United States has normally about 30% (plus or minus 5%?) of the world scientific activity in terms of authors and of papers published. In psychology the share seems to be about 75%, relatively oversubscribed by a factor of 2.5 for some reason. Does this mean that the United States values psychology so much more than do other nations that it provides so much of the world pool of knowledge? I think the answer must be that we value some service activity associated with psychology more than others and my guess is that the activity employing these psychologists (who also do the research) is college teaching. Quite possibly, the United States, with its highly democratized college education system, has a greater need for teaching of this subject; an alternative explanation would be some large need for a service function of social workers who had been taught this field. In the subject of economics an opposite pattern of behavior appears. Here the United States contribution seems low (about 20%) and that of "other" countries, very high. My suggested explanation here is that only a small part of economic literature actually constitutes an international corpus of knowledge to which all nations are adding and on which all draw. A considerable part of economic journal publication must consist of essentially domestic information of much more interest to one's own nationals than to those of other lands. In any such fields related more to domestic than international interests, one must expect a world distribution that approximates more a "per country" or "per capita" basis than to the less even spread of international knowledge.

THE STRUCTURE OF SCIENCE

At this point we have a general conceptual basis for scientific manpower and its distribution and change with time in each country and in the world distribution. Manpower has already been linked to the production of basic science by studies of production distribution and by the transience/continuance flow that seems to underlie it. We can now go further and show that the scientific papers themselves, produced by such manpower, form a system having a highly visible structure and, indeed, one that appears to be highly deterministic in its development. The studies of Small and Griffith (24–25) show that by using the properties of co-citation, which clean up most of the random noise in a citation network, the universe of scientific papers exhibits a clustering structure. The papers are, indeed, distributed in a space of surprisingly small dimensionality—most of the behavior can be accounted for by no more than the two dimensions of a common map such as we use in geography. Furthermore, the spread of the clusters seems to correspond remarkably well to entities that we intuitively feel to be the basic subfields of which science is composed. With time the old clusters change little, and only slowly do new ones appear near the outer boundaries of the map.

I suggest that one might combine the features of the Griffith-Small mapping with that of the uniform birthrate that produces exponential growth of the international knowledge system. What emerges is a concept that might be called the Jigsaw Puzzle Model. Since scientists act as if they are discovering knowledge rather than generating it *sui generis,* we take a (probably infinite) map that is a picture map of the entire knowledge system, past and present, and divide it into individual puzzle pieces whose size and boundaries correspond to those of the atomic entity of the scientific paper. Pieces may vary in area with subject—for example, pieces in the continent of chemistry may be generally smaller than those in particle physics—and each worker has a fairly constant rate of laying down area of new pieces. The central parts of the puzzle began to be systematically pieced together in the seventeenth century, and we have by now a rather large assembled section of somewhat irregular outline. Perhaps it still contains a few holes and a few places in which pieces have been incorrectly joined, thereby holding up the solution.

At any given time we may take as axiomatic that all the loose pieces that are relatively easy to join to the main body have been so joined. There seem to be no worthwhile scientific contributions that are just lying there for the taking. All the time the world's research scientists are standing around picking up likely pieces, examining sections of the front, and trying to fit some new piece to the completed corpus. From time to time someone succeeds where all others concerned with that area have failed. At this point the puzzle is changed by the addition of the new piece; circumstances and patterns exist that did not before and therefore the newly located piece has a large chance of becoming the locus for further successful items of solution by others. The cooperative efforts therefore have some tendency to produce action wherever action has been and in general, over the average of the entire outer boundary, knowledge produces new knowledge at the constant *ca.* 7% per annum birthrate that characterizes the growth of systematic knowledge. Locally, there is clearly some tendency to fashion as the puzzle grows by pseudopods, having a high growth rate by virtue of strong emerging patterns that enable pieces to be laid down quickly. Locally, too, the team of workers occupied with such a pseudopodal area will constitute an invisible college, strongly relating to each other in the process of rapid advance, while exhibiting some detachment from the rest of the workers elsewhere in the puzzle. From time to time, a breakthrough will produce a new pseudopod, or two pseudopods will be joined to enclose an isolated hole that may rapidly be filled in to produce a major consolidating advance.

As another feature of the model, we may see overlapping discovery as the tendency for more than one of the puzzle solvers to put their hands on the next piece to be fitted; the statistical distribution of such overlappings shows the Poisson behavior that results from virtually independent random chances that a given piece will be picked up. Lastly, but not quite so confidently, I suggest that the effect of new craft of experimental science, new instruments and techniques, might be viewed as a process making visible whole new groups of detached pieces that, having been invisible before, were not available for fitting to the body of the puzzle. Each time I have tentatively suggested this model, people have been concerned with the question of whether the puzzle is in fact infinite in extent. Having no answer for this question, I prefer to adopt a policy of "wait and see" since it makes no difference to the mechanics of the model. Individual areas certainly get completed from time to time, and perhaps the pre-Maxwellian feeling that everything was done except to look for an extra decimal place is a consequence of this. I also feel certain reservations in accepting Kuhn's idea of some major "paradigm shift" as a major feature of the historical unfolding of the jigsaw puzzle solution. Certainly, there exist major completions of sections, major movements of interest when a strategically important pseudopod has been made to erupt, but I do not see that this is necessarily something identifiable as a "paradigm" rather than just another newly visible part of the general pattern.

AN INTEGRATED SYSTEM

With the conceptual basis now covering the universes of researchers and the papers they produce, we may turn to the last part of the system that links these with the economic data. For most of the world's countries, particularly for the United States, there is a large body of fiscal data of money budgeted and expended in various ways in connection with the research scientists and their output of papers. Unfortunately, for our purposes such fiscal accounting is almost always complicated by money spent on this primary product being inseparably conflated with that spent on the incidentally produced goods and services produced by the same people. Almost certainly, the most useful general concept here is that of the Cost-of-Research Index introduced by Milton and others (26–29). In principle the index is nothing more than a computation of the total research expenditure per person engaged in the research or per paper produced from it. Such a calculation is given, for example, in Table 27, p. 120 of *SI-72*, where the present cost per person is shown to vary from $5600 for mathematics and computer science to $22,600 for the biological sciences.

It should be possible to regularize this index by introducing suitable definitions and concepts and by studying why it has the variations it does by field, by country, and with time. Clearly, a great deal of variation results from the enormous systematic differences between fields and between countries in the categories of spending included in "research expenditure." In some areas it includes the salaries of all research workers; in others, only those of the juniors (i.e., graduate student research assistants); and in others, again, salaries are included for only technicians, secretaries and the like. Again, in some areas the costs of major items of equipment, installations, and subsidiary services, such as accelerators, radio telescopes, computers, data systems, libraries, and office buildings are included; while elsewhere some of these items might be part of other fiscal systems, such as those of higher education or civil service administration. For the United States, the data given by *SI-72* (Table 60, p. 135), show that almost exactly half of the research expenditure is in salaries and wages (the table is badly arranged, but apparently 20% goes for junior salaries and the wages of aides, and 30% goes for senior salaries) and the other half is for equipment, indirect costs and the like. Studies in the United Kingdom, Canada, and the Soviet Union agree in this splitting down the middle of the average research budget. We can guess, therefore, that to a first approximation research cost may be computed by doubling the average scientist's salary. Multiplication by the equivalent full-time number of research scientists enables us to calculate the research budget of an institute or of a nation.

In the context of my stated goal of a search for indicators that relate to others and follow simple and "explainable" laws, perhaps the most helpful data in this area have been provided by a study made by OECD (30). Their statistics indicate that for 20 of the more highly developed nations, there is a linear relation with little scatter giving the equation:

$$\frac{GERD}{GNP} = k \; \frac{R\&D \; Manpower}{Population}$$

where GERD is the gross expenditure of R&D and k has a value of about 4.3, so that an outlay of 1% of a nation's GNP on research and development corresponds to a deployment of 1%/4.3, or 23 R&D workers per 10,000 inhabitants. The fundamental equation may be rearranged into the form:

$$\frac{GERD}{R\&D \; Manpower} = k \; \frac{GNP}{Population}$$

which states that the average cost of R&D per worker is a fixed multiple (with these definitions, 4.3) times the average per capita income of the country. If the cost of all R&D is similar to that of basic research in having about half the outlay go into salaries, this principle states that the salaries of such workers around the world tend to be a fixed multiple of the average per capita income of the nations. It seems reasonable for an overall average, but the fiscal definitions are hard to interpret, so I find it difficult to get much further here. The less developed countries will probably show a higher parameter for this equation, since the per capita income of the total population is low compared with that of their middle-class professional urban population in which the researchers find their economic context.

A particular criticism of *SI-72* is that neither for the total "science and engineering" activity nor for the "basic scientific research," which is most adequately dealt with, are there sufficient data of expenditure and population to make calculations of this sort. Then again, the entire area of "applied research" has been consistently neglected so that, although it seems to be consuming much effort (we have 0.4% basic research, 0.6% applied research, and 1.7% development, as a fraction of GNP), one has no idea who is paying or what products, services, or publications are coming out. Table 1, p. 102, combined with Table 2, p. 103, gives perhaps the best data linking R&D expenditures with those of manpower. Unfortunately, the complete ranges of countries and of dates are not exactly the same in the two tables but pooling the information, we can compute the parameter k. Although it must be drawn from quite different definitions from that of the OECD data (it is of order of magnitude 10 rather than 4.3) we note that it has been falling steadily in all countries. In France and Germany it is perceptibly higher than in the United States; in Japan and in the USSR it is perceptibly lower. Being an absolute number, free of currency units, k is insensitive to forces of inflation so that the general fall in value and the difference from country to country probably reflect the appropriate levels of support in a real way. Again, there are not enough detailed facts to see exactly what is happening. The lack of correlation between Tables 1 and 2 indicates that they both have merely "Gee Whiz" value. Another small trouble that should perhaps be recorded to save others confusion is the fact that the important Table 14d, p. 100, in *SI-72* has been badly misprinted with columns mislabelled or out of order. The first two columns should be headed "Basic Research" in current and then in 1958 dollars; similarly; the second pair is "Applied," the third pair, "Development." Except for this table, there is no breakdown between the R costs and the D costs by source or by performer (let alone with the two taken together) so the data cannot be used to make estimates of the separate jobs of production, services, and the performance of research. It would be valuable to seek out such data in future annual analyses so that we may have a first approximation to a national balance sheet of our purchases and priorities and the ways that these measure up to our place in the international system. With clear and well-defined cost-of-research data, it should be possible to link all expenditures with numbers of people and, therefore, also with the numbers of research papers produced. We would then know the cost in manpower and money of the goods and services we purchase at various priorities for various purposes.

REFERENCES

1. H. Mineur, *Histoire de l'astronomie stellaire jusqu'a l' époque contemporaine*, Hermann & Cie, Paris, 1934.

2. H. Mineur, "Progrès économique et progrès astronomique," *L'Astronomie*, **63**, 294–300 (1949).

3. E. W. Hulme, *Statistical Bibliography in Relation to the Growth of Modern Civilization*, Butler and Tanner, London, 1923; and D. de S. Price, *Science Since Babylon*, Enlarged Ed. Yale University Press, New Haven, 1975, p. 170, n. 3.

4. J. D. Bernal, *The Social Function of Science*, Macmillan, New York, 1939.

5. D. de S. Price, "Quantitative Measures of the Development of Science," *Archives internationales d'histoire des sciences* **14**, 85–93 (1951); and *Actes du VI Congrès International d'Histoire des Sciences*, I, Amsterdam, Hermann & Cie., Paris, 1951. pp. 413–421.

6. D. de S. Price, *Science Since Babylon*, Enlarged Edition, Yale University Press, New Haven, 1975, Ch. 8.

7. G. N. Gilbert and S. Woolgar, "The Quantitative Study of Science: An Examination of the Literature," *Science Studies*, **4**, 279–294 (1974).

8. H. Krauch, "Priorities for Research and Technological Development," *Research Policy*, **1**, 28–39 (1971).

9. B. Barber, "Resistance by Scientists to Scientific Discovery," in B. Barber and W. Hirsch, Eds., *The Sociology of Science*, Free Press, New York, 1962, pp. 539–556.

10. R. K. Merton, *The Sociology of Science*, N. W. Storer, ed., University of Chicago Press, Chicago, 1973, ch. 13; N. W. Storer, *The Social System of Science*, Holt, Rinehart & Winston, New York, 1966.

11. D. de S. Price and S. Gürsey, "Studies in Scientometrics I: Transience and Continuance in Scientific Authorship" and "Studies in Scientometrics II: The Relation Between Source Author and Cited Author Populations," *International Forum on Information and Documentation*, Moscow **1**, 2, 17–24; **1**, 3, 19–22, both 1976.

12. National Science Foundation, *American Science Manpower, 1964, A Report of the National Register of Scientific and Technical Personnel*, National Science Foundation, Washington, D.C., 1966. Data calculated from Table 11, p. 60.

13. OECD, Directorate for Scientific Affairs, "Frascati Manual," *The Measurement of Scientific and Technical Activities*, OECD, Paris, 1970.

14. C. Freeman, "Research and Development: A Comparison between British and American Industry," *National Institute Economic Review*, No. 20, May 1962, 21–39, especially p. 30, Chart 2, and p. 38, Appendix Table 9.

15. Lord Rothschild, "The Organization and Management of Government R. & D.," *A Framework for Government Research and Development*, Her Majesty's Stationery Office, London, November 1971 (Cmnd. 4814).

16. R. K. Merton, *The Sociology of Science*, The University of Chicago Press, Chicago, 1973, Chs. 12–14.

17. OECD, "The Overall Level and Structure of R. & D. Efforts in OECD Member Countries," *International Statistical Year for Research and Development*, **1**, Paris, 1967, especially p. 33, Graph F. (Better data can be found in *UNESCO Statistical Yearbook 1970*, Table 3.6.)

18. D. de S. Price, "Citation Measures of Hard Science, Soft Science, Technology and Non-Science," in C. E. Nelson and D. K. Pollock, Eds., *Communication Among Scientists and Engineers*, Heath, Lexington, Mass., 1970. pp. 3–22.

19. R. A. Fairthorne, "Empirical Hyperbolic Distributions (Bradford-Zipf-Mandelbrot) for Bibliometric Description and Prediction," *Journal of Documentation*, **25**, 319–343 (1969).

20. D. de S. Price, "Measuring the Size of Science," *Proceedings of the Israel Academy of Sciences and Humanities*, **4**, No. 6, 98–111 (1969), and "Some Statistical Results for the

Numbers of Authors in the States of the United States and the Nations of the World" (with Suha Gürsey), preface to ISI's *Who is Publishing in Science 1975 Annual*, Philadelphia, PA, 26–34, 1975.

22. R. E. Lapp, *The Logarithmic Century*, Prentice Hall, Englewood Cliffs, N.J., 1973.

23. R. L. Merritt and S. Rokkan, Eds., *Comparing Nations*, Yale University Press, New Haven, 1966.

24. H. Small and B. C. Griffith, "The Structure of Scientific Literatures I: Identifying and Graphing Specialties," *Science Studies*, **4**, 17–40 (1974).

25. B. C. Griffith, H. G. Small, J. A. Stonehill, and S. Dey, "The Structure of Scientific Literatures II: Toward a Macro- and Microstructure for Science," *Science Studies*, **4**, 339–365 (1974).

26. H. S. Milton, "Cost-of-Research Index, 1920–1965," Combat Analysis Department Technical Paper RAC-TP-209, Research Analysis Corporation, March 1966.

27. H. S. Milton, "Cost-of-Research Index, 1920–1970," Economic, Political & Social Sciences Technical Paper TP-430, Research Analysis Corporation, July 1971.

28. A. V. Cohen and L. N. Ivins, "The Sophistication Factor in Science Expenditure," *Science Policy Studies* I, Department of Education and Science, London, Her Majesty's Stationery Office, 1967.

29. National Science Foundation, "A Price Index for Deflation of Academic R&D Expenditures," NSF 72-310, Government Printing Office, Washington, D.C. 1972.

30. OECD, "Patterns of Resources Devoted to Research and Experimental Development in the OECD Area 1963–1971," SPT (74)12, OECD, Paris, p. 4, Graph B.

Models of Scientific Output

5

Manfred Kochen

Scientific output is usually assessed by processes of peer review or managerial judgment. Because science indicators are expected to play a role similar to that of social indicators (1) in helping participants to anticipate the consequences of rapid technological change, a need arises to provide better information systems for participants in this early warning and planning process.

An important step in that direction was *Science Indicators 1972 (SI-72)* (2). It has been used in nearly every hearing involving appropriations for the National Science Foundation, in Congressional hearings, by the President of the United States, and by many authors and public speakers.

At national and lower levels, concerned observers are asking how well our science-technology system is performing and whether it is getting better or worse, how it arrived at its present condition, and what can be done to improve it if that is necessary. The technology for continuously monitoring the state of science may soon be available, and the evidence indicates that it will be used to set up sophisticated information systems. Therefore it seems urgent to develop a rationale on which to base the design and use of such systems.

Wise planning of science and technology (S&T) requires some understanding of S&T processes and choosing among various points of view and values. Emphasis on quantitative indicators seems to proceed from a presumption of a completely objective rational basis for S&T planning and an assumption that a mechanistically based science of science is not only necessary but also sufficient. This assumption should be examined critically, as should other conceptual schemes for measuring scientific output.

The infant science of science (see, for examples, 3–7) cannot yet provide an adequate rationale for choice of science indicators, or even yet be judged as an aid. Still, the use of such indicators, even crude ones, may help to advance theories of science planning; at the least, they may increase the level of responsibility in criticism of science. The art of constructing science indicators, and the science of S&T and S&T planning shape each other as both become more sophisticated, and thus could even affect the growth of science itself. For, as Kenneth Boulding has observed, "Increasingly the scientist is creating the universe which he studies" (8).

As a step toward clarifying what constitutes a better information system to help in planning science, I attempt in this paper:

1. To clarify three conceptual issues central to evaluating such information systems.
2. To compare five models for S&T and S&T planning.
3. To evaluate various output measures for the five models.
4. To suggest priorities for further research in this area and for constructing new science indicators.

Some of the ideas presented here—particularly four of the models of science planning—are drawn from a paper by Harvey Brooks (9). The discussion of science indicators draws on the literature of economic and social indicators as well as on discussions with such authorities as A. Campbell and S. Seashore. A view of science as a humanistic learning process is proposed, with a suggestion that a theory explaining the growth of knowledge may emerge from that process.

CONCEPTUAL ISSUES IN PLANNING AND OUTPUT JUDGMENT

Concept of Science in S&T Planning

The movement to assess the state of science and to attempt to direct its growth with the help of indicators seems to be based on a mechanistic concept of science and S&T planning. Thus, the search for objective yardsticks by which to measure the performance of a given scientific research organization seems guided by extrapolation from situations where prediction is assumed to be possible through application of general laws—for example, in judging whether a chemical process is proceeding in the predicted direction. But we must distinguish between "the body of knowledge" represented by the above and "the image of knowledge" (10) which comprises sources, limits, and legitimizations of different kinds of knowledge. The images of knowledge need not correspond to the search for either paradigmatic cases or general laws.

The mechanistic conception is exemplified by a model making an analog of the growth of a scientific specialty as the spread of an infection and using concepts and techniques of mathematical epidemiology (11). Here the key variable is $A(t)$, the number of active contributors to a specialty, operationalized by the number of authors who have contributed at least one article listed in an authoritative bibliography of that specialty up to and including year t. The increase (rate of change) in the number of such active contributors is assumed to be proportional both to their number ("infectives") and to the number of potential contributors (those whose papers appear in the bibliography after year t, the "susceptibles"), less the rate at which active contributors withdraw.* There is thus a nonlinear differential equation in $A(t)$. Similar differential equations can be written for the rate of change in the number of susceptibles and removals. Even if this system of nonlinear equations cannot be solved, conditions for stability and times of reversal of epidemic states can be derived, and these have been fitted to data with good accuracy. Such a system of equations can be interpreted to describe a mechanistic dynamic system.

When the search for science indicators is viewed as more nearly similar to the search for economic or social indicators than to predictors of the course of a mechanistic process or system, the limitations of the mechanistic conception of science become readily apparent. For if a high-level decision maker were to use a mechanistic model to solve a complex problem—such as how much to invest in safety systems for a nuclear reactor—he might rephrase the problem, perhaps in terms of the values of lives lost or damaged by a very improbable nuclear catastrophe, and thus calculate the level of safety-equipment expenditure that optimizes cost-benefit ratio, subject to important constraints. But his results would be limited by the fact that the value of a life, as such, is not amenable to the methods of scientific inquiry as is the value of an energy source.

*A somewhat better explanation of the decline of a discipline is achieved by treating the susceptibles as analogous to subproblems rather than to potential contributors (12).

In conceptualizing how decisions which involve what has been termed a "trans-scientific question" (13) are reached, one must be able to take into account a "priceless good" such as "public confidence that the quality of individual life stands foremost as an aim of government" (14). As the word *priceless* suggests, quantitative indicators for this aspect of source output are hardly appropriate. And in another formulation of the nuclear reactor-design problem (15)—"Will the emergency core-cooling system work with a sufficient degree of reliability to provide reasonable and sufficient protection for the public?" —resolution may require legal or political techniques carrying the problem beyond science.

A new humanistic conception of science has been proposed (16) to deal with the problem of science as an activity of fully developed persons. It subsumes the mechanistic conception of science, going beyond that by admitting the investigation of idiographic instances (individual events, persons, states of consciousness) that happen only once and are not interchangeable. Hans Kammler (17) has shown that Maslow's investigation of unique instances demands terms referring to classes. It embraces the holistic approach, which would treat a person as one unit not analyzable by reductionist methods; it allows subjective reports as evidence; and it is vitally concerned with the relation between the knower and the known. In science, as in the act of discovery by an individual, a single critical event often triggers a major advance or insight. This is an unique event, not subject to investigation by the dominant contemporary techniques of experimental psychology. Explanations of such processes of discovery and invention have been attempted (18–20). But much more must be done before the potential of this seemingly promising area of investigation can be assessed (21).

Contrast this with the mechanistic conception reflected in the following quotation:"Physics has had enough of psyche, whereas psychology can never have enough of physics." Persons and societies are regarded as complex systems of a basically physical kind, a concept encouraging the hope that eventually quantum chemistry may explain some brain functions and even also enlighten humanity as to the nature of interpersonal relations. Psychology is irrelevant for explaining the behavior of atoms. "The role of the observer is to observe—and to plan and interpret observations—not to become the subject of physical theory" (22).

Some historians and humanists are also questioning the fruitfulness, even the possibility, of a scientific approach to science planning—of a science of science such as has been proposed (3, 5, 6, 23). They argue that generalizations based on the history of science break down when tested in individual cases examined in sufficient detail. Others feel that humanistic explanations of human events and activities are far more powerful than those furnished by much of science; and science planning, certainly a human activity, should thus be explained humanistically.

Whether science planning is to be based on a mechanistic concept, on traditional legal procedures and points of view, or on a new humanistic concept of science is an important issue. Our image of knowledge, and of science planning in particular, is a kind of theoretical knowledge. Theoretical knowledge, it has been suggested, is a key characteristic of a postindustrial society (one concerned with services rather than manufacturing or primary industries,—e.g., the United States) (24).Indeed, the idea that we can plan, the consciousness that we can bring knowledge to bear on shaping our future, is a major step in cultural evolution.

However, the choice of intellectual technology (e.g., linear programming) used to help in science planning depends on the underlying concept of science. Intellectual technologies may serve to increase the security and authority of the technostructure, which dominates much of the world where S&T is done; planning is directed toward the interests of the technostructure rather than the goals of the consumer or the public user, argues Galbraith (25). Planning has replaced the market, probably permanently. The very word *technostructure* suggests the extent to which the mechanistic concept, from which the present technology has emerged, pervades the contemporary establishment of science and S&T planning. Roszak (26) says it this way: . . .

> to the extent that their [scientists'] conception of what science is prevents them from seeking to join knowledge to wisdom, they are confessing that science is not gnosis [real knowledge that integrates, humanizes, supplies meaning, from which science emerged and divorced itself], but something far less. And to that extent they forefeit—deservedly—the trust and allegiance of their society.

It may be dangerous to proceed with a mechanistic image of knowledge. Considerable critical discussion should be devoted to this issue—and it must be done before committing major resources to information systems for science planning based on a mechanistic concept.

Judgment of Scientific Output

The output of scientific institutions is often judged for purposes of planning. But if information about scientific output is to enter the planning process, the question of who provides it becomes critical. For success in planning depends a great deal on who does it. Science has reached some of its peaks when individual scientists have been able to plan their own work freely or pursue their investigations without conscious plans (an arrangement that is increasingly rare).

Who Judges   (see also Ezrahi, this volume). Until the past decade it was generally assumed that only other scientists were qualified to pass judgment on the work done by their peers, both individually and in aggregate. Yet, the value of the output of organizations engaged in S&T has for some time also been judged implicitly when their supporters compare the output and needs of organizations competing for the same resources. Even when scientists advised those responsible for making plans and policies affecting science, the advice had to be based on conclusions that could not be derived by scientific methods alone; thus, the performance and needs of S&T organizations were in effect evaluated, like those of U.S. military organizations, by nonprofessionals. A further judgment factor comes from members of the general public who are (rightly) concerned with whether we are spending enough, too little, or too much on science by comparison with other societal goals and with whether our S&T institutions are organized appropriately for current and future challenges. Weinberg (27) approaches this problem by tying the level of decision making—and thus the judges of scientific output—to the level of the institutions responsible for social goals; thus, decisions affecting the improvement of health care, for example, would originate closer to those who would carry them out than would decisions involving the deeper, broader, or purer aspects of science.

For What Purposes?   The performance of an individual in an S&T organization is usually evaluated by a director, department head, or other organizational leader. In nonauthoritarian organizations the evaluator uses inputs from peer-review groups. He may also take into account such measurements as the number of publications, honors, and prizes and demand for the individual's papers and talents. A matter of considerable concern is the search for indicators of researchers' potential promise, to be used for recruiting, hiring, and awarding of special stipends. Measurements of quality have been used at various levels of aggregation, ranging in size from the output of individuals to the kind of global dimensions seen in *SI-72* and *SI-74*.

At the level of a research department, which is often associated with a university doctoral program, quality ratings have been developed for evaluating departmental graduate faculty and effectiveness of the doctoral program (28). Such measurements as number of refereed or reviewed publications, citations by others, invited talks at national or international congresses, monetary amounts of research grants, and membership in the National Academy of Sciences have been attempted; and the number of Ph.D.'s produced who have published at least five papers, have given invited addresses, and are in leading institutions have also been attempted. In mathematics, the most recent indicator of the quality of doctoral programs is the number of Sloan Fellowships awarded to graduates of a program (29). In the 1960s this was a good (if lagging) indicator of productivity and eminence in mathematics (correlated with winners of Fields Medals), but its usefulness may decline in the future with shifts of emphasis toward applied mathematics—a phenomenon that may also be illustrative for other fields and for higher levels of aggregation.

Such indicators may play a part in deciding how much to support some scientists in relation to others in an organization. The performance of organizations and of units within organizations is also generally judged by managers at higher levels for purposes of relative funding. Unfortunately, there is no assurance that performance evaluations are used to identify groups showing the most promise, so that special attention can be paid to their results.

Two main weaknesses of the current procedures are their dependence on talent in a managerial hierarchy and their almost exclusive reliance on budget allocation as a management and planning tool. Peer review works well for relatively autonomous, mature organizations whose scientists, no longer in the early stages, strive primarily for good conceptual schemes; strong managerial ability and judgment seem to work better for mission-oriented institutions. The effectiveness of peer review depends critically upon the procedure for choosing peers. In the development of the managerial hierarchy in a given S&T organization and of the criteria for evaluating such managerial abilities, political processes provide the determinant. These hierarchies are often less mature and effective than their non-S&T counterparts because S&T is probably more difficult to plan and manage than other activities for which executives and managers are trained (usually in schools of business administration); non-S&T organizations, centuries old, have profited by the possibility of codifying much more of the managerial know-how and experience; and the supply of talented and experienced S&T managers is probably smaller than the demand—though that imbalance may now be reversed.

Thus arises the desire for an algorithm to allocate resources and support on some objective basis, using the values of quantitative indicators as input, and so apparently to reduce the critical role of judges of performance. However, because the assessment of quality probably cannot be completely objectified, much of it may have to remain subjective. Even in business, the problem of investment and resource allocation cannot be isolated from the total problem of running a firm; no set of quantitative indicators is by itself sufficient for decision, although that is not to say that such indicators are not necessary or helpful.

If measurements or evaluations of scientific output are to be used for planning, who controls the indicators and who uses them in evaluation may matter more than how the measurements are made. Planning is largely a political process; it works best with the participation of those affected. But we cannot assume that those responsible for or contributing to the maintenance of an indicator will not be self-serving—for example, that a government agency responsible for an indicator referring to it, to a larger organization of which it is a part, or to a part of itself, would willingly supply data that make it appear incompetent or unworthy. The wisdom of disqualifying institutions for the responsibility of judging their own output is clear.

Application of Output Judgments

Measurements of scientific output have been applied to specialties nationally, such as mathematics in the United States; to specialties within institutions, such as mathematics in the Mathematics Department of the University of X; to individuals, such as Lord Kelvin and all his works; and to institutions, such as the IBM Research Center. The importance of this issue is that the units to which budgetary allocations are made are the *loci* of power in the S&T establishment. It is they who control the reward system, and any policy must be executed and enforced and its results interpreted through them.

In universities, for example, the basic units are primarily the discipline-oriented departments, where much power is concentrated. Professors employed by these departments make their living by teaching but make their reputations by research. The prevalent mode of research is still to select a problem which fellow specialists recognize as important—a selection usually stimulated by an article in a specialized journal, lecture, or personal conversation. The solution of the chosen problem is directed back to the specialized journal or the fellow specialist; graduate students are trained to follow the example of their mentors. All this activity encourages extreme specialization and fragmentation and fails to foster attitudes of sharing. Shifts toward different organizational units, and structures occur that, in time, erode the common good.

For this paper, however, it seems best to focus on types of S&T institutions, the most natural categories for output measurement. Research institutes less than 10 years old, industrial research laboratories, government laboratories, and independent intellectual centers should probably be the *foci* of science planning and of the study of its growth. Much smaller levels of aggregation are of value for planning only if they are deviant exceptions to be flagged; much larger aggregations do not appear to be emerging as viable budgetary units.

MODELS OF SCIENCE FOR PLANNING AND STUDY OF GROWTH

Types of S&T Institutions: Independent Variables

Institutions engaged in S&T may be classified along numerous dimensions. These include diversity. Major innovations seem to occur when scholars are recruited for their individual promise, by (or near) a large diversified center (30–31). Giving freedom and autonomy to younger people also seems more important than stardom at the top for high-quality output and productivity. Three main dimensions are therefore considered most relevant to information systems for science planners.

Size.   Assuming that size is correlated with diversity, the most obvious way to measure size is by relative annual budget or man-years. Small laboratories are typically organized around a principal investigator in a university, supported by research assistants, graduate students, and technicians. They are often engaged in trailblazing, in extending the frontiers of a specialty, and creating new conceptual schemes.

The annual budgets of large organizations are usually of an order of magnitude exceeding 100 man-years. They may differ from small organizations in the ratio of support personnel to senior staff, in overhead rates that reflect large equipment installations, and in time horizons. They naturally work on projects requiring large teams and resources.

Programmaticity.   Some institutions are highly programmatic, having clearly defined goals and matching strategies, and organizational structures and managerial procedures and abilities that reflect them. Managers tend to run relatively "tight shops" and to allow relatively little freedom to subordinates or autonomy to subunits.

At the other end of the scale are relatively unstructured institutions with good scientists who, though also highly disciplined, enjoy considerable opportunity for adaptability; they can shift direction to pursue whatever leads appear promising, even if this diverts them from the chosen direction. They are recruited primarily for their individual promise and only secondarily to match a job description.

Highly programmatic institutions tend to structure programs first and then select people who fit them. Thus, the quality of their output is more sensitive to the quality of the programs and the abilities of those who plan and manage them than to those who carry out the programs. Their work is more likely to be safe, sound, and mediocre than risky, bold, and great.

Internal/External Dimension.   This may be measured by the ratio between the extent to which an institution exploits—that is, uses without improving—science and the extent to which it adds to science. (These terms are intended to apply to research strategy and behavior rather than to motivation). Whether a field is "ripe for exploration" and whether its practitioners prove sufficiently responsible and competent are internal criteria. A scientist may decide which experiments to do next according to the logic of his path of inquiry, constrained and guided only by the basic concepts, methods, and standards of his discipline; or instead, he may be responsive to the interests and needs of those paying his salary or to the potential beneficiaries of his work. Internal scientific factors such as the available concepts, techniques, and technologies constrain the scientist's freedom to choose problems (wisely). Internal criteria for significance also include the illumination that certain results provide for a variety of other scientific problems—which, if solved, would in turn remove barriers to solving further problems and lead to increasing codified understanding.

The internal/external dichotomy has been discussed extensively in the literature and has been criticized by Elkana (32). Institutions with directions influenced primarily by internal factors tend to do pure science; they are programmatic and tend to explore systematically and chart a disciplinary specialty. Institutions with directions more responsive to external conditions tend to do applied research and development—on the technology side of the science-technology couple. Whether a proposed experiment is likely to further technology or some goal such as improved health care or national prestige is an external criterion. Clearly, when support for S&T becomes increasingly dependent on a larger public, a shift toward the external criteria seems likely (see Ezrahi, this volume).

Individuals ranked toward the internal pole of this scale tend to have considerable autonomy in choosing and solving problems. Their choices are guided by internally generated ideas rather than by external stimuli, demands, or data, and they tend to be more intuitive and insightful than those at the opposite end of the scale.

To operationalize the programmaticity dimension or scale, we could measure the fraction of time individual scientists in an organization spend on problems of their own choice, those falling outside any program, and then aggregate this over the entire organization. Another measure of this dimension could be the fraction of scientists hired, not because their qualifications match the job descriptions of a program but primarily because of their promise as creative individuals.

A possible measure of the internal/external dimension could be the fraction of a scientist's time charged to the account of some client or sponsor who expects a specified return on his investment, aggregated over all scientists in the institution. Another measure may be the extent to which the institution must do what it is doing in order to continue receiving the support of clients or sponsors interested in results rather than the promise represented by creative individuals.

The cube in Figure 1 spans the three scales mentioned above, showing the types of institutions occurring at the corners. However, the three "dimensions" are not exactly independent variables; size and programmaticity may be correlated; nonprogrammatic groups driven primarily by internal factors tend also to be autonomous.

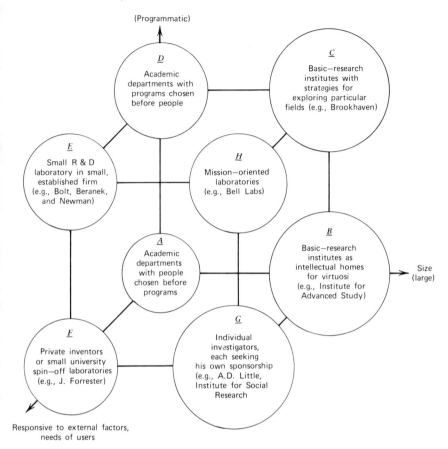

Figure 1. Some extreme instances of institutions (by size, programmaticity, and internal/external dimension).

Institutions engaged in S&T are, of course, distributed throughout the cube. The distribution is probably shifting so as to become more concentrated in the large, programmatic, externally responsive region near corner *H*. In the interior of the cube is the kind of institution (e.g., American Cyanamid) that hires scientists in areas likely to yield new processes or products for some client; it encourages these scientists to move with their ideas right from their inception in basic science through development to marketing, use, and providing services. Also in the interior of the cube are specialized institutions within large high-technology corporations that form links in the hoped-for chain of transfers from basic research to the development, engineering, manufacture, marketing, and servicing of a product.

Questions about the distribution of national science resources—whether it is good enough, or how far it is from variously defined optima—could be posed by imagining it also distributed over this cube.

Many other independent variables could, of course, be considered in clarifying the effect of an information system on science planning. Also, the circles at the corners of the cube in Figure 1 are not exactly opposites in all specific cases; most academic departments choose faculty *both* for their individual promise and their fit into a program and expect mutual shaping and enrichment of the program and their personal growth. Institutions that fit at one point of the cube are embedded inside institutions that fit elsewhere.

This entire approach to models for science planning has a social-science flavor and is not entirely consonant with the warning above about accepting a mechanistic image of knowledge and its growth. Nevertheless, this is a useful vehicle for specifying the kind of image which might supplant or complement the one now most applicable and in use. It also clarifies the implications for measuring output.

Types of Research, Development, and Utilization in S&T: Dependent Variables

The purposes of the S&T planner are related to the purposes of the institutions doing the research in S&T and the effects of the research. If the effects of work done in some institutions are far-reaching in the sense that they profoundly affect many people over a long period of time, planning such work should have a correspondingly long time-horizon and be done at a high enough level to ensure the purpose. The comparison of different conceptual models for measuring scientific output should take into account not only types of organizations classified along the three dimensions described in the preceding section but also, as two more dimensions, aspects of the purposes of those organizations interpreted as dependent variables.

Considered first is "consumer-reach" of the organization—defined as the number of people to be affected by its output, weighed by the influence each science-user exerts and by the depth of the effect on him. The concepts of relativity theory had a profound effect on physicists, whose own ideas also had fairly profound effects and a philosophical impact on vast numbers of people in walks of life far removed from science and scientists. The creation of conceptualizations, world-images, and discoveries with such far-reaching effects can hardly be predicted, much less planned. Probably no indicators exist that would have enabled us in 1900 to predict Einstein's 1905 accomplishments, or even to pinpoint Einstein, without a very high rate of false positives.

The notion that parity conservation might not hold was also profound. But it affected only the thinking of physicists rather than that of all scientists, and the thinking of the general public not at all. Nonconservation of parity is perhaps a slightly more predictable discovery in the short run than relativity, although it probably could not have been planned either.(Knowledge that some of the principals involved in the former discovery were raised in a different culture might prove to be useful in making such predictions (33). Technological development, manufacturing, and marketing are obviously far more amenable to planning; but anyone with experience in creating market plans in the United States will testify to how far even these are from being an applied science.

A control variable that can be termed "degree of controllability" is closely related to consumer-reach. Controllability ranges from zero—where output cannot even be predicted—through various degrees of predictability in which output can be nurtured, to work that can be programmed and even planned. Research on the cure for a disease that afflicts only a dozen people, where the research may or may not shed light on more fundamental aspects of disease processes (with associated possibilities for transferring ideas, techniques, or results to the cure or prevention of other diseases), is at the lower end of the consumer-reach scale. In the mid-range is the work of the scientific assault troops who advance the frontiers of a specialty; control and management of their activity can probably be nurtured or developed at a natural growth rate. The outcome of such work is predictable but probably not subject to planning either. The work of the scientific equivalent of military policemen or mop-up troops, those who explore a paradigm or work out special cases of general laws in the region behind the frontier of a specialty, is perhaps not only more predictable but also more amenable to planning and programming.

Five Models of Science Planning

Science as an Autonomous Self-Regulating Mechanism. In this view, the image of science emerges as either a "delicate plant" (34) or a"vigorous weed" (9, p.48), with S&T seen as growing according to autonomous laws like those governing large and complex organisms. The study of multiple independent discoveries (35) supports the view that science is governed by internal forces more than by policies responsive to external conditions. When technologies, conceptualizations, or methods necessary for certain discoveries become sufficiently advanced, someone will use them to make the expected discoveries. Thus the discoveries of Planck, Crick and Watson, and others, occurred because science has reached the appropriate stage, with individual scientists seen as the substitutable discoverers; if they had not been there, someone else would have made these discoveries around the same time.

Plans that do not harmonize with these autonomous laws are more likely to harm the growth of S&T than to help it; at best, they have no effect. The most that planners or scientists interested in discovering the laws of this growth can hope to do is to predict. If they can then recommend information systems based on an understanding of these laws, that would perhaps lead to an acceleration of the unfolding of S&T. The primary goal of science, according to this view, is the creation of conceptual models that enlighten. Technology is set apart from science by the fact that the former lends itself more readily to planning than the latter.

Mechanistic models, such as the one (described previously) for explaining growth in the number of active contributors to a specialty, fit with this view. Typically, a specialty is initiated by the creator of a new paradigm or the discoverer of a trail-blazing finding—for example, Newton, Darwin, Einstein—through the vehicle of a seminal paper, which then both stimulates further work and makes it feasible. The conceptual schemes thus created and developed are often far-reaching. The models of Goffman (11) or Nowakowska (12) do not necessarily capture the full essence of this picture of sequential development from a discovery, but their image of science as a self-regulating autonomous organism is nevertheless limited by its exclusion of relations to other types of knowledge, culture, and wisdom and thus is still no more humanistic (as opposed to mechanistic) than currently prevalent images of biology.

Science as a Supporting Service (27).   This model too fits the more mechanistic conception of science by being based on the assumption that socially desirable goals can be attained, with science as a means; S&T is thus justified to the degree that it contributes to the attainment of social goals (see Ezrahi, this volume). Study of basic problems of fertility is now supported primarily because it is believed necessary for population control and, only secondarily, because it is possible and interesting. Mathematics is supported in accordance with how much it is believed to have enriched other branches of science and to have offered many diverse areas of application in the past. Health care delivery is viewed as a science-based social industry (36–37). Challoner (38) has argued that "given the continuation of the current approach toward the managing and funding across the spectrum of disciplines basic to health, the country is headed for stultification and possible social disaster."

Institutions responsible for effecting change in the state of some portion of the world—for example, reducing the level of phosphorus pollutants in Lake Erie to a specified level by 1980—allocate some of their resources for the underlying basic research. But a large corporation with a project in this area may be concerned primarily with the cost to its stockholders or with its competitive position in the industry. Basic research so sponsored is often hastily executed, not generally applicable, and frequently "semi-duplicated" as a repetition of similar work. A typical instance of impact-inspired industrial research is seen at IBM, which has several operating divisions, each with an attached research and development laboratory responsible for providing scientific and technological support for the division's products and operations. Each laboratory is funded by a fixed proportion of the division's overhead. In addition, a separately funded corporate research center ensures a strong technological base for the company. But budgetary arrangements tend to further hostility and rivalry between the research center and divisional laboratories, and the two problems considered to be industry's most difficult—transferring research outputs to where they contribute to the marketing goals of a corporation and stimulating central research laboratories to be more responsive to the needs of the operating divisions—(39) remain unsolved. Once again, we see that attempts to delineate, restrict, and confine the image of science may distort it to the point where unplanned science is preferable to such counterproductive planning.

The Socialist View.   Marx has been interpreted as emphasizing that the interrelatedness of various strands of social behavior, historically generated by needs, results in a total system of relations not admitting of fragmentation. His dialectic may thus be seen as humanistic. Lenin said, ". . . the teaching of Marx is all-powerful because it is true. It is complete and symmetrical, offering an integrated view of the world. . . .It is the legitimate inheritor of the best that humanity created in the 19th Century in the form of German philosophy, English political economy, French socialism. . ." (40).

The S&T policies recently in force in the USSR seem to have been shifting in the same direction as those in the United States (41). Both bureaucracies confront the same dilemma: One horn is a laissez-faire relation between science and government, at the cost of an efficient chain of transfer from science to consumer; the other horn is increasing the intervention of government and the technostructure into selection, training, and employment of scientists and the imposition on science of ideologies or conceptualizations at the cost of stifling innovation and creativity.

Societal needs, political processes, and S&T become increasingly interrelated as S&T grows and becomes more visible to a public that may be increasingly concerned about its impact. But if science is to be viewed as essentially free inquiry, no nonscientific ideology or conceptualization can be forced upon it without destroying its essential character. Thus, a science that helped (willingly or not) to create a society like the one described in George Orwell's novel *1984* would be contributing to its own destruction.

The Marxist model, with its stress on economic and technological determinants of S&T can lead to excesses that fetter the spirit of inquiry characteristic of science, whether mechanistic or humanistic; in conspicuous contrast is the view of science as a social order "in which investigation, criticism, intellectual revolution are *always* possible and *always* welcome" (42). Little reliable information about the Chinese version of a Marxist state is available, but science seems to be required to serve the interests of the state first; one direction of emphasis in technology can be deduced from the rapid development and recent testing of H-bombs.

Science as a Focus of Postindustrial Society (43). From this viewpoint science is seen not in terms of its products or outputs but as a *process*, a new nonproductive tertiary industry. The view of science as an autonomous self-regulating system was intended, in part, to counter the Marxist view. Similarly, the view of science as a supporting service or overhead was countered by the view of science as an essential part of culture or high civilization; the latter view implied that society should support an elite of S&T special talent, despite the lack of immediate or apparent economic function. Toulmin suggested adding to the classical economic functions of production and consumption a third basic notion—*employment*—which would encompass activity in sectors such as radioastronomy, archeology, and pure mathematics. Thus the creation of theoretical knowledge, the signpost of post-industrial society, would be included among the society's economic functions. According to this view, scientists are resources with potentials that must be realized as fully as possible—including those stemming from the increased leisure technology has brought to post-industrial societies, giving rise to spiritual and intellectual demands and new values.

One way of conceptualizing science as a process is as follows (23): At any given time the corpus of facts, findings, and techniques, representing what is known to be known in a specialized field of knowledge, is part of a larger set of recorded propositions that represent the outstanding questions or the body of what is known to be unknown in that special field. Both what is known and what is known to be unknown characterize, in part, the specialized field of knowledge. Both are limited by the predicates and the conceptual vocabulary specified by a paradigm that governs the field, if it is a paradigmatic field. The process by which knowledge grows is assumed to be one in which an investigator selects a problem in the region of outstanding problems and then, in solving it, extends the frontier of what is known to be known to include what was previously an unanswered question. In so doing, however, he may also extend the frontier of the *set* of these outstanding questions; for each problem he solves, he or someone else may now think of one or more new questions that no one before had thought of asking or had been able to ask. The key criterion for advance, according to this concept, is the rate at which new and good questions are posed, rather than the rate at which established questions are settled. Support for research, according to this view, is justified in accord with the degree of uncertainty removed and the extent and quality of enlightenment produced.

Of course, good work may increase rather than decrease uncertainty. Research may be convergent or divergent. In convergent research an investigator recognizes a problem, creates a well-defined problem statement, analyzes it into solvable sub-problem statements, and creates a solution. In divergent research, an unsolvable problem is replaced by several unsolvable ones. The very features that ensure its stability may divorce the scientific community from the rest of society and threaten its long-term survival, unless the society at large learns to value astronomy, pure mathematics, and so on as part of general culture.

Science as Learning.   This view stresses the benefit to future generations. It resembles the model just outlined, except that it has an unlimited time horizon. It stresses the rate of learning rather than learning designed for the achievement of some objective. Investment in S&T is justified by the extent to which it contributes to the rate of learning.

Science differs from other forms of higher learning by offering a distinctive proven path to valid knowledge. Replicability of experimental findings, amenability to codification into abstract generalizations, unity in diversity, logical consistency of theories, and communicability of valid knowledge and codified understanding are some of the criteria at the heart of science. Can a humanistic extension of science conform to these criteria? Can the acquisition of knowledge of a unique whole person be replicated and can the result even be shared? At first glance, it hardly seems likely; if we abandon the criteria of science, we should perhaps no longer call what is left "science."

Yet, the following remarks of Rabi and Einstein support the view that even if our concept of science cannot be revised in the direction of humanism, it must be fused with a humanistic view of social change.

What the world needs is a fusion of the sciences and the humanities. The humanities express the symbolic, poetic and prophetic qualities of the human spirit. Without them we would not be conscious of our history; we would lose our aspirations and the graces of expression that move men's hearts. The sciences express the creative urge in man to construct a universe which is comprehensible in terms of the human intellect. Without them, mankind would find itself bewildered in a world of natural forces beyond comprehension, victims of ignorance, superstition and fear.

I. I. Rabi

Real human progress depends not so much on inventive ingenuity as on conscience.

A. Einstein (44)

Common to these several views is the concept of science as a system of inquiry (45). Tasks and opportunities are not only certain kinds of states of the external environment, they are also special states of internal representations as well—for instance, contradictions and gaps in knowledge. Science, the process, is an outstanding example of learning—which may be viewed as the formation, revision, and use of representations and systems of representation for recognizing and coping with an increasing variety of opportunities and traps (21). We may be creating needs, even dependencies, with our scientific progress; certainly, new relations are added. Investment in S&T is justified by the rate at which we learn to recognize and cope with opportunities and traps and to readjust our representations and systems of representation to comply with changing external demands. One hopes that this will occur at least as speedily às the problems that threaten life or quality of life are generated.

The distinction between opportunities and traps depends critically on a system of values, and these may themselves be the result of natural, evolutionary processes (46). If this is so, value concepts become objects of scientific study in themselves—as do states of consciousness, the bases for transculturation,* and unique persons. Western cultural evolution has now probably reached the stage of awareness of its image of knowledge and the import thereof and a consciousness of the potential for shaping it—and, thus, for planning.

Overall Consideration of the Five Views.   Each of the five views has validity. They diverge when applied to the day-to-day behavior of S&T workers and planners. If a scientist requests funding from a large segment of the public or its representatives, he is less likely to succeed with arguments based in a view of science as high civilization than in a view of science as a service. An argument from science as an autonomous living system hardly abets requests for public support, for why should the taxpayer shift funds from areas of immediate concern to an activity that will grow regardless of his support? The taxpayer may trust in the judgment of the expert and defer to his advice, but if too much time passes without output that seems to the taxpayer to justify his trust, it is likely to decrease. If liberal education results in a new generation of rebels who seem to add more to social costs than to social values—as judged by the proponents of liberal education—then this

---

*This means recognizing and articulating the purposeful explanations of the world—whether in terms of magic, religion, or practical wisdom—that every culture is hypothesized to have; creating analogies of these with Western science; and analyzing problems indigenous to Western culture (32). Elkana confined this to explanations of the physical world. But it would seem important to include the human world as well, particularly if we are to fulfill the promise of his approach for steering between the West's Scylla (decline amidst shared starvation and failure) and Charybdis (destruction by the revenge of S&T receiver-countries, followed by their destruction of one another).

Table 1 Science as Service: Research, Development, Testing, Evaluation, and Utilization

Kind of Scientific Activity	Quality	
	High, Good	Lower, Poorer, Less Mature
"Basic" (change internal representation of world or state of science)	Progress toward problems of increasing depth (47); illuminates and deepens understanding of neighboring and embedding fields	Merely accumulates observations; over-specializes; chooses unrealistic problems; produces divergent research
"Applied" (change state of external world to cope with recognized problem)	Investigates the fundamental problems thoroughly; selects solvable problems of social value and uses relevant science validly; likely to effect change for the better	"Quick and dirty" approaches to impress; crash programs "covering the water front"—may effect change for the worse

support will decline. Science as a service is now realistic, moved by social as well as realistic factors.

The relation of science to society is somewhat parallel to the relation of mathematics to the sciences: Is science queen or servant? If the question is still meaningful, the obvious answer is "both." Queens in ant colonies, like queens in empires, both rule and serve; they cannot rule long without serving, and they cannot serve in the needed role of leadership without ruling.

The issue is not whether science rules or serves, is "on top or on tap," is adding to or effecting needed changes in the world—but whether it does well in any of these roles and functions. Some of the criteria for determining this are sketched in Table 1.

Appropriateness of Models to Various Types of Institutions

The judgment of how appropriate a model is can be thought of as an intervening variable. The number of people affected by science is a dependent variable to be optimized. Different kinds of institutions have impact on different segments of the population, present and future, in diverse ways but with considerable overlap. If resources are allocated to an institution according to a model judged to be appropriate to that institution—for example, "science as a service" —this affects reach or impact; choice of another model would affect reach or impact differently.

The choice of model also affects controllability, another dependent variable. The judgments about both these dependent variables (shown in Table 2) are based on what is judged to be the most appropriate model. The degree to which the performance of an institution in a category could be directed and planned by managers and the extent to which it can only be predicted by investigators who are content merely to study S&T institutions have been combined. The alleged unpredictability of some S&T has been used to argue for its autonomy—for example, perhaps maser technology could not have been forecast from research on the microwave spectrum of ammonia, but conceivably, the start of maser or laser technology could have been triggered by research on something else—perhaps the search for optical storage devices or further theoretical work exploring the implications of D. Gabor's work.

In Table 3, which summarizes the attempt to compare and evaluate the five models, the first criterion assesses relative realism or range of applicability. The third criterion asks: "How well does the model lead to appropriate formulation of problems?" and the fourth poses these questions: "How well does each model account for the phenomenon of excessive fragmentation and overspecialization?" and "Do the explanations suggest any remedies?"

'"Appropriate" formulation means taking account of second-order effects or interactions among variables when they are important (e.g., safety features in the nuclear reactor, pollution in car design). The Marxist model, as used in much of the Socialist world, tends to stress primary factors and to exclude secondary ones even when it is known that considering them at the outset may later prove most cost-effective.

OUTPUT MEASUREMENTS IN RELATION TO SCIENCE MODELS

Criteria for Evaluating Measurements

For each model of science planning, at least two criteria of scientific signifi-cance can be used in judging the adequacy of various measurements of output. An indicator is scientifically significant if its behavior can be ade-quately explained. The first criterion is the extent to which the output meas-ure helps to assess the achievement of the major purpose of S&T assumed by the model. The second criterion is the extent to which the output measure helps to assess errors of omission and commission.

The purpose of science—according to the autonomous or self-regulating system model, for example—is to produce conceptual schemes, theories, and codified understanding. The first criterion then asks of each output to measure the extent to which it helps assess the quality of a conceptual scheme. According to this model, S&T institutions aim to reward the crea-tion of high-quality conceptual schemes and informative assertions of max-imum content. They may fail by rewarding people whose output turns out not to be a good conceptual scheme, in the sense that it is not rich in implications, is inelegant or unrealistic, or is not fruitful in stimulating re-search. The institutions also may fail by not rewarding the creation of a rich, elegant, fruitful, conceptual scheme.

A widely held current view of what constitutes a good conceptual scheme and how it is arrived at emphasizes the consensus-reaching process which may be interpreted as a reward mechanism. (see Cole et al., this volume.) Thus the development of a "good" conceptual scheme accompanied by failure to reach consensus may be interpreted as an error of omission. Seri-ous questions exist about the diversity of viewpoints and cultural values that must be represented among those whose consensus is required, how large a group it should be, to what extent the consensus-reaching process involves forcing compliance rather than explanatory persuasion, and so on. Errors of omission are probably more frequent and less serious than errors of commis-sion. The second criterion asks, for each output measure, the extent to which it helps detect these two kinds of errors.

Table 2. Judgment of Appropriateness of Models, by Dependent Variables

Institutions Shown in Figure 1		Three Independent Variables Describing Institutions		
Point in Cube	Description	Size[b]	Program-maticity[c]	Internal/External
A	Academic departments, individual promise-oriented	Small	Nonprogrammatic	Factors internal to science predomina[te]
B	For example, Institute for Advanced Study	Large	,,	,,
C	For example, Brookhaven	,,	Programmatic	,,
D	Academic departments program-oriented	Small	,,	,,
E	For example, BB&N	,,	,,	Responsive primarily to non-scientific criteria
F	Private inventor	,,	Nonprogrammatic	,,
G	For example, A.D. Little	Large	,,	,,
H	For example, Bell Labs	,,	Programmatic	,,

[a] Other models apply secondarily; possibly "Science as Learning" applies somewhat in all cases.

[b] Correlates with diversity of resources.

[c] Nonprogrammatic—personnel autonomous; programmatic—administrative policy less favorable to individual autonomy.

[d] This seems to hold when small autonomous groups, containing young "stars" recruited for their promise as scientists, are in or near large centers.

Table 2. *(continued)*

Judgment about Appropriate Model that Applies Primarily [a]	Dependent Variables	
	Reach or Significant Impact	Controllability and Predictability
Autonomous, self-regulating mechanism	Furthest philosophical[d] influence (e.g., relativity), impact on all intellectuals	Unpredictable, except some discoveries made with help of planned research technologies
"	Impact on scientists only, mostly one discipline (e.g., nonconservation of parity)	Slightly predictable
Science as tertiary industry or science as service	Explore a paradigm: impact on fellow specialists, government, industry; reward = eponym	Can be developed or nurtured, predicted (with great effort)
Science as learning: investment in future (training in disciplines)	Impact on graduate students, disciplines	Same (with less effort)
Science as service	Sustained effort: impact on professions, government, industry	Controlled by market and technostructure; predictable
"	Patents and profits, impact on industry	"
"	Reports to clients; often wide impact through key clients	Short-term planning possible, early-warning and technological forecasting possible
"	Industry-wide, may change lifestyle of many people	Long-term planning may be possible

119

Table 3. Comparative Evaluation of Five Models

Criteria	Models				
	Autonomous, Self-Regulating Mechanism	Science as Support-ing Service	Marxist	Science as Tertiary Industry	Science as Learning
1. How much of the variance in behavior of U.S. institutions does it explain? Where is it appropriate?	Probably a good description of the few top academic departments and basic-research institutions that account for most of the basic progress	Probably a good description of most large S&T organizations dependent on public or industrial support, where most of the money is spent	Probably S&T institutes in Communist countries are planned thus	Urban areas based on science— (e.g., Rte. 128)	In a secure strong society with faith in its future (rare)
2. Utility for S&T planners	Little for basic science; some for science based on research technology; more for technology	Increases as organizations become more dependent on public support	Not in United States	Possibly for new-town planning— 2 to 3% of new growth	None

3. Maturity level of development, sophistication	Promising, (e.g., work of Price, Goffman)	No quantitative models, but most promising	?	?	Some analytical work
4. Explanatory power, richness of deductive structure	Fair, by counts of works of synthesis, (e.g, good theory, books, review papers); emergence of "world brain" as new social organ (48)	Good, via study of relatedness, cohesion, cross-illumination among fields. (Even mathematics should not be judged by mathematicians alone)	Low	Good	Good
5. Potential fruitfulness; Goodness in generating good questions, ideas, research	Moderate	Good	Low though Very influential	?	Good
6. Humanistic?	No more than biology	No—science seen as source of human values, not vice versa	No	Increasingly	Increasingly

The five models sketched above thus give rise to 10 such criteria. Beyond these ten, there are general, seemingly model-independent criteria such as: (a) Does the output measure identify major strengths of, and in, the total scientific enterprise? (b) Does it identify weaknesses? (c) Does it identify utilization of science output of any kind?

Global measures of a and b were reported in SI-72, but they depend crucially on what are defined as the strengths and weaknesses of science, and those in turn vary with images of science and models of science planning. Significant increases in unemployment of natural scientists in 1972, would be considered a major weakness according to the "science as a tertiary industry" model, but this is not dramatized as such on pages 51 and 59 of SI-72, where this kind of unemployment rate is compared with that for all workers. The occurrence of significant conceptual advances in 1972 which would be highlighted according to the autonomous-growth model, were not dramatized either (see Holton, this volume). Such global measurements as percent of GNP by country suffer from the fact that they measure input into science rather than output of science (applications for research grants follow the dollars and the dollars go where the results are—a self-perpetuating circular process), that GNP is not as good as a measurement of economic welfare as, say, "mean economic welfare," and other comparably well-known defects.

Another possible global measurement of utilization is the ratio between the estimated total number of all kinds of uses of scientific publications coming out in one year and that year's number of pages of publication (49). Lack of use of scientific output may of course be preferable to misuse. But science, as such, is ineffective if its outputs are not used even to advance science for its own sake. Science indicators are, at best, likely to serve clearly specified purposes for which there is an appropriate model of science planning into which they also fit. A good example of such output measures is furnished in the early-warning reports based on an analysis of patients (50). They may help in science planning, but they play only a limited role in the process.

Further criteria of a methodological nature specify the properties desirable for any indicator (51). An indicator is required to correspond faithfully to the trends it is intended to measure or forecast—for example, the number of publications must correspond roughly to increases or decreases in scientific productivity. Statistical adequacy is another property required to provide assurance that the numbers can be relied on in the future as well as in the past. That assurance depends on a good data-reporting system; good coverage of the population for a substantial time-period and if complete coverage is not feasible, a good sampling procedure; seasonal adjustments; and error estimates (see Kruskal, this volume). Yet another required property is sensitivity to changes, at least to trend reversals or significant shifts. Additional criteria include currency (which is met if values of the indicator are available promptly), smoothness of the time series, efficiency of measurement (it should not require an extremely costly massive effort), and ease of interpretation.Few indicators can meet all criteria to the same extent. Clearly, complete coverage, or even large-scale sampling and low cost are incompatible; so are the joint requirements for currency of information and low cost. Obviously, trade-offs must be arrived at.

In Table 4 a small sample of indicators, or measurements of scientific output with respect to some of these criteria, are compared. Most of the methods for measuring output are well known, except perhaps for rate of change in the "quality of life." One method of measuring quality of life (52) has been to interview people, asking questions such as these: "Suppose that a person who is entirely satisfied with his life would be at the top of the ladder, and a person who is extremely dissatisfied with his life would be at the bottom of the ladder. Where would you put yourself on the ladder at the present stage of your life in terms of how satisfied or dissatisfied you are with your own personal life?" The responses form a "self-anchoring striving scale." This work is being advanced by A. Campbell at the University of Michigan.

Suggested Modifications and New Measures of Output

Peer Review Plus Delphi Plus Improved Peer Selection. The peer-review method is fairly well suited to measuring the output of academic science. Its effectiveness has recently been studied to determine how well senior scientists can judge the methodology of applied social science projects (54). Noble maintains that peer review by itself is inadequate for quality control. To overcome two major shortcomings—possible biases of judges and errors in selecting judges—I propose that peer-review panels engage in some kind of teleconferencing, using the Delphi procedure for canceling out biases. The use of information technologies such as computer networks with store-forward-edit-retrieve capabilities may enable panelists to observe and perhaps to accelerate the process by which consensus is reached. The panelists could also be chosen from a larger pool of potential peers and chosen with greater sensitivity, according to techniques now under experimentation (55). This would permit representation of a greater variety of viewpoints, matching the panels to increase the likelihood of success (using the FIRO three-way measurement of personality, for example) and to permit more alteration and revision of the questionnaire. Different experts give different weights to diverse criteria; to avoid the confusion that could result, the problems would be structured to make the criteria explicit; and the panelists would thus be helped to realize the trade-offs they must make.

Table 4. Appropriateness of Methods of Measuring Output.
(1 = Overall Judgment of Good to Fair; 0 = Poor, Remote, Irrelevant)

Criterion Method Helps to Assess	Sample Methods of Measuring Output							
	Change in "Quality of Life" (52)	Frequency of Citation Counts	Occurrence of Discoveries and Inventions [a] (53, 30)	Share of Total Talent Utilized	Peer Review (Refereeing) by Specialists	Population of Active Contributors	S&T Contribution to GNP	Publication Volume
Quality of conceptual schemes (autonomous model)	0	1	1	1	1	0	1	0
Errors in rewarding work of developing conceptual schemes	1 [b]	1	0	0	1	0	0	0
Attainment of social goals (Weinberg model)	1	1	1	0	0	1	0	0
Errors in rewarding attainment of social goals	1 [c]	1	0	0	0	0	1	0

Ability to cope with spiritual, intellectual needs, to specify new goals, aspirations	1	1	0	1	0	0	1	0
Future social productivity; learning rate	0	0	1	1	0	1	1	1
Strength of science as a whole	0	1	1	0	1	1	0	0
Weaknesses of science as a whole	1	1	0	0	1	0	0	0
Sensitivity to coming shifts	1	0	0	1	0	1	0	1
Totals	6	5	5	4	4	4	4	3

[a] Good but lagging indicator.
[b] Assumption is that reward for scientists who do good work changes "quality of life."
[c] Assumption is that managers are the ones rewarded

125

Understandable presentations, given by a few experts, would provide the panelists with a common knowledge base.

These techniques and technologies are being applied to refereeing manuscripts, evaluating the performance and potential of individuals being considered for new positions or for promotion, and assessing the output of groups engaged in basic research where subjective evaluations of quality (e.g., of conceptual schemes) are more reliable than most quantitative objective measures.

Citation Counts Plus Content Analysis (see Zeisel, this volume). The number of times an article has been cited in other publications is a fair measure of attention paid to it. A scientist who publishes little and is not cited but lectures a lot may be popular in an irrelevant sense while scientists whose papers are much cited would scarcely be considered unpopular in the sense of being neglected or ignored. Popularity, however, is not per se an important property of science output. What does matter is its quality and impact, and a substantial literature now exists that critically explores the use of citation-based assessments of "importance," "quality," and impact. For example, a good correlation exists between the number of citations to scientists and their honors, number of publications, and other common measures of excellence (56). Virgo (57) has also shown that citation frequency consistently predicts which of two papers will be judged the more important, this being the more frequently cited paper.

Ordinary citation counts do not measure the extent to which important scientific output is communicated to and used by the authors it might help. To remedy this fault, I suggest that each passage in a citing paper that refers to a prior publication should be analyzed for the presence of predicates indicating how much the reference has contributed to the citing paper.* This would involve the search for words that, in context, carry positive or negative evaluations of the references to which they refer, using a special context-sensitive grammar and thesaurus. For example, if a citing paper states that it relies on a result in some cited paper in a way that is essential for the main point of the citing paper, that is given high rank. Although this would require a sophisticated content analysis involving inference, it might be simplified if the paper contained a good abstract resembling an aphorism. The sentence responsible for the reference would be identified as antecedent to the main conclusion of the citing paper.

---

*This idea occurred to me prior·to and therefore independently of my learning about similar projects already under way. Several independent approximations to what I have in mind involving the qualitative content analysis of citations have since appeared in print; see the papers by S. Cole, Moravcsik and Murugesan, and Chubin and Moitra (58). See also Cole, Cole and Dietrich; and Garfield, Malin and Small, both in this volume.

Second-highest rank might be given when a cited paper is criticized in a citing paper. Perhaps a computer-implementable algorithm could be found that ranks the references in a citing paper on this score. Also, since the various papers citing the same reference differ in their merit and influence, influence could be measured by the same kind of modified weighted citation count being proposed here; and the result of peer-review judgment could be used as a measurement of merit. Each cited paper, then, would be assigned a score composed of the number of articles that cite it, each weighted for the quality of the citation as well as for its merit and influence. Thus, a paper that contributes greatly to several highly meritorious and influential citing papers obtains a high score and is considered influential. A paper that is widely cited but has little direct influence on the citing papers—and where the citing papers are not especially meritorious or influential either—would get a low-to-medium score. If citing papers criticize or fault a paper, that need not detract from the cited paper's contribution to the citing paper; the contribution is large if the citing author's view that it was erroneous or objectionable provided the stimulus or even the idea for his own paper.

A problem arises if the author of the citing paper fails to acknowledge intellectual influence or debt properly. But the reward system of science usually applies severe sanctions for such error or deceit, and these sanctions are applied through the peer-review system.

The proposed output measure would be feasible if scientific publications or manuscripts were widely available in machine-readable form. The lowered cost of source-recording terminals and computer-based editing is making this increasingly cost-effective for a large number of scientific institutions. The resulting indicator can be used as an objective complement to the peer-review results in all applications involving the peer-review method. It can also be used for evaluating the performance of research institutes with scientific missions (59) and of basic-research labs in organizations with a mission related to social goals.

Measures of Interrelatedness.   An analog of Leontief's input-output analysis appears to be feasible and fruitful. Suppose that publications can be classified into disciplines or missions $1,....,n$—such as 1 = mathematics, 2 = physics, ... 20 = crime prevention, 21 = antiballistic defense, ... n—according to whether they appear in a journal affiliated with a national (or international) mathematical or physical society, or authors' and readers' professional affiliations or primary areas of concern. (As in economics, there is a problem of how many sectors at what level of aggregation to deal with, i.e., whether to lump all the life sciences into one or to treat separately specialties such as paleobotany.) Let a_{ij} be the average number of inputs to produce one unit of output. Let p_j be the number of 1975 publications cited per year in discipline j. Then $\sum_j a_{ij} p_j$ is the total number of uses of 1974 papers in discipline i that were used to produce a 1975 paper in some other discipline. Let k_i be the type-token ratio between the number of papers of field i so used and the total number of uses, or the reciprocal of the average number of times one 1974 paper in field i is cited per year. The $k_i \sum_j a_{ij} p_j$ is the average number of 1974 papers in discipline i that were used in one year. This could be revised when 1976, 1977 . . . papers that cite the 1974 papers appear. This total is some fraction f_i of p_i.

The resulting system of equations has certain characteristic values for f_i / k_i that can be compared with the estimates to monitor deviations from equilibrium and trends. If 1975 articles in field i cited only 1974 articles in their own field, then the matrix (a_{ij}) is diagonal and the characteristic values of f_i / k_i are those diagonal elements.

In some sense, this kind of inbreeding is incestuous, and if it is already excessive and becomes more so, it may indicate danger for the health of science communication. Some function of the characteristic values of the matrix (a_{ij}) could be used to monitor the extent of cross-fertilization among missions and disciplines. Higher powers of the matrix may be usable for tracing some of the antecedent influences of work done several generations of cited publications before selected works in an area of current importance.

Another use of some of these citation-based techniques is to measure the speed with which a country recognizes new disciplines or problem areas and the speed with which it abandons useless but traditional notions. These factors were suggested by Ben-David (60) as critical in accounting for the fast growth of German science in the middle of the nineteenth century and of American science after 1880.

Manpower-Related Indicators. The number of key people—for example, associate or full professors in a department—making lateral moves is often viewed as indicative of deterioration in the institution losing them. At a higher level of aggregation, the fraction of the labor force that is professional/technical and the rate of increase in the fraction of the labor force employed in services indicate to some degree how far a country or institution is along Bell's scale, which ranges from preindustrial to postindustrial societies (61). This is, presumably, an indirect effect of S&T output and its utilization.

Another promising method of measurement for assessing social change is the use of time (62–63). This involves a breakdown of the categories on which scientists in responding to interview questionnaires, could say reliably how they spend their time. Shifts could be quite revealing.

Publication-Related Indicators. Suppose the indexes of two widely used textbooks, in the same field but a decade apart, are compared. Certain authors and subject headings absent from the earlier text will appear in the later one. The institutions where these authors were employed and where the new topics were developed would be noted, and their appearance in standard texts could be used as a measure of intellectual impact. Content analysis of publications (in machine-readable form) could be used to measure the ratio of theory to empiricism and its growth rate. Theory is indicated by the presence of abstract symbols that codify knowledge and illuminate or enlighten many varied circumstances. This would be correlated with the ability to plan, to control growth, and to utilize innovations. If Bell is right, it would also be correlated with position on the pre- to-postindustrial-society scale.

Measures of Fragmentation. One vehicle by which the process of S&T copes with its large output of specialized publications is the the review paper. A good review paper shows how numerous new findings fit together not only with each other but also with older findings and with possible applications and extensions. Good review articles are difficult to write, and the incentives are not great. Fragmentation uncompensated by efforts at evaluation and synthesis, such as in good review papers, is deplored. Improving incentives and aids to competent authors might remedy the situation. This is an area worthy of further study. In the meantime, it would be desirable to know whether such uncompensated fragmentation is improving. A possible measure is the percentage of publications that are eventually cited by some review paper (23). This measure does not discriminate review papers by quality, but the measure may perhaps be extended to take this into account. The idea for doing so is to examine the passages for concentrations in the review paper that use a cited paper. In an extremely poor review paper, there may be as many separate and unconnected passages as there are references, one for each. In a good review paper, each passage may connect several references and the passages may be themselves interconnected. An algorithm may possibly be developed on which content-analysis by computer can make the measurement of an indicator feasible.

Topic-Related Indicators. The proliferation of specialties, indicated perhaps by the numbers of separate registers and subject headings, is a measure of the health of S&T if it is interpreted in conjunction with the number of works of synthesis. Fragmentation in excess of attempts to interconnect the fragments is counterproductive. According to the self-regulating model, the natural laws governing the growth of S&T are such that if the excess becomes too extreme, compensating forces come into play; they restore equilibrium, and the entire process is then stable. According to the cultural-evolution view, planners are nature's instruments for monitoring the appropriate indicators and making sure that stability and equilibrium are maintained.

Other Suggestions.   An important possible indicator is the rate at which new questions are being formulated, compared with the rate at which outstanding problems are being worked on. The ratio of divergent to convergent research is yet another important indicator, as is the fraction of scientists doing consistently good work who can do what they think is important. The portion of an institution's scientists working on weekends or after business hours may indicate the scientific vigor and promise of that institution. The growth in the portion of young high-quality investigators can be another indicator.

The gap—or lack of realism—between the time horizon of a project that in retrospect seems most appropriate and the one that was actually used is indicative of the quality of scientific management. Another indicator is the extent to which higher-order and interactive effects are anticipated in time to take them into account. For example, consider the question: "How poisonous will plutonium in large quantities be, and can it be reliably sequestered?" Consider first when the question *should* be asked (perhaps when nuclear reactors are first being designed) and next when it *is* asked. The lag between these two times is indicative of how sensitive S&T organizations are to the larger social problems and their inevitable relationship to modern science. This sensitivity amounts to foresight. A measure of the vigor and precision of foresight about where social sensitivities will be focused in the next decade would be one of the most important indicators we could hope to find.*

Suggestions for Further Work

The more valuable the output of an institution engaged in S&T, the more it seems to resist early identification. The obviously significant outputs may not be detectable in citation counts, simply because their value is so obvious after it has quite suddenly become so. A more promising line of inquiry into the relative merits of different measurements of scientific output might begin with a significant end-product and trace its intellectual antecedents (64–65).

*Essential notions in this paragraph were included in a private communication from Joshua Lederberg, November 22, 1974.

Extensions of techniques for citation analysis may be helpful, as may the material reported by Lipetz (66).

Further critical studies along some of the lines suggested in the previous section—measures of interrelatedness and fragmentation, modified peer review, and citation counts—may prove to be fruitful. Simulations of the effects of science planning (67), possibly connected with the data of science mapping (see Garfield, Malin, and Small in this volume) and with the choice of science indicators as independent variables (used in the planning simulation), may be a fruitful research approach. At the same time unusual opportunities exist for semiempirical testing of concurrent indicators in institutions in fields of science now being shaped.

The theory of fuzzy sets and its applications may be a promising subject of research. It started in about 1956 with systems engineers who felt that the design of a control system to park a car or tie a knot, for instance, should be based on totally different concepts from those based on classical set theory, where an element either does or does not belong to a set. The theory of fuzzy sets deals with sets such as large numbers and great distances and it assigns a grade of membership to elements in such sets. It is now being vigorously developed at certain universities in Japan, Rumania, France, and the United States. Concurrent indicators might be used to predict the short-term course of this field and to test it by comparison with how it develops.

Quantitative and qualitative analyses of science and S&T planning models, plus empirical testing, should be given priority. A modest experimental science-information system should then be created for a more detailed analysis of specific indicators and the problems associated with their monitoring and use. Research on this type of information system can proceed in three stages. The first stage simply involves examining the existing time series that measure scientific output for auto- and cross-correlations. The second is to search for causal relations, testing hypotheses derived from the more quantitative models of S&T; and the third is to investigate the use of the information system for planning and policymaking. The last stage involves testing hypotheses, such as the following:

The use of the information system increases the quality of criticism and debate about science policies, provides the general public with "better" orientation in and attitudes toward science, improves the choice and formulation of issues, and affects the conduct of science.

To illustrate a more detailed hypothesis about how an information system affects the conduct of science, consider the following:

If the statement "Copper will soon become rare; we should look for substitutes or recycling methods, even though the copper industry opposes this" is brought to the attention of institutions responsive to copper demand, it is more likely to lead to increased research for copper substitutes than if such a statement is not brought to the attention of appropriate institutions.

Action toward better science planning and the creation of supporting information systems is not likely to wait for these research recommendations to be implemented and successfully concluded. Since we are beginning to recognize that knowledge of ourselves and our environments is necessary for the attainment of a higher quality of life, legislation like the National Development Act suggested recently by Haggerty (68) may be enacted. This possibility has implications for indicators to monitor the effectiveness of deliberately seeking, diffusing, and applying knowledge as an integral and continuing part of government policymaking. These implications should be investigated by those concerned with science indicators.

CONCLUDING COMMENTS

The development of information systems to help science planners requires stronger scientific underpinnings than are now available. A comparison of five approaches toward conceptualizing science planning processes suggests that the view of science as a service to support the attainment of social goals seems to account for the behavior of most scientific institutions in the United States. That view has increasing utility for science planners and shows promise as a basis for further development of the needed scientific underpinnings. Yet it does not reflect the potential of science for helping to shape a more viable image or for selecting the most significant social goals in the first place. Both the limitations and potentialities of science and scientific attitudes in what may be science's most important role are not fully captured by that view. The view of science as a learning process helps more in fusing scientific and humanistic attitudes and images of the world.

The choice of quantitative indicators, which information systems to help planners would monitor, depends on the view of science and science planning. The whole notion of monitoring the state of science by quantitative indicators takes on less importance under a view that is more humanistic and less mechanistic. Still, certain indicators, if used with an awareness that they are but one of several inputs to a science planner, may be valuable. Rate of change in quality of life, if developed into an adequate indicator, has more promise than the most common alternative indicators, according to several criteria. Measures based on modified citation counts and peer-review techniques also appear to be promising.

We should be aware, in considering the preceding discussion, that science indicators are only one of several inputs to decision makers, whether they are voters or legislators. Quantitative output measures usually capture only approximately the full essence of what they are intended to indicate. They are subject to a variety of interpretations, often based on many diverse values. Major decisions within S&T institutions, or affecting them, are seldom carried out unless there is consensus among the key people involved. It would be naive to expect that even the most rational decision makers will be compelled or persuaded to agreement by objective indicators. But they can help increase awareness that perhaps the disagreements boil down to value differences along only a *few* dimensions, and thus they can raise the intellectual level of their debates. This may point to responsible paths toward agreement.

REFERENCES

1. R. A. Bauer, Ed., *Social Indicators*, MIT Press, Cambridge, 1966.

2. *Science Indicators 1972; Report of the National Science Board*, Government Printing Office, Washington, D.C., 1973.

3. D. J. De Solla Price, "The Science of Scientists," *Medical Opinion Review*, 1, 88–97 (1966).

4. D. J. de Solla Price, *Times Literary Supplement*, London, 28, July, 1966, pp. 659–661.

5. M. Bunge, *Scientific Research*, Vol. I, Springer, New York, 1967.

6. Polish Academy of Science, *Problems of the Science of Science*, Ossolineum, Wroclaw, 1970.

7. H. W. Menard, *Science: Growth and Change*, Harvard University Press, Cambridge, 1971.

8. K. E. Boulding, "Dare We Take the Social Sciences Seriously?" *American Psychologist*, 22, 879–887 (1967).

9. H. Brooks, in *Problems of Science Policy*, OECD, Paris, 1967, pp. 1–20.

10. Y. Elkana, "The Problem of Knowledge in Historical Perspective," *Proceedings of the 2nd International Humanistic Symposium*, Athens, 1974.

11. W. Goffman, "Mathematical Approach to the Spread of Scientific Ideas—The History of Mast Cell Research," *Nature*, 212, 449–452 (1966).

12. M. Nowakowska, *Language of Actions and Language of Motivations*, Mouton, The Hague, 1972. Also *Theory of Actions: An Attempt At Formalization* [in Polish], PWN, Warsaw; through *Behavioral Science*, 18, 393–416 (1973).

13. A. M. Weinberg, in *Civilization and Science*, Ciba Foundation Symposium, North-Holland, Amsterdam, 1972, p. 113.

14. J. Lederberg, "A Freeze on Missile Testing," *Bulletin of the Atomic Scientists*, 43, 4–6 (1971).

15. S. Toulmin, in *Civilization and Science*, Ciba Foundation Symposium, North-Holland, Amsterdam, 1972, p. 116.

16. A. H. Maslow, *The Psychology of Science*, Harper & Row, New York, 1966.

17. Hans Kammler, "A Set-Theoretical Explication of Causal Relations," unpublished research note, private communication to author, 1975.

18. H. Poincaré, *Science and Hypothesis*, trans. by J. Larmor, Dover, New York, 1952.

19. J. Hadamard, *The Psychology of Invention in the Mathematical Field*, Princeton University Press, Princeton, 1945.

20. G. Polya, *Mathematics and Plausible Reasoning*, Vols. I and II, Princeton University Press, Princeton, 1954.

21. M. Kochen, in J. Belzer, H. G. Holzman, and A. Kent, Eds., *Encyclopedia of Computer Science and Technology*, Vol. 3, Marcel Dekker, New York, 1975.

22. M. Bunge, in I. Lakatos and I. E. Musgrave, Eds., *Philosophy of Science*, North-Holland, Amsterdam, 1968, p. 147.

23. M. Kochen, *Integrative Mechanisms in Literature Growth*, Greenwood Press, Westport, 1974, pp. 44–63; 90–137.

24. D. Bell, "Measurement of Knowledge and Technology," in E. B. Sheldon and W. E. Moore, Eds., *Indicators of Social Change*, Russell Sage Foundation, New York, 1968, pp. 145–246.

25. J. K. Galbraith, in B. R. Williams, Ed., *Science and Technology in Economic Growth*, Macmillan, London, 1973, pp. 39–58.

26. T. Roszak, "The Monster and the Titan: Science, Knowledge, Gnosis," *Daedalus*, **103**, 17–32 (Summer 1974).

27. A. M. Weinberg, "Science, Choice and Human Values," *Bulletin of the Atomic Scientists*, **22**, 8–13 (1966).

28. D. Roose and C. J. Andersen, *A Rating of Graduate Programs*, American Council on Education, Washington, D. C., 1970; A. M. Cartter, *An Assessment of Quality in Graduate Education*, American Council on Education, Washington, D. C., 1966.

29. R. D. Anderson and W. H. Pell, "A Measure of Quality of Doctoral Programs," *Notices of the American Mathematical Society*, **21**, 260–262 (1974).

30. K. W. Deutsch, J. R. Platt, and D. Senghass, "Conditions Favoring Major Advances in Social Sciences," *Science*, **171**, 450–459 (1971).

31. R. Hahn, *The Anatomy of a Scientific Institution*, University of California Press, Berkeley, 1971.

32. Y. Elkana, "The Problem of Knowledge," *Studium Generale*, **24**, 1426–1439 (1971).

33. M. Kochen, in M. Kaplan, Ed., *Isolation or Interdependence? Today's Choices of Tomorrow's World*, The University of Chicago Press, Chicago, 1975.

34. M. Polanyi, *Personal Knowledge*, The University of Chicago Press, Chicago, 1958.

35. R. K. Merton, *The Sociology of Science*, The University of Chicago Press, Chicago, 1973, pp. 343–382.

36. President's Science Advisory Committee Report, *Improving Health Care Through Research and Development*, The White House, Washington, D. C., March 1972. Also *Scientific and Educational Basis for Improving Health*, HR 5640, HR 5948, Government Printing Office, Washington, D. C., 1973.

37. S. P. Strickland, *Politics, Science, and Dread Disease*, Harvard University Press, Cambridge, 1972.

38. D. R. Challoner, "A Policy for Investment in Biomedical Research," *Science*, **186**, 27–30 (1974).

39. B. Moores, *Management Studies*, **11**, 163–167 (1974).

40. M. Eastman, Introduction to Karl Marx, *Capital, Communist Manifesto and Other Writings*, Modern Library, New York, 1932, p. xxi.

41. H. Brooks, "What's Happening to the U.S. Lead in Technology?" *Harvard Business Review,* **50,** 114 (1972).

42. A. Rapoport, *Science and the Goals of Marx,* Harper, New York, 1950, p. 232.

43. S. Toulmin, "The Complexity of Scientific Choice II: Culture, Overheads of Tertiary Industry," *Minerva,* **4,** 155–169 (Winter 1966).

44. *A Treasury of the Art of Living,* Sidney Greenberg, Ed., Wilshire, Hollywood, Calif., 1967, pp. 256–257.

45. C. W. Churchman, *The Design of Inquiring Systems,* Basic Books, New York, 1971.

46. B. Glass, *Science and Ethical Values,* University of North Carolina Press, Chapel Hill, 1965.

47. K. Popper, *Conjectures and Refutations: The Growth of Scientific Discovery,* Basic Books, New York, 1963, p. 222.

48. M. Kochen, Ed., *Information for Action: From Knowledge to Wisdom.* Academic Press, New York, 1975.

49. M. Kochen, *Principles of Information Retrieval,* Melville/Wiley, Los Angeles, 1974.

50. U. S. Department of Commerce, *Technology Assessment and Forecast: Early Warning Report of the Office of TAF,* Government Printing Office, Washington, D. C., 1974.

51. G. H. Moore and J. Shiskin, *Indicators of Business Expansion and Contractions,* Columbia University Press, New York, 1967.

52. H. Cantril, *The Pattern of Human Concerns,* Rutgers University Press, New Brunswick, 1965.

53. J. Ben-David, *The Scientist's Role in Society,* Prentice-Hall, Englewood Cliffs, N.J., 1971.

54. J. H. Noble, Jr., "Peer Review: Quality Control of Applied Social Research," *Science,* **185,** 916–921 (1974).

55. M. Kochen, "Quality Control in the Publishing Process and Theoretical Foundations for Information Retrieval," in J. Tou, Ed., *Software Engineering,* Vol. 2, Academic, New York, 1970, pp. 119–154; also, "Improving Referee-Selection and Manuscript Evaluation," in *Proceedings of the First International Conference of Scientific Editors,* Jerusalem, April 24–29, 1977.

56. A. E. Cawkell, "Search Strategy: Construction and Use of Citation Networks, with a Socioscientific Example: Amorphous semi-conductors and S. R. Ovshinsky," *Journal of the American Society for Information Science,* **25,** 123–129, (1974); J. Cole and S. Cole, "Measuring the Quality of Sociological Research: Problems in the Use of the Science Citation Index," *The American Sociologist* **6,** 23–29 (1974).

57. J. A. Virgo, "A Statistical Measure for Evaluating the Importance of Scientific Papers," Ph.D. Dissertation, University of Chicago Graduate Library School, December, 1974.

58. S. Cole, "The Growth of Scientific Knowledge: Theories of Deviance as a Case Study," in L. A. Coser, Ed., *The Idea of Social Structure,* Harcourt Brace Jovanovich, 1975, pp. 175–220; M. J. Moravcsik and P. Murugesan, "Some Results on the Function and Quality of Citations," *Social Studies of Science,* **5,** 86–92 (1975); D. E. Chubin and S. D. Moitra, "Content Analysis of References: Adjunct or Alternative to Citation Counting?," *Social Studies of Science,* **5,** 423–440 (1975).

59. A. J. Matheson, "Centres of Chemical Excellence?" *Chemistry in Britain,* **8,** 207 (1972).

60. J. Ben-David, "Scientific Productivity and Academic Organization in Nineteenth-Century Medicine," in B. Barber and W. Hirsch, Eds., *The Sociology of Science,* Free Press, Glencoe, 1962, pp. 305–328.

61. D. Bell, *The Coming of Post-Industrial Society*, Basic Books, New York, 1973.

62. A. Campbell and P. E. Converse, Eds., *The Human Meaning of Social Change*, Russell Sage Foundation, New York, 1972.

63. A. Szalai, Ed., *The Use of Time: Daily Activities of Urban and Suburban Populations in 12 Countries*, Mouton, The Hague, 1972.

64. E. T. Crawford and A. D. Biderman, Eds., *Social Scientists and International Affairs*, Wiley, New York, 1969.

65. IIT, *Technology in Retrospect and Critical Events in Science*, IIT Report to NSF, December 16, 1968.

66. B. A. Lipetz, *A Guide to Case Studies of Scientific Activity*, Intermedia, Inc., Carlisle, 1965.

67. M. E. Beres, B. M. Koehler, and G. Zaltman, "Communication Networks in a Developing Science: A Simulation of the Underlying Socio-Physical Structure," *Simulation and Games*, **6**, 3–38 (1975).

68. P. E. Haggerty, "Science and National Policy," *Science*, **184**, 1348–1351 (June 28, 1974).

PART **II**

PARTICULARS

Taking Data Seriously

6

William Kruskal

This essay deals with themes suggested, but not limited, by *Science Indicators 1972* (1), the springboard text for our conference. My training and experiences as a statistician, a growing concern with the interrelationships of statistics and public decisions, and the stimuli of the Conference on Science Indicators together form the background for this set of reactions, reflections, sorrows, and suggestions. I do not attempt tight synthesis, although there are many connections among the sections to come.

My major theme is a plea in context that data be taken more seriously than they now generally are. Thus, I examine the way data are documented and treated in *SI-72,* in the publications on which it rests, and, to an extent, in social research more generally. With some exceptions, I

Preparation of this paper was supported in part by the National Science Foundation under Grant NSF GS 31967. I am indebted for many helpful suggestions and criticisms to Norman Bradburn, O. D. Duncan, Stephen Fienberg, A. L. Finkner, Denis F. Johnston, Joseph B. Kruskal, Norma Kruskal, Stanley Lebergott, Saunders MacLane, Margaret E. Martin, Robert K. Merton, Claus A. Moser, Frederick Mosteller, John W. Pratt, Margaret Reid, David L. Sills, Walt R. Simmons, Judith Tanur, and Edward R. Tufte. My thanks to these kind colleagues is muted only by my realization that I have not adequately attended to their numerous comments.

conclude that the data going into *SI-72* were not treated seriously enough in a number of senses. First, documentation of their modes of genesis is weak. Second, there is remarkably little study or consideration of the limitations of the data used: arbitrariness or vagueness of definition, degree of random error components, errors of bias, and so on. Third, there are some more technical statistical problems.

I point out that these deficiencies are by no means unique to *SI-72* but are widespread in economic and social science literature. There is discussion of the dangers in disregard of error structure, of the existence of relatively good treatments, and of public standards for proper treatment of data.

Other aspects of taking data seriously discussed here include the question of clear ascription of responsible authorship in government reports, pre- and postpublication review of those reports, and our common lack of fundamental understanding about graphical methods for presenting quantitative information, tabular methods for the same purpose, and indeed even about the straight prose with which we describe that information. I also discuss the status of what are called trend data, and questions of prediction and its accuracy.

A so-called Delphi experiment is reported in *SI-72,* and I close by commenting on it, primarily on the lack of documentation, of validation, and of replication. These deficiencies appear widespread in Delphi trials, yet still more widespread in other modes of extracting, refining, and summarizing expert opinion.

As the exposition moves along, I make some suggestions, beyond mere exhortation, towards taking data more seriously . . . that is, toward higher statistical standards. I must emphasize that I found *SI-72* and the Conference focussed on it to be highly stimulating. If I criticize *SI-72,* my remarks are meant to be friendly comments and suggestions, and I hope that they will be so understood. I am aware, in particular, that the social and bureaucratic history of *SI-72* may explain many of its characteristics; a critical look at those characteristics, therefore, will perhaps be helpful for the future.

GENERAL COMMENTS

Authorship and Review

SI-72 is a remarkably corporate document. It has no personal author to receive, as an individual, praise or blame. The closest indications of authorship are tucked away inconspicuously on the inside back cover—the membership of the National Science Board, the name of its Executive Director, and the name of the Staff Director for the National Science Board Report (of which *SI-72* is the 1973 avatar). The Chairman of the National Science

Board also appears in the prefatory material as signer of the letter of transmittal to the President of the United States.

Such diffusion of authorship for *SI-72* is more pronounced than in other NSF publications to which I was led by an interest in degree of attention to error structure. For example, *Research and Development in Industry, 1970* (2) has a signed foreword by the Director of the Division of Science Resources Studies and a prefatory paragraph listing five NSF authors and four cooperating individuals in the Bureau of Census.

The underplaying of personal authorship in government publications has not, to my knowledge, been studied systematically, and I suggest that such a study would be worthwhile; it surely relates to science policy and to government policy generally. One might conjecture that the stronger the professional staff of a government agency, the more frequent is clear ascription of personal authorship, although I expect that causation runs in both directions at once. One might also expect that pressure for personal authorship is a force towards recruitment and retention of better civil servants, and surely personal authorship fosters increased personal responsibility. In any case, such questions deserve study.

The germane tensions within the agencies may be a little like the arguments among collaborators in scientific publication generally or perhaps like the more colorful arguments about credit lines on theatre posters and cinema front material. Pauline Kael, in a moving recent discussion of the relative roles of producers, directors, and actors of motion pictures says

> In no other field is the entrepreneur so naked a status seeker. Underlings are kept busy arranging awards and medals and honorary degrees for the producer, whose name looms so large in the ads that the public—and often the producer himself—comes to think he actually made the pictures (3).

A stimulating study of authorship name orderings in scientific papers is provided by Harriet Zuckerman (4).

Related to the question of individual authorship is that of *review* in the usual channels of scientific discussion. Government reports like *SI-72* do not as a rule* go through the normal scientific publication procedures of prepublication refereeing and postpublication review and abstracting by scientists outside the government agency. Presumably, intra-agency reviews occur, but that may be an entirely different matter. Since, in addition, kudos for publication in standard scientific journals is absent or weak in many government agencies, one might fear the growth of self-perpetuating enclaves of

---

*A conspicuous exception is the very publication you are reading, a publication in which *SI-72* is examined in the broadest critical sense.

second-rate scientists, only weakly stimulated and sharpened by encounter with the turbulent, brawling, changing, growing extramural world of science. Able administration can mitigate this problem, and I know that the Civil Service Commission, for example, has programs of interchange between agencies and universities and, perhaps, also among different agencies. Nevertheless, cases of parochial isolation, both in and out of government, are not hard to find.

I have no special reason to think that the staff of the National Science Board suffers from scholarly isolation, but the surprising paucity in *SI-72* of references to the relevant outside literature gives the appearance of such isolation. For example, I found no references to the abundant literature of the broader social indicator movement, not even to the then-forthcoming publication by the Office of Management and Budget *Social Indicators, 1973*, an important, widely discussed book with a long gestation period (5). Again, the Delphi material cites none of the extensive literature —sympathetic, neutral, or negative—about the Delphi movement. A third instance may be found in Appendix A of *SI-72*, a presentation of about 68 tables giving the detailed numerical information that is summarized earlier in the report. Those tables cite as documentation by my count 18 published sources of which eight are NSF documents, four are published by other U.S. government agencies, three are from international organizations, and three are from private organizations. One does get an impression of considerable endogamy.*

One might say that *SI-72* is not, after all, a scholarly monograph, but rather a crisp, uncluttered summary of important findings. Nevertheless, documentation for those findings is important—as a matter of self-discipline and conscience for the staff, as an addition to the persuasiveness of the message from *SI-72*, and as a model for other government reports. Some publications meet this problem by having two physically separate volumes, one a relatively brief summary of context and conclusions, the other a much more detailed discussion of procedures and details of definition, design, and analysis. In *SI-72*, the device of self-contained appendices was adopted.

TECHNICAL PROBLEMS OF STATISTICAL EXPOSITION

Graphical and Tabular Materials; Statistical Prose

SI-72 may properly be praised for the high professional levels of its graphical and tabular presentations. Nor is that faint praise. One need only look briefly

---

*The impression is softened by the citation of five private publications elsewhere in *SI-72*. But the impression is stiffened if we note the absence of citations of well-known, widely discussed, and relevant private publications, for example, Morgenstern (6).

at otherwise excellent books and journals to find examples of incoherent, tendentious, or nearly fraudulent graphical and tabular presentation. Examples of misleading graphical work may be found even in the Office of Management and Budget's *Social Indicators, 1973,* as is trenchantly pointed out by Stephen E. Fienberg and Leo A. Goodman (7).

One merit of the graphical material in *SI-72* to which I can testify on a personal wavelength is the cleverly modest, but effective, use of highly contrasting colors so that persons like myself with color-deficient vision might better see the patterns. Graphs and tables are well placed, neither too large nor too small, uncluttered, yet with reasonable titles, legends, and citations.

Among readily available expositions of current standards for graphical and tabular materials one may cite, respectively, Calvin F. Schmid and James A. Davis—Ann M. Jacobs (8). It must be said, however, that we are in a primitive state of knowledge about graphic and tabular presentation, primarily a state of poorly organized craftsmanship, conventional dicta, and rules of thumb. So far as I know, psychologists studying perception have been little interested in so mundane an application as statistical presentation, and statisticians (and geographers, demographers, meteorologists, etc.) who carry out direct experiments (e.g., on density of contour lines, utility of color, bar charts vs. circles) work on restricted topics without the benefit of proper psychological background. Even such imaginative recent innovations as the Tukey hanging histogram, the Tukey stem and leaf tallying approach (9), and the brilliant Chernoff (10) use of schematic faces to express numerical vectors, even these, to my knowledge, have not been properly examined in carefully designed psychological experiments. These innovations do, however, signal a return to graphics of some measure of scientific respectability, so we may see before long an informed joint attack by psychologists, statisticians, and others.

There is, to be sure, a literature of empirical investigation into a variety of graphical methods; F. J. Monkhouse and H. R. Wilkinson, for example, report and cite journal articles from many sources (11). In a more recent article H. Wainer does deal with Tukey's suggestions, but avowedly as a preliminary sortie only (12). Current interest also extends to graphical methods extended into the time dimension by television, motion pictures, and so on; this has been called "kinostatistics" by one of its more vigorous investigators, Albert D. Biderman (13). Such an inevitable extension of statistical graphics will make still knottier those questions of perceptual psychology—avoidance of misleading or biased patterns, emphasis upon patterns for insight, memory, and discovery—that are still largely unexplored for old-fashioned statostatistical graphics.

Perhaps this gap between infrequent, relatively superficial empirical work on graphs and fundamental psychological investigations of perception is perfectly natural and understandable. There are also signs that it will be bridged. For example, Naomi Weisstein and Charles S. Harris describe a perceptual experiment that may lead towards fundamental understanding of graphics (14). Perhaps not irrelevantly, that work was done at a laboratory in which statisticians and psychologists are in nearby offices.

The ups and downs of respectability for statistical graphics would themselves form an interesting chapter in a history of scientific method, and perhaps—broadly interpreted—a chapter relevant to science policy. To what extent do leaders of policy—science policy and other—base their judgments on graphical presentations of the underlying facts, or presumed facts? To what extent do national leaders distrust graphs, and even numbers, as compared with verbal summaries? Do graphical presentations permit readier expression of numerical uncertainty than do other modes? Does not increased use of graphical method permit readier communication from government agencies and executives to an educated public?

Around the turn of the century, Karl Pearson, an almost elemental force for more and better statistical thought in all areas of life, including, with gusto, matters of public policy, was thinking and lecturing about graphical methods. But later in Pearson's life, and certainly in the careers of R. A. Fisher and the other great statistical minds of the first half of the century, there was a falling away of interest in graphics and an efflorescence of devotion to analytical mathematical methods. Indeed, for many years there was a contagious *snobbery* against so unpopular, vulgar, and elementary a topic as graphics among academic statisticians and their students. The recent advances by Tukey, Chernoff, and others appear to be bending that snobbery towards a more balanced approach. High time, too, with the increasing availability and use of computer-generated graphic methods!

Along with difficulties of communicating statistical analysis and conclusions by graphical and tabular methods, there are at least equally serious difficulties of clarity, honesty, accessibility, and so on in the ordinary prose used for statistical exposition. The balance, for example, between soporific dullness and suppression of full detail is never easy, although many cases of statistical prose are inadmissible in the sense that they could readily be improved simultaneously towards greater liveliness and greater precision. One can find in textbooks and journals a few suggestions towards better exposition in plain language, and there are examples of masterfully written statistical explanation on whose excellence almost all will agree. Yet I know of no empirical, metastatistical study of how well various prose devices perform. Even in poetry this empirical study has had its beginnings, although perhaps not yet statistical ones: Consider I. A. Richard's famous 1929 *Practical Criticism* (15).

There are many problems of statistical prose. One of them, discussed elsewhere in this essay, is how to state clearly the meaning for nontechnical readership of a confidence interval (or for that matter, of a fiducial interval or a Bayesian interval). A special set of problems arises for a conscientious head of a government statistical group that publishes many reports with figures, tables, and explanatory text. How is one to maintain expository and technical standards without the imposition of mechanical—and hence ultimately deadening—rules? I know of at least one U.S. federal statistical agency where this question is under active investigation (see Monroe G. Sirken et al. [16]).

Underlying all of these communication problems is heterogeneity of audience. What is clear to the Colonel's lady may not be to Judy O'Grady, and vice versa. Much of the statistical prose of the other essays in this volume seems exemplary to me, but my mother might find it overly abstract or pointless. We do not seem to have good methods for dealing with heterogeneous audiences, and it is hard for us to remain sensitive to the problem so that we use what weak methods we know, mainly sequential recapitulation in terms addressed first to the peanut gallery, then to the parterre, and then to the orchestra.

There exists, of course, a literature on statistical prose and related matters. For example, there have been studies (but all too few) of the effects of wording on response in opinion surveys (see, e.g., Elisabeth Noelle-Neumann [17]). The U.S. Bureau of the Census—and, no doubt, Census Bureaus in other countries—have carried out investigations of question wording. I suppose that there has been work in the fields of marketing, advertising, and communications. My impression, however, is that most of this research and writing has been episodic, noncumulative, not leading to development of useful theory, and that in any case, it is more concerned with the wording of questions than with exposition. May I be forgiven for gross ignorance if I simply do not know of relevant research and publication that fill the gap deplored in these paragraphs.

Problems with Statistical Conceptualization

I noted two elementary errors of statistical concept as I read SI-72. The first, on page 10, is in the context of a discussion of citations in scientific literature. It is remarked as a limitation of the citation approach that important articles may fail to be noticed because of their language or the journal in which they appear; on the other side, poor articles may be frequently cited as whipping boys, and articles presenting only minor improvements in technology may come to be frequently cited as convenient statements of a procedure or a result. Then the report says that such "... limitations of the [citation] indicator, however, are minimized by the extremely large number of citations involved in the present case." In other words, a possible source of bias is not worth concern because of the large sample size.

In general, of course, the situation is just the opposite. Under common circumstances, bias is unaffected by sample size, but plays a larger *relative* role as sample size increases because then sampling and measurement *random* errors have a reduced effect on averages or proportions. In the present case the situation is not so simple because the sampling method is apparently not a probabilistic one but nonetheless, I see no reason to expect bias to decrease as sample size increases.

Second, a survey-based discussion of public attitudes toward science includes the statement that the ". . . chances are 95 to 100 that the survey results do not vary by more than 2 percent (plus or minus) from the results that would be obtained if interviews had been conducted with all persons in this population" (p. 96).

This sounds at first blush like a Bayesian statement, but it seems clear from context that it is rather a confidence interval statement that is simply wrong in an all-too-frequent way. The point is that, once the survey is made, a reported percent either is or is not within 2% of its population value. We do not know which is the case. What we do know, however (everything else *comme il faut*), is that the *method* of procedure is one that, under hypothetical repetitions of the survey, followed by calculation, provides intervals that include the population value with probability .95. There are various clear ways of saying this, and the use of a vulgar misstatement makes me suspicious of the statistical bona fides of this part of *SI-72*.

It is not so much that the misstatement will trouble many readers—most of them will not notice it or be sensitive to the distinction—but the evidence does suggest insufficient participation in the study of the well-trained statisticians in, or available to, the Foundation.

My unease is extended by two other technical problems with the quoted statement. First, the $\pm 2\%$ is apparently an upper bound, appropriate when the (unknown) true percent is 50%. The actual length of a particular confidence interval would be proportional to $\sqrt{\hat{p}\,(1-\hat{p})}$, where $\hat{p}$ is the estimated true percent divided by 100; that square root has its maximum when $\hat{p} = 0.5$. (I stay away here from issues of adequacy of the asymptotic normal approximation implicit in the prior discussion.)

Second, the interval of the quoted statement and its corresponding 95% probability apply to a single reported percentage from the study. There are, however, many reported percentages, and it would indeed be surprising if some of them were not outside of the $\pm$ 2% range. This is a form of the familiar so-called multiple comparison problem, one to which the writer of this part of *SI-72* appears to be insensitive. (See Rupert G. Miller, Jr. or Peter Nemenyi [18]). So far as I know, there is no good, generally applicable solution to the problem, and perhaps there cannot be one. This aspect of the transition from numbers to conclusions presents one of the major open problems of statistics. An interesting exposition of it in the context of employment statistics is given by Geoffrey Moore (19).

I close this section by drawing back. Only two errors of statistical concept in 139 pages is a good score in most leagues of social statistics or, indeed, of any statistics.

DATA LIMITATIONS AND ACCURACY

A widespread problem in the analysis of data is lack of attention to how the data were generated, to their limitations, and in general to the error structure of sampling, selection, measurement, and subsequent handling. This ubiquitous—but by no means universal—inattention to error structure of data can occur even when there is widespread, if amorphous, cynicism about the data.

Consider, for example, the so-called body counts and pacification statistics in the Vietnamese War. One source for information about the distortion of pacification statistics is the U.S. Senate (20). Distortions of body counts have been widely reported—for example, Alvin Shuster in a *New York Times* article, quotes the following court martial testimony of William L. Calley, Jr.: "It was very important to tell the people back home we're killing more of the enemy than they were killing us. You just made a body count off the top of your head. Anything went into the body count. . ." (21). In my opinion, it is to the shame of American statisticians, including myself, that there was little or no professional outcry against the misuse of statistics on Vietnam. That outcry might have been made, whatever one's views at the time about the war itself.

In *SI-72*, a profusion of data is presented via tables, graphs, and summaries in the text. There is almost no commentary about limitations and error structure of the data presented.

Supporting Documents

Detailed tabulations of the economic, demographic, and educational data summarized in the main text are presented in "Appendix A—Indicators" of *SI-72*. Each table lists its supporting documentation, and it seemed worthwhile to me to look at that documentation, insofar as I conveniently could, for discussion of data limitations and error structure. Even if *SI-72* lacks examination of its data, perhaps the more nearly primary sources would contain such examination. Regrettably they do not.

Some supporting documents were cited many times, others only once. Some tables had a single citation, others several. Counting, as usual, leads to difficulties, ambiguities, and arbitrary decisions. Without worrying the details here, I counted seven citations to unpublished sources (e.g., for several tables beginning with Table 52, "Special tabulation by National Science Foundation based on data supplied by the U.S. Office of Education").

Then I counted 18 apparently published sources and I examined the 11 of those that I could find, without extensive research, in the library of the University of Chicago. The joint distribution by kind of source and findability follow.

Table 1. Published Sources Given in *SI-72* As References for Its Data.

	Found	Not Found[a]	Total
NSF	5	3	8
Other U.S. government agencies	4	0	4
International organizations	1	2	3
Private	1	2	3
Total	11	7	18

[a] Of the seven not found, one was listed in the card catalogue, but was mysteriously missing from the shelf, and has not been discovered. The other six were not in the card catalogue according to my amateurish eye. No doubt, with professional help from a librarian, some of them would have been turned up.

The selection process by which I did or did not find a source may lead to bias. For example, publications not owned by the library may, on the average, show less attention to data errors than those that are owned. In the other direction, a source I could not find because it was missing might well, on average have been more careful about data examination. With the time and energy I had to prepare this essay, however, I saw no ready way of solving this possible problem, and besides, there is this evidence that the unfound sources would not materially change the conclusions: Documents I found immediately did not differ in level of data examination from those I found after more extended effort.

Now what came of my examination of the 11 documents at hand? Discouragement. The levels of concern about data accuracy, completeness, and so on seemed to me unhappily low. I would grade them C to F on an absolute scale and B to D on a scale relative to ambient current practice. The most common problems were inattention to measurement error (e.g., classification in the trichotomy: basic research, applied research, development); inattention to conceptual ambiguities (the same example serves); inattention to nonresponse or noncoverage; and absence of ancillary studies to get at some of these problems.

To illustrate, I describe in a little more detail three of the documents that seemed better than others in terms of concern for quality of data or that seemed more interesting for other reasons.

The first is *Unemployment Rates and Employment Characteristics for Scientists and Engineers, 1971* (22). It is cited for Tables 48 and 49. This report, responsive to a wide concern about unemployment of scientists and engineers, is a discussion of two separate surveys. The first, about scientists, was based on an essentially complete mailing to all 305,402 persons who had reported to the 1970 National Register of Scientific and Technical Personnel. Its usable response rate was 83%. The second survey was based on a one-in-five systematic sample from a joint mailing list of 23 major engineering societies, a list comprising about 40% of the nation's engineers. The number of questionnaires mailed was 94,720; the usable response rate was 63%. No discussion is given about overlap of the two populations, about scientists who did not respond in 1970, or about the 60% of engineers not on the joint mailing list.

The first survey was analyzed without any discussion of sampling or measurement error. (There was, after all, no sampling in the usual sense.) The second was analyzed with estimated standard errors for dichotomous traits via the conventional formula for random sampling without replacement.

Both surveys asked for information about employment and unemployment. The questionnaire forms are reproduced in the report, and it deserves a gold star for that. (But the earlier 1970 form was not reproduced.) The engineering survey form contained many forced polytomies (geographical, engineering specialty, job function, etc.), and one might worry about respondent despair followed by arbitrary choice or chucking the form into the wastebasket. I noted no discussion of this problem.

I found little discussion or expressed concern about accuracy of response or about difference in response rate between those employed and those unemployed. There seemed to be little concern about the individuals outside the two universes. If 60% of the nation's engineers are absent from the mailing list and usable responses came from 63% of those on the list, then usable responses came from .63×40=25% of the nation's engineers. That figure seems to me a small fraction, especially in the light of plausible differing employment characteristics on which to base the comprehensive conclusions given. For example, it seems plausible that an unemployed engineer is less likely to receive a questionnaire, but more likely to respond than one who is employed.

At least, qualifying discussion should be prominent. It is not; what I did find was a brief paragraph on page 114, a sentence or two on page 116, and a paragraph on page 117; these pages are near the beginning of "Appendix A. Technical Notes." The discussion of sampling error on pages 116 and 118 is brief, strange, and incomplete. Its nomenclature is not standard, and a procedure (presumably to obtain confidence intervals for small probabilities with sample sizes too small for the normal approximation) is mentioned but not described. There are several variant approaches to this problem in the statistical literature.

The central point, of course, is that there was no reported serious effort to check on sources of error in the original NSF report. Then *SI-72* used some of that report's data without indicating their uncertainties or the uncertainties *about* those uncertainties. No doubt later reports will quote some of these numbers from *SI-72*, without indication of uncertainty. Thus we see a special case of secondary use of poorly documented data.

The second document on which I report, *Research and Development in Industry 1970* (23) is cited for Tables 19, 20, 34, 65, 66, 67, and 68 in *SI-72*. This document describes a stratified sample survey of industries. The stratum of large manufacturing industries was fully sampled, and smaller manufacturing industries and nonmanufacturing ones were sampled at varying rates depending on industry and employment magnitude. The large manufacturing industries (plus those in a few other fully sampled strata) are said to account "for almost 95% of the total R&D performance funds" (p. 26).

I found no detailed discussion of nonresponse, but it states on page 26: "the very few large companies that did not reply were mailed the census mandatory form MA-121. Less than 1 percent of total R&D was were [sic; the reader may choose to taste] obtained in this way and included in this report." Form MA-121 is not further described, but the basic questionnaire form (RD-1) *with* instructions is, to the report's credit, given in full.

The methodological part of the "Technical Notes Appendix," on which most of my comments are based, was prepared by the Bureau of the Census, which was "the collecting and compiling agent" for the study. Page 26 includes one semistandard Census discussion of confidence intervals; unfortunately, it contains wording that may mislead in a way similar to that described earlier for page 96 of *SI-72* and that had begun to become fixed in Census terminology. Page 28 gives a table of sampling standard errors. There are brief, general remarks about nonsampling sources of error on pages 26 and 27 of the "Technical Notes" and even (another gold star) on page 4 of the "Summary," where noncomparability with Bureau of Labor Statistics surveys is briefly discussed.

I must mention, however, that I have a sense of discomfort when I read so general a statement as this one on page 26: ". . . in addition to the sampling errors as measured by the standard error, the estimates are subject to errors in response coding, processing, and inputation [sic] for nonresponse." It is of course better to list such problems than to maintain the more usual silence. Yet it would be better still to be in a position to say something about the magnitude of these nonsampling sources of error. That they do not materially increase the overall random variability is implicit, but there is no supporting evidence whatever; in addition, there is possibility of bias from these sources.

The NSF report contains the running trichotomy: basic research, applied research, and development. I conjecture that this trichotomy must be highly uncertain, and my sympathy goes out to the chaps in the comptroller's offices of General Electric, American Telephone and Telegraph, and elsewhere, who must allocate their R&D funds among the three categories on the basis of the instructions reproduced on pages 107–109 of the report. The distinctions are doubtless important and clear-cut in many cases, but surely are fuzzy in many others. One may fear arbitrary and shifting methods used by the respondents on the basis of superficial job titles, division names, and fiscal reporting conventions. I looked in vain for any mention of such problems—for example, of exploratory visits to a few large respondents, or of ancillary surveys about which kind of corporate employees did the classificatory work and how they did it. One might also look at whether responses from a given company changed over time in synchrony with changes in the identity of the responsible employee.

How can such sources of error resulting from shifting definition affect *SI-72*? I can only conjecture, but let us suppose that the changing climate of opinion and of government pressure made *basic* research, compared with applied research and development, less attractive as the decade of the 1960s ran its course. There must be many industrial research efforts that are marginal as to whether they are basic or applied; the distinction, after all, is to a large extent one of the investigator's state of mind. If more and more of this marginal research were classified as applied rather than basic, then the movement of basic research in industry might be considerably distorted. In *SI-72* we find graphs on pages 44–45 about the movement of industrial basic research; for example, such research in physics-and-astronomy has gone from almost $150 million in 1967 (current dollars) to less than $100 million in 1970. How much of that substantial drop is real and how much a shift in definition following a shift in fashion?

I must add that pages 77–78 of *SI-72* include some healthy cautionary remarks about the accuracy of industrial R&D statistics.

Finally, I noted with some surprise that the R&D summarized in this study excludes "research in the social sciences or psychology, or other nontechnological activities or technical services" (p. 107). Also excluded are quality control, market research, product testing, and some other activities. The rationale for these exclusions—especially the social science-cum-psychology one—is not made clear. Indeed, one has to read with care to discover what is excluded. I am surprised that R&D social scientists have apparently not been complaining for years about this official snub. It would not be easy to remedy it retroactively, but what snub is?

My third example presents difficulties of a related, but somewhat different, sort. Tables 13a and 13b of Appendix A present summary statistics about the U.S. trade balance in technology-intensive products by year for the period 1966–1971. Table 13a gives imports, exports, and balance by broad category of product; Table 13b gives analogous totals by country and by class of country. Both tables give as source a serial publication of the Bureau of International Commerce in the Department of Commerce, *Overseas Business Reports*. For Table 13a, the source issue is 72-005, April 1972; for Table 13b, it is 72-001, May 1972. (The latter identifying number is apparently a misprint for 72-011, surely a venial misprint.)

When I examined these source materials, however, I found myself unable to obtain many of the numbers given in Tables 13a and 13b. As one example of difficulty in attempting to reconcile the numbers in Appendix A to *SI-72* and the numbers in the cited supporting documents, consider that part of Table 13a which lists annual imports of chemicals to the United States for 1966–1971. I compare next those import totals with the totals from the cited supporting document from the Department of Commerce, OBR 72-005,

Table 2. Stated Imports of Chemicals (Millions of Dollars)

	1966	1967	1968	1969	1970	1971
Table 13a, App. A						
SI-72	910	896	1058	1146	1328	1499
Cited Source	955	958	1129	1228	1450	1612
Difference	45	62	71	82	122	113

April 1972. There is an unexplained, substantial, and shifting difference between the two sets of numbers. In the Department of Commerce source, the total for chemicals is subdivided into eight subcategories, two of which are further dichotomized. I am tempted to play numerical detective in an effort to understand the above differences as omissions of some subcategory or subcategories of chemicals. In 1966 the smaller subcategory totals, starting from the smallest and, in order of size, are 27, 50, 53, 60, 61, 66, . . . None of them (nor their sums) fits the 45 discrepancy above, even allowing for rounding. In 1971 the corresponding ordered sequence begins 85, 106, 116, 119, 122, 127, . . . Just possibly, omission of the 116 (dyeing, tanning, and coloring materials) might explain the 113 discrepancy after considerable rounding, but that explanation does not hold for 1970, 1967, or 1966. Curiously, it *does* hold *exactly* for 1968 and 1969, when the dyeing etc. totals were 71 and 82, respectively. Might fluctuating categories have been omitted in different years?

The analysts preparing Table 13a may have had access to more detailed material than the cited Department of Commerce source, and categories may have been rearranged and recombined to fit the technology-intensive purposes for which Table 13a was prepared. Discussion of such more detailed data and consequent computations is, however, absent, and one cannot help questioning the level of statistical practice and exposition.* Surely, good practice calls for enough explanation so that the persevering reader can reconstruct for himself from the cited source what appears in the derived table. This is, in a way, analogous to replication of experiments or to repetition of calculation: It gives science its central *public* character.

Quite possibly I missed clues in *SI-72* or in the Commerce reports that would have permitted the reconciliations I sought. If so, I regret my obtuseness, but I do not blush too deeply, for the public or semipublic nature of the corrections I receive may be salutary beyond the bounds of my own skull.

The two Commerce reports do not themselves say anything about error structure for the reported numbers, but they refer for "further information regarding coverage, valuation, and compilation" to Census reports, FT 135 for imports and FT 410 for exports. These are serial publications, respectively,

U.S. Foreign Trade. Imports, Commodity by Country

U.S. Foreign Trade. Exports, Commodity by Country

*One of the sharp-eyed, sharp-witted correspondents listed in my introductory footnote of acknowledgment has suggested what may well be the explanation of the discrepancies in the table above. If one looks, not at *Overseas Business Reports* issue 72–005, but at issue 72–011, in particular at its Table 2 on page 19, one finds a column of total imports by years that is the same as the 955, 958, etc. row of our table above. But if one adds the imports as subdivided into countries and regions (Canada, Japan, 19 American republics, etc.), one finds the *smaller* numbers given by Table 13a of *SI-72*! Those totals have to be struck by the reader; they are not given as such in Table 2. Footnote 1 to Table 2 (on p. 25 of 72–011) explains that the global totals include trade "with other countries in the Western Hemisphere, in Eastern Europe, and in Asia." There is no explanation in 72–011 of why those countries are separated out, but the demerits of Table 13a are graver—inaccurate citation and the use of a subtotal that is neither described, explained, nor motivated.

My correspondent, while on the track of the above explanation, found the spoor of a second infelicity: On p. 19 of Table 2 of 72–011, the total chemical imports by year are subdivided into three kinds of chemical products. The sum of the three subtotals by kind is properly the same as the grand total for each of the years 1966–1971, *except* for 1969, where there is a discrepancy of $4 million—small from a percentage standpoint but perhaps troublesome in some analyses. No explanation seems forthcoming for what may well be a typographical error. There is not space here for discussion of the poorly understood ubiquitous problem of numerical and logical inconsistency in large statistical tabulations.

I have examined the discussions of accuracy and error in these publications, an easy task because almost nothing is said (except in connection with a sampling method used on shipments of low value). For reasons I do not understand, however, the discussion for exports is superior to that for imports. There is more detail and more frankness. For example, in the tricky issue of accuracy of valuation, the export volume says: "The extent to which the statistics reflect the stated value definition depends largely upon the accuracy of reporting by shippers on their export valuations " (24). The import statement is still more bland. It says: "Value information is required to be reported on import entries in accordance with Sections . . ." (25).

Now surely the Census statisticians know of alleged gross misreporting of valuation for customs purposes in international trade. Oskar Morgenstern discusses it (6). The recent articles by Richard Barnet and Ronald Muller refer to many studies and allegations of misreporting to minimize taxation or for similar reasons; for example, on page 114 of the second part of their article, they speak of a Rand Corporation study that ". . . concluded, because of the widespread use of transfer pricing and the importance of intracorporate transactions, Department of Commerce balance-of-payment statistics on foreign trade . . . are 'totally unreliable' " (26).

In addition to valuation, classification of commodity and country of destination or origin are subject to error. The Census materials mention these possibilities, but there is no evidence about magnitude and there is no indication that the amounts of error may be very large.

The absence of careful discussion of error in these Census foreign trade publications is in sharp contrast with the extensive attention given to error structure for the Census of Population and for the demographic work done by the Bureau of the Census. It would be an interesting study to examine the history and sociology of circumstances whereby one part of a large, technical, important government agency lives in so different a culture from that of another part. Perhaps it is a matter of frustration: one can really do something with and about errors in counting the population; errors in export statistics may, in contrast, be very difficult to understand and mitigate; and errors in *import* statistics may be quite impossible to handle. Yet surely that is too simple a conjecture. At least as important must be views in and out of government about the relative importance of different kinds of statistics.

Everyone Does It.

One response sometimes made to concern about inadequate attention to the quality of data is that everyone commits the same sin. The state of the art, so to speak, discourages attention to accuracy. To be up and at the substantive analysis is much more exciting.

It is true that the prevailing atmosphere of data analysis in economics, demography, and other related disciplines is not uniformly one of sharp attention to data defects. A comprehensive discussion for economics is that by Oskar Morgenstern (6), where many examples are given of economic data riddled with error and—much worse—not examined for error by their producers and most users. Some of the data he criticizes appear in SI-72, for example, gross national product and import-export statistics. Morgenstern says that it is desirable

> ... to stop important government agencies, such as the President's Council of Economic Advisors, the various government departments, . . . [etc.] from presenting to the public economic statistics as if these were free from fault. Statements concerning month-to-month changes in the growth rate of the nation are nothing but absurd and even year-to-year comparisons are not much better. . . . It is for the economists to reject and criticize such statements which are devoid of all scientific value, but it is even more important for them not to participate in their fabrication (p. 304).

Morgenstern tilts his lance with special emphasis at international comparisons such as those made near the start of SI-72. He discusses national income statistics on pages 276–282 and unemployment rates on pages 240–241. Some thoughtful writers believe, however, that Morgenstern pushes his argument too hard; for example, see Raymond T. Bowman, Morgenstern's reply, and Julius Shiskin (27).

So in a sense, it is true that, if not everyone, at least many people do neglect problems of data accuracy. But is this part of the intrinsic human condition, a sort of brand of Cain at which we can only howl? Of course not! There is wide variability in the degree to which data error structures are studied, and many authors do creditable work indeed. In On the Accuracy of Economic Observations Morgenstern refers to the pathbreaking attempts by Simon Kuznets to study errors of the U.S. national income statistics (pp. 254ff.). The Bureau of the Census, especially in its demographic sections, brings excellent minds to bear on the estimation of error via ancillary studies, theoretical analysis, and ransacking of prior data. (I mention, in particular, its program for studying underenumeration as a function of age, sex, and race. A recent report on one important phase of this was recently written by Jacob S. Siegel [28].) Indeed, the Census has recently published a report to instruct its analysts and authors about good ways to describe error structure in Census publications (29). This report is generally excellent, but its initial form contained the often misleading wording about confidence intervals to which I referred earlier.

Also worth citation as an example of error reconciliation and removal is a recent joint study by the U.S. Census and Statistics Canada to bring into statistical balance hitherto conflicting export and import accounts (30). Other examples of careful attention to error structure by government agencies may be found in the work of the National Center for Health Statistics and of the Statistical Engineering Laboratory of the National Bureau of Standards. These examples are, of course, not inclusive; a reader from an agency not mentioned here should not assume lack of approbation.

Attention to data accuracy outside government agencies also has, of course, wide variation. One can easily find examples of concern for data at levels far more careful than those of documents supporting Appendix A of *SI-72*. For example, I have been reading in Margaret G. Reid's *Income and Housing*, and I find intense concern with measurement error (31). One may point to many publications of the National Bureau of Economic Research as examples of reasonable attention to data accuracy. In short, good examples exist and are not difficult to find. The argument that everyone else is inattentive, whatever merit it may have or lack in principle fails because its hypothesis is wrong.

It is not only in economics that we find recognition of these problems of data quality and understanding of error structure. In psychology, for example, I have recently noted an apt paper by Donald W. Fiske (32) and in anthropology, an apt book by Raoul Naroll (33). No doubt, many other examples may be found. An excellent survey of nonsampling error problems is given by Frederick Mosteller (34).

Students of statistics in government and of statistics for public policy discussions have, at least since Quetelet, fretted about data accuracy. A nice discussion in the 1920s, for example, is given by Arthur Bowley (35). For a recent treatment, see the 1971 *Report* of the President's Commission of Federal Statistics (36).

A more recent example gives me an opportunity to quote myself, and better still, to present a striking quotation from Walter Lippmann embedded in my prose.

There are two extreme attitudes that can be taken towards statistical assertions or analyses about matters of public importance. Sometimes these attitudes appear to be held almost simultaneously.

The first is the naive attitude of credulity, of accepting asserted numerical facts without serious question. There is a mystique about numbers that lends itself to the rhetoric of propaganda: 99 percent pure, the parade was viewed by 9,672 people, 487 motorists will be killed next July 4. These are familiar cases in which everyone recognizes difficulties of meaning, accuracy, or of tendentiousness, yet they are assertions that apparently carry much more uncritical interest and persuasiveness than they deserve. A Vietnamese general is quoted as saying:

Ah, les statistiques! Your Secretary of Defense loves statistics. We Vietnamese can give him all he wants. If you want them to go up, they will go up. If you want them to go down, they will go down.[2]

At the other extreme we find the cynical attitude that damns all statistical arguments, on the grounds that a given set of observational data can be twisted toward almost any conclusion by a sufficiently clever and malicious analyst. This is the attitude that gave rise to the famous canard, "Lies, damn lies, and statistics." Such criticism of statistics is also naive; it is like criticizing language itself because it can be used to tell untruths, to persuade hypocritically, and to lead men astray. Further, the criticism is fatuous, for many arguments have to be statistical, in the sense that the only way many important questions can be answered is to gather numerical data and to analyze them.

Both of the above naive attitudes lead to troubles that can be mitigated by concentrating on the uncertainties of quantitative data, that is, on the idea of data fallibility or data errors. By carefully studying the error structure of the numbers in a statistical analysis, we are at once largely protected from credulity and from the mischief of distorted analysis. We may also be able to use the study of data fallibility to make at least partial corrections; for example, there has been much attention paid to errors in stated ages of people for censuses, and methods for mitigating the effects of those errors are known.

Unfortunately, there is not nearly enough emphasis on data fallibility in statistical arguments related to public matters, and we are therefore often led to premature or wrong decisions, and to fruitless argument. Neither, to be sure, is there adequate consideration of error in many investigations of basic science. But these are likely to be self-correcting, without the haste and pressure of public policy issues, and without the dire consequences that accompany wrong conclusions in the public arena.

Some readers may think that this stress on data fallibility is quibbling, unconstructive captiousness, but that is not the case. Errors will always be present and must be evaluated; we cannot simply pull the blanket of ignorance over our head and try to neglect the sound of things going wrong in the dark. To omit an investigation of error is to assume—often tacitly—that error is negligible. Walter Lippmann said in 1922 that

> the study of error is not only in the highest degree prophylactic, but it serves as a stimulating introduction to the study of truth. As our minds become more deeply aware of their own subjectivism, we find a zest in objective method that is not otherwise there.[3]

The systematic study of data fallibility is one function of the professional statistician, who should of course be knowledgeable about specific subject matter areas in order properly to examine errors in those areas. The absence of a careful study of error structure is good evidence that a statistical analysis is inadequate;

[2]Quoted in review of To Move a Nation by Roger Hilsman, The New York Times, June 9, 1967.
[3] Public Opinion (New York: The Macmillan Company, paperback edition, 1961), pp. 409–10.

unfortunately, some statistical analyses for public issues lack careful study of errors. It is that lack which we deplore and hope to remedy (37).

Finally, it is worth recalling that the social indicator literature includes some cautionary discussions about error structure. I note in particular the article by Amitai Etzioni and Edward W. Lehman (38).

Are There Standards?

Are there published standards for proper concern with error structure? Of course there are. They will not agree exactly, and there are many open questions, but standards exist and agree in large measure.

First, there are the standard textbooks for survey research. Second, there is a journal literature and some monographs like Morgenstern's. Third, there is a literature within federal agencies. I have already mentioned the Census publication on modes of error presentation. A more comprehensive one is Circular No. A–46 of the Office of Management and Budget. A relatively late revision was issued on May 3, 1974, and it is, in part, reprinted in the July 1974 issue of the *Statistical Reporter* (Statistical Policy Division). From that reprinting I especially note the following topics from "Standards for Statistical Surveys":

Purpose of the survey
Relation to other surveys
Target populations and extent of coverage
Sampling (includes detailed description, controlling and measuring non-sampling error, etc. as standards)
Method of collection (including pilot studies)
Consideration of sampling and nonsampling errors
Allowance for pretests
Questionnaire and instructions

and the following from "Standards for the Publication of Statistics":

Label data and define terms
Describe the survey design
Appraise the data

There are other government publications, for example, one published by the U.S. Department of the Army that was based on a predecessor to the OMB Circular (39).

Expense

One might say that to study error structure is too expensive. Even if this were true, it does not preclude discussion of possible errors, but I think that the argument of cost is seldom strong. Although careful examination of measurement bias may indeed be expensive, diversion of only a small fraction of money from the field operations of a survey will often suffice for a reasonable first attack at error. The real question is that of allocation of resources between the substantive study itself and the metastudy of its accuracy. In my opinion, almost always the allocation could be increased for the understanding of data accuracy. That opinion may of course represent a professional leaning, and further careful study of the question would be worthwhile. On what grounds can such issues be sensibly argued?

Utility

Does knowledge of data accuracy structure make a difference? Should it make a difference? One may, in part, answer these questions by asking a more fundamental question: What is the utility, or supposed utility, of the indicators in SI-72, even if they were error free? A modestly witty further question, such as "What is the utility of a newborn baby?" is not enough. Choices must be made, yet no one seems to have any clear idea of how to compare utilities for statistical programs. We are at the level of generalities and anecdotes.

In general terms, knowledge of data error structure has two basic functions. First, it helps decide when the difference between two related indicators reflects an underlying real difference. For example, SI-72 has many key statements like this one: "The fraction of total Federal outlays devoted to [research and development] fell from 12 to 7 percent between 1965 [and] 1972. The decline was due in large part . . ." (p. 20). The statement and explanation presume that there really was a decline from 12 to 7%. But perhaps there were changes during that period in definitions and measuring techniques that accounted for part of that decline, or even showed it to be greater. Perhaps some of the change results from sampling fluctuations in surveys on which the conclusion is based.

The second way in which knowledge of data error structure plays a role is in redesign of the data collection system and in deciding where best to put resources to use towards reducing errors. Are we, for example, better off interv.ewing 10 scientists in each of 50 research laboratories, five scientists in each of 100 laboratories, or what? To answer such design questions requires knowing relative error magnitudes. Even more serious design questions relate to potential sources of bias, for in general, investigations of such matters require serious side investigations to get at magnitudes of bias and methods for reducing it. For example, there is almost no empirical information about whether differing confidentiality arrangements will affect responses to a survey; the Bureau of the Census is planning a sequence of special investigations to throw light into this shadowy corner.

Yet perhaps the most direct questions are: Do data errors really have an effect? Are we only playing with statistical elegance? Isn't crude information enough, as it is enough for most personal decisions of daily life?

The last form of the question is the most provocative, for crude information is not enough in daily life. Witness the sick man who has the initiative, time, and money to get more than one independent medical opinion and, thereby, makes a better decision. Or witness the buyer or seller of a house or automobile, or whatever, who takes the trouble to examine the market rather than to rush impetuously ahead.

Some errors make a difference and others do not. Difference for what? Recent newspaper articles describe a recently discovered error in the Consumer Price Index (40). Apparently, the National Automobile Dealers Association's "Blue Book," a source of used car price information for the CPI, changed its basis for price quotation in the Spring of 1974 by including air conditioning for the first time. No one noticed the change until December and accordingly, the CPI was too high. For example, the October value of the index, 153.2, should apparently have been 152.9. That is in a sense a small error—only 0.3 points out of some 150—yet it will have a substantial effect on labor agreements, government pensions, and other economic arrangements linked to the CPI.

Franklin Zimring points out how two government agencies arrived at vastly different numbers of guns imported into the country, although both agencies presumably based their totals on the same basic documents (41). The number of guns imported relates directly to the legislative discussion of gun control.

Morgenstern gives many examples of the effect of data error (6,27).

Errors in the census of population affect congressional redistricting, revenue sharing, and many federal programs (see Siegel [28], U.S. Bureau of the Census [42] and, as yet unpublished work by I. R. Savage and Bernard M. Windham). Errors in import-export statistics affect legislation and public opinion about foreign trade and international relations (e.g., see Barnet and Müller [26]). Errors in income data affect estimated tax revenue. Errors in traffic counts have mightily affected the financial structure of some toll roads. The error structure of seismographic data has appeared centrally in negotiations about the underground testing of atomic bombs.

The British coal mine strike of the winter of 1973–1974 was said to have been exacerbated by a statistical error; according to the *New York Times* of February 23, 1974, because of a misunderstanding about handling vacation time, ". . . instead of seeming to be better off by 2 per cent, the miners are 8 to 10 per cent worse off . . . Prime Minister Heath denied that his Government had made a 'ghastly mistake' . . . Harold Wilson accused the government of incompetence." I have been informed by an official of the British statistical ministry that the mistake or misunderstanding was on the part of the press, not of the government. The example is apt in any case.

Geoffrey Moore says

> . . . in a time of rapid inflation it is easy to imagine ways in which inaccuracies [in price indexes] . . . could make a great deal of difference in the derived output estimates [of real GNP]. . . . I can't help thinking how ironic it would be if by overdeflating the current dollar figures, in a well-intentioned effort to get rid of the effects of inflation, we should have talked ourselves into recession (43).

Edward Denison gives a stimulating discussion of how errors in estimating the gap between actual and potential GNP may have seriously affected national economic policy in the 1960s (44). Rosanne Cole deals in a different, but equally stimulating, way with the effects of data errors on GNP business forecasts, and other aggregate economic quantities (45). Allan H. Young also treats these problems (46).

Many of the above examples are cases in which errors became known. What of the errors that are never known? They always exist. The central questions are those of *magnitude* of error levels. Perhaps fairly crude measurements will do for, let us say, the balance of trade. How do we know whether the actual error level is within that permissive interval of crudity without having quantitative evidence about the actual error levels? Perhaps the data are too precise, and we could save money by relaxing standards. More likely the errors are at a level far cruder than we are prepared to tolerate.

"Trend Data"

In *SI-72* and in most of the other NSF publications discussed earlier, there are many time series of economic, demographic, and other data, neatly tabulated and plotted. These are called "trend data" in some of the NSF publications, and they are clearly regarded as important.

Why are they important? Presumably, at least for one or both of two reasons. First, retrospective examination of these time series might lead to fundamental relationships among the variables; second, extrapolation —more or less formal—might permit prediction of the future behavior of the series. Although the NSF publications do not say why the time series data are regarded as important, a fair guess is that predictive extrapolation is the major motive. One uses the series, apparently, to make an informed guess about their future course.

It is remarkable that so broadly used an approach—not only in NSF publications, but in all sorts of social, financial, economic, and other contexts—has been so little studied as a method.* How accurate is intuitive extrapolation under these or those circumstances? What are its variations among individual extrapolators? How strongly is it affected by different modes of tabular or graphical extrapolation?

When one recognizes that the simple plotting of a time series followed by intuitive extrapolation is for many an admirable and advanced form of quantitative inquiry, it seems strange that there has been so little discussion of it from either a psychological or statistical viewpoint. Yet the far more developed discipline of demography has been in a similar position. Until recently, so far as I know, demographic predictions of population have been unaccompanied by probabilistic statements of accuracy. Many demographers have argued, and continue to argue, that such statements are impossible or, at least, inadvisable. Other demographers—for example, Nathan Keyfitz (49)—look forward to development of a calculus of probabilistic uncertainty for demographic predictions. Other interesting, broadly based statements have been written by Otis Dudley Duncan; and Denis F. Johnston (50).

The central difficulty, of course, is that of the shocks of changing social conditions on the usual demographic models. One insufficiently used device is to judge a prediction method by trying it on past data, independent of the data generating the method. Sometimes that is not possible, but often it is; for example, discussions are given by W. H. Williams and M. L. Goodman; James E. Hacke, Jr.; and Byron W. Brown, Jr. and I. Richard Savage (51). P. Newbold and C. W. J. Granger are also worth citing for discussion of semimechanical time series extrapolation methods; their article is followed by a sharp, illuminating discussion (52). A clear treatment of relatively recent predictions, and of ones found erroneous by hindsight, appears in *The New Yorker* (53). Albert Wohlstetter uses retrospective evaluation prediction to examine accuracy in American predictions of Soviet strategic arms (54).

Related methods of using internal variability to obtain probabilistic predictions include the jackknife approach (Tukey [54a]) and so-called cross-validation and predictive inference (Geisser, Stone [55]). Frederick Mostel-

---

*There are some related studies, for example, those of Victor Zarnowitz about the economic predictions of businessmen (47). A brief, tantalizing reference to empirical intuitive extrapolation trails is given on page 208 of a 1959 essay by Paul A. Samuelson (48); he informs me, however, that the details of his observations have not been published.

ler describes the jackknife approach (34); good entries to the literature in this area are provided by S. Geisser and M. Stone (55).

Predictions with associated error bands might well have been used for at least the more important of the many time series plotted and tabled in *SI-72*. Without that analytical help, the reader is left with only his eyes and general experience to guess even the near-term future. It might seem presumptuous to give predictions for error, and it might also be depressing, for realistic prediction error bands are usually saddeningly wide. Yet is it not better to do something of this kind than to maintain the customary overoptimistic silence?

It is healthy and sporting to look at older predictions and to see how wide of the mark they have been. Such a game has been standard in demography for many years. More qualitative predictions of social, scientific, and technological events are also frequently in error and make fair sport. Two such recent critiques are by Garrett Birkhoff of 1968 predictions by Herbert Simon and Allan Newell about progress in high-speed intelligent computation, and by *The New Yorker* (53) of predictions by Herman Kahn and Anthony Wiener about the economy and society.

It seems to me, however, that the game of swatting the intrepid predictor is too simple to be more than sport. But if predictions were made with associated sensible error bounds, then the future critic, looking back with hindsight, would only criticize fairly if the eventuality lay outside those bounds. But the wider the bounds at the time of prediction, the less credence, utility, newsworthiness, and public interest in the prediction. How much space, for example, would newspapers give to a political poll leading to a conclusion like this: The proportion of electoral votes for candidate D will, taking bias, sampling error, and random measurement error into account as best we can, lie with high confidence between 45% and 59%?

This is, of course, an example of the familiar statistical tension between wanting to say something precise and wanting to say something correct. The tension will never be resolved, but it can be explored and analyzed.

THE DELPHI STUDY

An interesting part of *SI-72* is its Delphi study. This example of the so-called Delphi method began with six panels of participants having experience and knowledge in scientific and social affairs; the panel members were asked a structured set of questions, summary information about the answers was fed back for a second round of questioning, and then the second round of replies was summarized.

Some panels were asked *normative* questions—for example, desirable criteria for allocation of future scientific funds. Some panels were asked *predictive* questions—for example, the extent to which science and technology could ameliorate various problems of public concern, aside from the question of whether there should be increases in scientific and technological funding. Other panels dealt with *interpretations* of past events—for example, the effects of past R&D funding changes. The names and professional affiliations of the panelists are given, but we are not told which panelists were on which panel. Surely that is an unfortunate omission.

The Delphi method is a way of obtaining coordinated views from a panel of respondents (often, but not necessarily, experts) without having them meet for joint discussion. There are three essentials of the Delphi method as it is usually presented,

1. Anonymity of response
2. Statistical summarization of the distribution of responses, and communication of that to the respondents
3. Iterative questioning, with continued statistical feedback

The advantage of Delphi over ordinary committee or group discussions include anonymity, less dependence on rhetorical power and personal influence, and greater flexibility in committee formation because its members need not be brought physically together. Some of those advantages might be regarded as disadvantages—for example, the direct interaction of committee members in a meeting room might be a great aid to wise conclusion.

There are many varieties of the Delphi method. Panel selection and instruction, question choice and refinement, mode of feedback, number and nature of iteration and so on all permit a wide scope for variation. Indeed, why should one consider so formless and ad hoc a procedure? The reason, I think, is not hard to find. Institutions and governments of all kinds have constant needs and desires for advice, usually, but not always,* supposedly expert advice. How should the prince, the politician, the executive find his advisors, arrange their deliberations, and summarize their recommendations? Insofar as advice is more than decoration or a stalling device, the problem of arranging for effective advice must be felt in almost all societies at all times. To my knowledge, most of the literature on this topic has been descriptive and anecdotal. A recent study, itself by a committee, that goes

*Nonexpert advice may be sought for a number of reasons. For example, the U.S. Bureau of the Census has received help from groups of nonexpert citizens on what aspects of a census form might be confusing. (Of course one might say that a nonexpert is, paradoxically, an expert on the reactions of at least one nonexpert.) Public opinion polls might be regarded as advice from nonexperts. Panels from the general population have been widely used for marketing studies.

beyond anecdote to serious general consideration deals with advice for science policy (57).

The Delphi method is of interest because it is a relatively new way of structuring advice, and a way that can in principle be studied because it permits specification and independent replication. To be sure, committee advice in principle also permits specification and independent replication —indeed, *The Science Committee* (58) suggests just that—but I see no movement in that suggested direction. Yet there has been some attempt to specify, replicate, understand, and validate the Delphi method.

The most complete discussion of the Delphi method known to me, along with an extensive annotated bibliography, is by H. Sackman (59). Sackman's report is highly critical of Delphi, but I think, by standards that are inappropriately or anachronistically high, even for most run-of-the-mill social science and surely for studies of the advisory process.

Yet it is apparently the case that most Delphi studies are poorly documented and do not use the opportunities for evaluation replication that they have. At any rate this seems to be the case for the Delphi study in *SI-72*. We are told nothing in detail about how the panel was selected, nothing about the number of refusals, and little in detail about the instructions to the panels. (How does one convince busy panelists to take difficult, vague questions seriously?) Only sketchy information is given about the nature of feedback, and there was no replication in an attempt to get at inherent variability. That is a pity because, in the Delphi context, it is relatively easy to replicate with effective independence, especially in comparison with conventional advisory committees where independence would generally be very difficult to attain.

As I read the Sackman report and look at some of its references, however, I sense that the *SI-72* Delphi study may not have been so very atypical. In general, the Delphi standards for documentation, replication, together with sampling procedures, pilot studies, and so on, seem to leave a great deal to be desired. That impression has been confirmed by reading a Delphi study, unlisted by Sackman, that I discovered in an unlikely place, a House of Representatives document (60).

I close by repeating that, unsatisfactory as is the state of validation and understanding for Delphi, it is far more unsatisfactory for ordinary advisory committees, ungraced by a glowing Greek name. Perhaps the work of social psychologists on problem solving by small groups will eventually have some applicability here. (One attractive entry to that literature is by Elliot Aronson [61]). Meanwhile, I recommend to the NSF that it encourage empirical, theoretically based, systematic study of the role of science advisory committees.* It has many of those committees in its own house and it might wish to begin by arranging for modest formalization and replication of some of their activities.

REFERENCES

1. *Science Indicators 1972: Report of the National Science Board, 1973,* National Science Board, National Science Foundation, Government Printing Office; Washington; D.C., 1973.

2. National Science Foundation, *Research and Development in Industry, 1970,* Government Printing Office, Washington D.C., 1971.

3. P. Kael, "On the Future of Movies," *The New Yorker,* **50,** No. 24, 15–59 (August 5, 1974). Page 47.

4. H. A. Zuckerman, "Patterns of Name Ordering Among Authors of Scientific Papers: A Study of Social Symbolism and Its Ambiguity," *American Journal of Sociology,* **74,** 276–291 (1968).

5. U.S. Executive Office of the President, Office of Management and Budget, *Social Indicators 1973,* Government Printing Office, Washington, D.C., 1973.

6. O. Morgenstern, *On the Accuracy of Economic Observations* 2nd ed., Princeton University Press, Princeton, 1963.

7. S. E. Fienberg and L. A. Goodman, "Social Indicators 1973: Statistical Considerations," in R. Van Dusen, Ed., *Social Indicators 1973: A Review Symposium,* SSRC, New York and Washington, 1974, pp. 63–82.

8. C. F. Schmid, "Graphic Presentation," *International Encyclopedia of the Social Sciences,* Macmillan and Free Press, New York, 1968; J. A. Davis and A. M. Jacobs, "Tabular Presentation," *International Encyclopedia of the Social Sciences,* Macmillan and Free Press, New York, 1968.

9. J. W. Tukey, *Exploratory Data Analysis,* Addison-Wesley, Reading, Mass., 1977.

10. H. Chernoff, "The Use of Faces to Represent Points in k-Dimensional Space Graphically," *Journal of the American Statistical Association,* **68,** 361–368 (1973).

11. F. J. Monkhouse and H. R. Wilkinson, *Maps and Diagrams, Their Compilation and Construction,* 3rd ed., Methuen, London, 1971.

12. H. Wainer, "The Suspended Rootgram and Other Visual Displays: An Empirical Validation," *The American Statistician,* **28,** 143–145 (1974).

13. A. D. Biderman, "Kinostatistics for Social Indicators," *Educational Broadcasting Review,* **5,** 13–19 (1971).

14. N. Weisstein and C. S. Harris, "Visual Detection of Line Segments: An Object-Superiority Effect," *Science,* **186,** 752–755 (1974).

15. I. A. Richards, *Practical Criticism: A Study of Literary Judgment,* Harcourt, Brace & World, New York, Paperback 1956. First published 1929.

*These studies are, of course, not unknown. A recent example, sponsored by the Department of Health, Education, and Welfare, is described by John H. Noble, Jr. (62). Allan Mazur has suggested similar experiments to throw light on resolving disputes among experts; in particular, he sketches an experiment to investigate the importance of oratorical ability in adversarylike hearings about public policy questions with major scientific aspects (63). See also Alan Porter and Jacques Vallee et al. (64). An example of experimentation on persuasion within small groups is described by Harvey London et al. (65).

16. M. G. Sirken, B. I. Shimizu, D. B. Brock, and D. K. French, "Statistical Standards for Analyzing Data Based on Complex Sample Surveys," *Proceedings of the American Statistical Association, Social Statistical Section*, 1973, pp. 1–7.

17. E. Noelle-Neumann, "Wanted: Rules for Wording Structured Questionnaires," *Public Opinion Quarterly*, **34**, 191–201 (1970).

18. R. G. Miller, Jr., *Simultaneous Statistical Inference*, McGraw-Hill, New York, 1966; P. Nemenyi, "Linear Hypotheses: III. Multiple Comparisons," *International Encyclopedia of the Social Sciences*, Macmillan and Free Press, New York, 1968.

19. G. H. Moore, "Response to Commissioner Shiskin," *New York Times*, December 8, 1974.

20. U.S. Senate, 91st Cong., 2nd Sess., 1970, *Refugee and Civilian War Casualty Problems in Indochina*, Committee on the Judiciary, Subcommittee to Investigate Problems Connected with Refugees and Escapees, September 28, 1970.

21. Alvin Shuster, *New York Times*, 1 April 1971.

22. *Unemployment Rates and Employment Characteristics for Scientists and Engineers, 1971*, NSF 72-307, National Science Foundation, Government Printing Office, Washington, D.C., 1972.

23. *Research and Development in Industry 1970*, NSF 72-309, National Science Foundation, Government Printing Office, Washington, D.C., 1971.

24. *U.S. Foreign Trade, Exports, Commodity by Country*, FT 410, (March 1972), p. v.

25. *U.S. Foreign Trade, Imports, Commodity by Country*, FT 135, (December 1973), p. vi.

26. R. Barnet and R. Müller, "Global Reach," *New Yorker*, **50**, 41–42, December 2 and 9, 1974); and in book form by Simon & Schuster, New York, 1974.

27. R. T. Bowman, "Comments on 'Qui Numerare Incipit Errare Incipit' by Oskar Morgenstern," *The American Statistician*, **18**, No. 3, 10–20 (June 1964); O. Morgenstern, "Fide sed Ante Vide, Remarks to Mr. R. T. Bowman's 'Comments'," *The American Statistician*, **18**, 15–16 (October 1964); J. Shiskin, "Signals and Noise," "Letter to the Editors," *Fortune*, **69**, 66, (February 1964).

28. J. S. Siegel, "Estimates of Coverage of Population by Sex, Race, and Age: Demographic Analysis," in U.S. Bureau of the Census, *1970 Census of Population and Housing, Evaluation and Research Program*, PH C (E)-4, and *Demography* **1**, 1–23 (1947).

29. U.S. Bureau of the Census, *Standards for Discussion and Presentation of Errors in Data*, Technical Paper 32. See the revision, M. E. Gonzalez, J. L. Ogus, G. Shapiro, and B. J. Tepping, "Standards for Discussion and Presentation of Errors in Data," *The Journal of the American Statistical Association*, **70**, 5–23 (September 1975). Part II.

30. U.S. Bureau of the Census and Statistics Canada, *The Reconciliation of U.S.-Canada Trade Statistics, 1970*, 1973.

31. M. G. Reid, *Income and Housing*, University of Chicago Press, Chicago, 1962.

32. D. W. Fiske, "The Limits for the Conventional Science of Personality," *Journal of Personality*, **42**, 1–11 (1947).

33. R. Naroll, *Data Quality Control–A New Research Technique*, Free Press, Glencoe, Ill., 1962.

34. F. Mosteller, "Errors: Nonsampling Errors," *International Encyclopedia of the Social Sciences*, Macmillan and Free Press, New York, 1968.

35. A. Bowley, "Some Tests of the Trustworthiness of Official Statistics," *Economica*, **8**, 253–278 (1928).

36. President's Commission on Federal Statistics, *Federal Statistics*, Government Printing Office, Washington, D.C., 1971.

37. W. Kruskal, "Statistics, Public Policy, and Data Fallibility," in W. Kruskal, Ed., *Mathemat-*

ical Sciences and Social Sciences, Prentice-Hall, Englewood Cliffs, N.J., 1970, Chapter 4, pp. 49–51. Part of the Survey of the Behavioral and Social Sciences.

38. A. Etzioni and E. W. Lehman, "Some Dangers in 'Valid' Social Measurements," *Annals of the American Academy of Political and Social Science,* **373**, 1–15 (September 1967).

39. U.S. Department of the Army, Headquarters, *Statistics/Federal Statistical Standards,* Pamphlet 325-5, (March 17, 1965).

40. *New York Times,* 8 December 1974; *Washington Post,* 7 December 1974.

41. F. E. Zimring, "Firearms and Federal Law: The Gun Control Act of 1968," *Journal of Legal Studies,* **4**, 133–198 (1975).

42. U.S. Bureau of the Census, *Coverage of Population in the 1970 Census and Some Implication for Public Programs,* provisional version, 1975.

43. G. Moore, "On the 'Statistical Significance' of Changes in Employment and Unemployment," *Statistical Reporter* (March 1973), reprinted in *How Full is Full Employment?* American Enterprise Institute for Public Policy Research, 1973.

44. E. F. Denison, *Accounting for United States Economic Growth 1929–1969,* The Brookings Institution, Washington, D.C., 1974.

45. R. Cole, "Data Errors and Forecasting Accuracy," in J. Mincer, Ed., *Economic Forecasts and Expectation,* National Bureau of Economic Research, New York, 1969, pp. 47–82.

46. A. H. Young, "Reliability of the Quarterly National Income and Product Accounts of the United States, 1947–71," *The Review of Income and Wealth,* **20**, 1–35 (March 1974); reprinted under the same title and with substantial additions in Bureau of Economic Analysis Staff Paper No. 23, Social and Economic Statistics Administration, U.S. Department of Commerce, July 1974.

47. V. Zarnowitz, "Prediction and Forecasting, Economic," *International Encyclopedia of the Social Sciences,* Macmillan and Free Press, New York, 1968.

48. P. A. Samuelson, "What Economists Know," in D. Lerner, Ed., *The Human Meaning of the Social Sciences,* Meridian Books, World, Cleveland and New York, 1959, pp. 183–213.

49. N. Keyfitz, "On Future Population," *Journal of the American Statistical Association,* **67**, 347–363 (1972).

50. O. D. Duncan, "Social Forecasting—The State of the Art," *The Public Interest,* **17**, 88–117 (1969); D. F. Johnston, "Long Range Projections of Labor Force," in J. P. Martino and T. Oberbeck, Eds. *Long Range Forecasting Methodology,* Second Symposium, Alamogordo, New Mexico, U.S. Air Force, 1967, pp. 97–125,; and Discussion, pp. 125–129; D. F. Johnston, "Forecasting Methods in the Social Sciences," *Technological Forecasting and Social Change,* **2**, 173–187 (1970).

51. W. H. Williams and M. L. Goodman, "A Simple Method for the Construction of Empirical Confidence Limits for Economic Forecasts," *Journal of the American Statistical Association,* **66**, 752–754 (1971); J. E. Hacke, Jr., "Anticipating Socioeconomic Consequences of a Major Technological Innovation," in J. P. Martino and T. Oberbeck, Eds., *Long Range Forecasting Methodology,* Second Symposium, Alamogordo, N.M., U.S. Air Force, 1967., pp. 131–141, and Discussion pp. 142–146; B. W. Brown, Jr. and I. R. Savage, "Statistical Studies in Prediction of Attendance for a University," in H. Correa, Ed., *Analytical Methods in Educational Planning and Administration,* McKay, New York, 1975. p. 171–198.

52. P. Newbold and C. W. J. Granger, "Experience with Forecasting Univariate Time Series and the Combination of Forecasts," *Journal of the Royal Statistical Society,* Series A, **137**, 131–146, and Discussion, 146–164, (1974).

53. "The Talk of the Town; Notes and Comment," *The New Yorker*, **51**, No. 1, 29–30 (February 24, 1975).

54. A. Wohlstetter, "Is There a Strategic Arms Race?" *Foreign Policy*, **16**, 3–20 (Summer 1974); A. Wohlstetter, "Is There a Strategic Arms Race? (II) Rivals but No 'Race'," *Foreign Policy*, **16**, 48–81 (Fall 1974). The two above papers, together with further material, are in preparation for book publication. The expected title is *Competition or Race: Innovation and Changing Size of Strategic Forces.*

54a. J. W. Tukey, "Bias and Confidence in Not-Quite Large Samples," *Annals of Mathematical Statistics*, **58**, 614 (abstract) (1958); see also F. Mosteller and J. W. Tukey, "Data Analysis, Including Statistics," in Gardner Lindzey and Elliot Aronson, Eds., *The Handbook of Social Psychology*, 2nd ed., Addison-Wesley, Reading, Mass., 1968, Vol. 2, Chapter 10.

55. S. Geisser, "The Inferential Use of Prediction Distributions," in V. P. Godambe and D. A., Sprott, Eds., *Foundations of Statistical Inference*, Holt, Rinehart & Winston, Toronto and Montreal, 1971, pp. 456–469; M. Stone, "Cross-Validatory Choice and Assessment of Statistical Predictions," *Journal of the Royal Statistical Society*, Series B, **36**, 111–133 (1974), and Discussion, 133–147 (1974).

56. G. Birkhoff, "Mathematics and Computer Science," *American Scientist*, **63**, 83–91 (1975).

57. D. W. Bronk, *The Science Committee*, National Academy of Sciences—National Research Council, Washington, D.C., 1972.

58. *Ibid.*

59. H. Sackman, *Delphi Assessment: Expert Opinion, Forecasting, and Group Process*, Rand Report R-1283-PR, 1974; said to be forthcoming in book form.

60. S. Enzer, P. de Brijard, and F. D. Lazar, "Some Considerations Concerning Bankruptcy Reform," in Part III of *Report of the Commission on the Bankruptcy Laws of the United States*, 93rd Cong., 1st Sess., House Document 93–137, Government Printing Office, Washington, D.C., 1973. The authors are identified with the Institute of the Future.

61. E. Aronson, *The Social Animal*, Viking, New York (hardback), and Freeman, San Francisco (paperback), 1972, p. 74.

62. J. J. Noble Jr., "Peer Review: Quality Control of Applied Social Research," *Science*, **185**, 916–921 (September 13, 1974).

63. A. Mazur, "Disputes Between Experts," *Minerva* **11**, 243–262 (1973).

64. A. L. Porter, "Letter to the Editor," *Science* **183**, 1142 (March 22, 1974); J. Vallee, H. Lipinski, R. Johansen, and T. Wilson, "Computer Conferencing," "Letter to the Editor," *Science* **188**, 203 (April 18, 1975).

65. H. London, P. J. Meldman, and A. van C. Lanckton, "The Jury Method: How the Persuader Persuades," *Public Opinion Quarterly*, **34**, 171–183 (1970).

Economic Problems of Measuring Returns on Research

7

Zvi Griliches

It should be emphasized from the start that the view of science in this paper is primarily utilitarian. It focuses on the measurement of the economic "product" of science and views research as a form of investment in economically useful new knowledge and technology. Much of the discussion of public support for science is couched in these terms though we all know that this is not all there is to science. In fact, it may not even be the major part of science. In economists' terms, much of science may be a form of public consumption and a complement to, and by-product of, our educational system. An interesting topic, which I do not pursue further here, is what science may "really" be about, what the people who engage in it are actually doing, what they are maximizing, and to what incentives they are responding.

From an economist's point of view, the main problem with measuring the "output" of science is that very little of it is sold on any market; from a statistician's point of view, the problem is that very little of it emerges in units that are compara-

In this paper I borrow heavily from my previous papers on this subject (1–2). I am indebted to the National Science Foundation for research support that made some of this work possible.

ble and countable. The two problems are not unrelated. To my knowledge, there are three types of direct measures (indicators) of scientific output: invention lists, patents, and publication counts. Invention lists, and associated attempts to evaluate their relative economic and scientific importance, suffer from selectivity and incompleteness of coverage. The population from which a particular list is drawn is not clearly defined: only those inventions are listed that succeed somehow in drawing someone's attention to them.

Patents have a specific legal definition and the incentives to patent may vary greatly over time and space. Patent statistics are produced in large numbers but have not been used extensively by economists. The only really good intensive study is that of Schmookler (3), which takes patents as a measure of scientific activity and focuses primarily on an analysis of their response to the economic forces of supply and demand rather than on tracing the impact elsewhere of such measured swings in scientific activity. Only a few, not very successful, attempts have been made to show a relationship of patent counts to any other subsequent measure of economic output or productivity. The link between patents and productivity or other economic growth measures has not really been worked out; more could be done in this area. Unfortunately, patent statistics have not yet been computerized and classified adequately to allow a serious study of them, either by industry of origin or by industry of use.

Publication counts, the third direct measure of scientific activity, have also been used primarily to analyze the internal workings of different scientific disciplines rather than to establish a relationship between such a measure of "science" and some other measure of socially valuable product. A recent study by Evenson and Kislev did succeed in finding a statistically significant relationship between publication counts in the relevant biological and agronomic literature and subsequent changes in the yields of corn and wheat in different countries and regions (4). This is an interesting line of research, particularly as it allows measurement of the spillover effect of research from one country to another. Presumably, it could be improved by weighting the various publications by their relative importance as reflected in the more recently available citation counts (see papers by Kochen, and by Garfield et al., this volume).

Most of the economic research in this area, faced with the endless difficulties of the direct measures of scientific activity discussed above, skipped over this whole stage and pursued what might be called "indirect" or "ultimate" measures of the economic impact of science. By observing differences in the growth of GNP or industrial output over time and space and relating them to differences in the intensity of scientific activity as measured by research and development (R&D) expenditures or the employment of scientific personnel—and adjusting for such other sources of growth as the size of the labor force and physical capital—one tries to estimate the contribution of research to the growth in the observed economic magnitudes. Though there are different ways of doing it and while the algebra can become quite involved (1), basically, the model is very simple. All that is required are two numbers, p and k, which when multiplied together, will give the contribution of research to the observed rate of economic growth. The first number, p, is the social rate of return on investments in R&D; the second, k, is the net R&D investment ratio. We can observe the gross R&D investment ratio (the ratio of total R&D expenditures to GNP), which has hovered around .03 in recent years, in the United States; therefore, "only" two additional numbers (in addition to p) are needed to arrive at k—the fraction of current and past R&D expenditures that can be thought of as affecting GNP as it is currently measured and the fraction of current R&D expenditures that constitutes net-investment rather than just replacement or duplication.

The evidence on p, the rate of return on R&D expenditures, has been surveyed recently by Mansfield and myself, (5, 1–2). It is based primarily on two sources: calculations of returns to individual inventions and econometric studies of productivity growth in specific industries and for particular firms. Although each of these studies is subject to a variety of separate reservations, together they all point to a reasonably consistent relation between productivity growth and research expenditures and to relatively high (30–50%) rates of return, on average, to both public and private research.

Given such estimates of p and notwithstanding all the reservations about them, the main problem in using them further—and the main point of this paper—is that such estimates cannot be applied directly to the reported total of all scientific and R&D expenditures to compute their contribution to economic growth. The reason is not that many of these activities are not socially productive but rather that their contribution may not be reflected in the standard measures of national economic success such as "real" GNP or the rate of growth in productivity (real GNP per man-hour). Turning to such "indirect" measures of scientific output, we end by looking for something that is largely not even there—for reasons having to do both with the nature of much of R&D and the nature of our GNP and productivity accounts. There are at least two problems here:

1. Most of the R&D "product" is sold to the public sector. It consists of research on defense and space exploration, for which no adequate market valuation exists. By accounting convention, these "outputs" are measured by costs, resulting in zero contribution to measured productivity, by definition.* For example, whether the moon landing was successful or not,

*This is true for the bulk of defense and space research, which is contract research with an end-product of blueprints, formulae, or prototypes. It is also largely true for the research components of the "hardware" sold to government, because of the lack of appropriate price indexes by which to deflate these expenditures.

the effect on measured GNP would be the same. The contribution of the space effort is measured by the inputs into it, the cost of man-hours and hardware, and not by the success of these efforts. Similarly, whether Phantoms can fly faster than earlier planes and whether our security situation has improved or deteriorated as the result of research expenditures on military technology will not show up in the growth of national productivity as it is currently measured.

Or, taking another example, consider public research expenditures on health that lead to increased life expectancy and a lower incidence of some disease. These also would not show up in the current productivity accounts. Currently, such an improvement would raise both measured GNP and the hours worked by the labor force, leaving "productivity" (output per man-hour) largely unchanged. The contribution of such expenditures could be computed if we evaluated the changes in the "productivity" of obtaining hours of work from a given population. But that kind of calculation is not within the framework of the national-income accounts as they are currently constructed.

2. Much of the product of private R&D investments is in the form of new commodities or improvements in the qualities of old commodities. Whether this product shows up in current-dollar GNP depends on the short- and long-run monopoly position of the firms engaging in research and on the fraction of the social gain (consumer surplus) appropriated by them. Whether this product shows up in "real" GNP (in constant prices) depends on what happens to the price indexes with which the current net output of the particular industry is deflated. If the price indexes were to recognize these improvements in "quality" fully, the resulting real-GNP measures would reflect the social product of this research. But that is unlikely. If the price indexes fail to reflect these quality improvements, only the private product will show up in the real accounts—to the extent that firms succeeded in appropriating it via higher prices for the newer higher quality products.

If some of these unmeasured improvements are attached to products used in turn as inputs into the production of other private products, their contribution will show up in the productivity measures of the industries that purchased them. Thus for example, though the contribution of research on improving the performance of farm tractors may not show up in the output-input account of the tractor industry, it will have an effect on measured productivity in agriculture. But many new products are sold directly to consumers, and the improvements associated with their use are not caught in the conventional measures of GNP or consumption per capita. Examples are time-savers like pocket minicalculators or the whole air-transport industry: The fact that one can get to the West Coast from the East Coast in half the time required 15 years ago leaves hardly a trace in the national productivity accounts.

In a previous paper (1) I tried to guess the extent of this problem. By looking in some detail at the sources of funds and the announced uses of them, I concluded that in 1970, out of the $27 billion or so identified as "research and development" expenditures, half could be attributed to defense and space activities. Of the remainder, about $11 billion could be thought of as having a potential impact on aggregate productivity. These $11 billion consisted of $9.5 billion spent in industrial research; about $0.3 billion in intramural federal research; and $1.3 billion in research in universities, institutes, and similar organizations.*

This still does not allow for any "spillovers" from defense and space research. An upper-limit estimate might be based on the assumption that 20% of such expenditures have the same effect and the same rate of return as private expenditures aimed directly at affecting private productivity. Table 1 summarized the rough allocation of various R&D investments as to their potential impact on measured productivity in the private sector, including an allowance for externalities generated by that part of R&D (such as the space effort) not deemed to have a direct effect on productivity. A little over half of total R&D expenditures is estimated to be "productive" in this sense.

To use the figures listed in Table 1 together with the earlier reported estimates of ρ, we have to make an assumption about their division into social and private components. To simplify matters, I make the following assumptions: private R&D expenditures earn a "normal" gross rate of return of about 20% (10% for depreciation, 10% net) and yield an *additional* 20% in the form of externalities in the private sector, whereas public R&D investments (direct and indirect) yield a gross social rate of return of 30% per annum.† These assumptions and the distribution summarized in Table 1 yield an average ρ of 0.366.

The other adjustment that remains to be made is to estimate the part of R&D investments representing "net" investment. The distinction between *gross* and *net* is important because research investments both depreciate and become obsolete. They depreciate in the sense that much knowledge would be forgotten and rendered useless without continued efforts at exercising, retrieving, and transmitting it; this is what much of higher education is about. Also, since some of the newer findings displace, make obsolete, and

---

*The National Science Foundation has also published an allocation of Federal R&D funds by function (6). While the numbers differ somewhat, the conclusions are roughly the same.

†These numbers are explained and defended in greater length in Ref. 1, where several alternative calculations are also presented. Different ways of cutting into this problem yield rather similar results.

Table 1.   Distribution of R&D Expenditures (U.S., 1970) by Potential Direct and Indirect Effect on Measured Productivity

	Percentage	
Effects	Total	Productivity-Related
Direct Industrial R&D		
Private	40	35
Public	30	1
Public R&D		
Federal intramural	14	1
Universities, research centers, and related institutions	16	4
Indirect (Externalities) [a]	0	12
Total	100	53

[a] Equals 20% of the remaining R&D expenditures not obviously productivity related. $0.2 \times (100-41) = 12$.

duplicate large parts of previously acquired knowledge, their net contribution is smaller than would appear at first sight. In short, a non-negligible rate of investment in research may be required just to keep us where we are and to prevent us from slipping back. I shall assume that only about half of the observed R&D investment rate represents net investment with the rest devoted to "maintenance and replacement." (This is a reasonable assumption for series whose growth rate is approximately equal to the depreciation rate of the stock; it is equivalent to assuming a 10%-per-annum depreciation rate on past R&D investments.)

We can now multiply: 0.03 (average gross R&D investment rate as a fraction of GNP for the past two decades) times 0.53 (fraction of R&D likely to show up in output as measured, from Table 1) times 0.5 (assumption: "net" as fraction of "gross") times 0.366 (average gross rate of return on past R&D investments, based on earlier case and econometric studies), yielding .0029, or about one-third percent per annum, as the contribution to the rate of growth of aggregate output in recent decades. Although this is not large, it is not negligible either. It is on the order of a quarter of the "residual" or "advances of knowledge" as recently estimated by Denison (7) and others.

The point of the above exercise is to indicate that the "indirect" measures of the contribution of science, such as its effects on GNP and productivity growth, are also not the answer to the "What is the appropriate indicator?" question. They are interesting measures but also quite flawed. They contain the output of science only to the extent that it is marketed and measured in the conventional accounts, and that is very limited.

Further progress will depend on reopening the whole framework of national accounts and devising a new set of measures that can capture better the contribution of science to our lives. Until then, the fact that the contribution of science to our *measures* of economic growth is rather small reflects less on science than on the current state of our national economic accounts.

REFERENCES

1. Z. Griliches, "Research Expenditures and Growth Accounting," in B. R. Williams, Ed., *Science and Technology in Economic Growth*, Macmillan, London, 1973, pp. 59–95.

2. Z. Griliches, "Productivity and Research," in *Conference on an Agenda for Economic Research on Productivity*, National Commission on Productivity, U. S. Government Printing Office, Washington, D. C.., 1973.

3. J. Schmookler, *Invention and Economic Growth*, Harvard University Press, Cambridge, 1966.

4. R. E. Evenson and Y. Kislev, "Research and Productivity in Wheat and Maize," *Journal of Political Economy*, **81**, No. 6, 1309–1329 (1973).

5. E. Mansfield, "The Contribution of Research and Development to Economic Growth in the United States," in *R&D and Economic Growth/Productivity*, National Science Foundation, Government Printing Office, Washington, D. C., 1972, pp. 72–303.

6. *An Analysis of Federal R&D Funding by Function*, National Science Foundation, Government Printing Office, Washington, D. C., 1972, pp. 73–313.

7. E. F. Denison, *Accounting for U. S. Economic Growth, 1929 to 1969*, Brookings Institution, Washington D. C., 1974.

Citation Data as Science Indicators

8

Eugene Garfield
Morton V. Malin
Henry Small

Attempts to appraise the condition of science as intellectual activity or social institution have involved data compilation—reports on amounts of money spent on scientific research, magnitudes of scientific manpower, number of students enrolled as science majors at universities, and number of scientific papers produced or number of patents issued. For a variety of reasons, these all fail to indicate the "condition" of science. Perhaps the problem is that such data are compiled and presented without regard to a specific set of questions or set of hypotheses; thus a coherent framework for estimating the social or intellectual condition of science is missing. *Science Indicators 1972 (SI-72)* is an improvement over mere data compilation, and it should be applauded as a step, however preliminary and tentative, in the right direction. The purpose of this paper is to suggest further indicators relevant for measuring scientific activity, in the hope that this will lead to a better estimate of the condition of science.

At the Institute for Scientific Information (ISI), we operate on the fundamental as-

We thank Professor Belver Griffith of Drexel University, who participated in the mapping-science project and has contributed to the development of these ideas. Research was supported by a National Science Foundation grant to the Institute for Scientific Information.

sumption that citation data can be used as indicators of present, past, and perhaps future activity in science (1). The validity of this assumption must of course be tested; but at present standards against which the validity of citation analysis can be measured do not exist. All that can be done, and perhaps all we can expect to do, is to compare the results of different methodologies and attempt to find significant correlations between them. Thus, for example, Hagstrom (2) found a significant correlation between citation analysis and subjective peer judgment in the Cartter report on quality of science graduate departments of American universities.

CITATION ANALYSIS

Citation analysis is a bibliometric method that uses reference citations found in scientific papers as the primary analytical tool. Bibliometrics can be defined as the quantification of bibliographic information for use in analysis (3). The literature of science lends itself to quantification because each source article, report, note, book, and so on contains such bibliographic elements as authors' names, addresses (country, state, city, institution, department), titles (words and phrases), journal titles, place of publication, volume and page number, and date of publication. All are keys that when properly organized, provide a base for extracting analytical data. However, the added element of reference citations in scientific and scholarly literature is most significant for citation analysis.

Sources

The manipulation of all these bibliographic elements for sociological, historical, and other kinds of studies became practical on a large scale only when the computer entered the picture. Consider, for example, the difficulty of creating the *Science Citation Index (SCI)* data base without the aid of a computer. Approximately 3.5 million source items and 41 million citations have been input into this data base for the years 1961–1973. Approximately 400,000 source articles and book reviews, covering both current and past years, are added to the *SCI* data base each year.

Citation Data

Using the bibliographic elements extracted from the journal articles being processed, the information can be sorted in a variety of ways to provide data for studying a variety of questions and problems (see Kochen, also Cole et al., both this volume). Using citation data, Jonathan and Stephen Cole examined the phenomenon of social stratification among scientists, taking citation counts as a rough measure of peer recognition and of importance (interpreted as utility) of a scientific work (4). Derek Price has pioneered in the use of citation data in science policy studies and has made an important contribution to our understanding of citation networks (5). An early study of the use of citation data in historical research was done by Garfield, Sher and Torpie (6). Other studies, too numerous to mention here, are regularly reported in the *SCI* itself (7).

Science Mapping.   Currently at ISI, citation data are being used to study the relative impact of scientific journals (8), the impact and quality of the research of individuals and institutions, and the specialty structure of science. The last-mentioned application, "science mapping," is of particular interest because the methodology extends the usefulness of citation data.

Certain presuppositions, both historical and sociological, underlie the idea of "mapping" science by identifying key papers and events through citation analysis. The basic unit of analysis in mapping is the highly cited document. The assumption is that these articles and books are markers for critical scientific ideas or events, taken in the broadest sense. This includes theoretical formulations, speculative hypotheses, experimental results, procedures or methods, and any combination of these. The fact that some documents have been highly cited within a specified time-period confers upon them a special status as providing important "ideas" in their respective areas or specialties. It should be possible to identify the corresponding cognitive components for each highly cited item either by examining the citing context or by querying the citing author.

The former was done for a sample of highly cited papers in chemistry (9). For each highly cited paper, a sample of citing papers was obtained. Each citing paper was examined to determine where in the paper the reference was made. The terminology used by the citing author in referring to the highly cited item was noted. In general, the more highly cited the item, the more uniform the terminology: terms such as *Hammond's postulate* or *atomic scattering factors* or *orbital symmetry rules* were invariably associated with certain works. These works had a clear conceptual identity for the specialists who cited them. Less highly cited papers, though not always achieving this level of eponymy, could nevertheless be associated with distinct conceptual entities. Our hypothesis is that most of, if not all, the scientific ideas that have been regarded as important or influential can be associated with one or more scientific works that are at *some time* highly cited. Sometimes recognition through citation frequency comes soon after publication, but a two- or three-year time-lag is the norm. This view does not rule out the possibility of a long delay before a work is recognized—perhaps exemplified by the Einstein 1905 paper on relativity (10)—nor does it rule out resurrection of older and perhaps long-forgotten works as the basis for new departures.

Definitions. The implication of the phrase *at some time* should be considered carefully. Papers containing important ideas will not necessarily continue to be highly cited for all time. Eventually, an idea or paper may become so widely known that citing its original version is unnecessary: Knowledge of this kind may be "tacit" in Polanyi's sense (11). Or a new paper will supersede the original one by reformulating the idea in more up-to-date terms; the newer paper then receives all the citations to the idea.

The hypothesis does not say that *all* highly cited papers contain important "ideas," in the narrow sense of the term. Clearly, papers can become highly cited for important methods, procedures, and data compilations; the fact that they are not "ideas" does not make them less important. The availability of a good table of nuclear masses, for example, is probably of critical importance to the advancement of nuclear physics. Finally, *important* should not be confused with *correct*, for an idea need not be correct to be important. This is evident in the recent polywater controversy (12). The high negative citation rates to some of the polywater papers is testimony to the fundamental importance of this substance if it could have been shown to exist.

In the preceding discussion we have left the term *important* undefined. Like *quality* of scientific work, *importance*, as perceived by a scientist, is undoubtedly a highly complex matter, even though scientists are constantly called on to make such judgments of the work of their peers. The intent here is not to define some absolute scale of importance as measured by citations, but to operationalize the notion so that we can compare citations and scientists' perceptions in terms of responses to questions such as: What are the most important advances in your specialty in the last five years? or How would you rate the quality of this paper on a scale from 1 to 10?

This view of the highly cited paper stresses its role as a marker of discovery or invention (see Kochen, this volume). That the number of such works is relatively small is demonstrable from statistics on highly cited papers. Even though the absolute number of items cited is high, only about 1% of the items cited in a year are cited 10 times or more. It follows that few such seminal works exist for any single specialty or research area; statistics on average cluster* size indicate there may be as few as five per area. The

---

*Typically, a cluster is a core of discovery papers surrounded by succeeding works built on the original discoveries (see next section).

usually high citation of these works over a period of years exhibits growth and decay characteristics that fit with intuitive feelings about the rate of obsolescence of knowledge or shifts in intellectual fashions. The set of highly cited works can even be seen as quite concretely representative of the paradigm for the specialty. Indeed, Kuhn (13) may have had this in mind.*

The normal turnover in highly cited works from year to year—that is, the appearance of new cited works and the disappearance of old ones—reflects, then, the rate of change of scientific conceptions about the world. However, change in documents need not be revolutionary, since one document may merely replace another by formulating the key idea in a more useful way. Generally, the more highly cited a paper is to begin with, the longer it continues to be highly cited. Methodologically important papers, like Lowry et al. on protein determination (14), are consistently the most highly cited papers in the SCI file. Kuhn's observation that such methodological tools are critical to the paradigm is certainly borne out by citation analysis.

If the Kuhnian paradigm is defined in terms of highly cited papers, the structure of the paradigm changes constantly—though slowly. A sample of highly cited works from a single source year identifies not only works that contain original discoveries but also a set of more recent derivative works built on the original discoveries that are more ephemeral. The totality of these founding and derivative works will represent the research front during the source year but not the entire history of a given subject. It is precisely this change, or turnover, in the set of highly cited papers which can become useful in studying the history of a specialty.

As an illustration, consider Bohr's well-known 1913 papers on atomic structure. The first of the papers (15) was only moderately cited (six times) in the 1973 SCI. But the failure to find Bohr's paper highly cited in 1973 means that the research front in atomic physics has moved on—presumably, Bohr was highly cited when his work set the framework for atomic theory in the early 1920s. The implication is not that Bohr's paper lacks "lasting value," but that in general, a paper or an idea may have critical importance for one historical period or environment and not for another. Citations are an "indicator" of a paper's importance or timeliness for a particular historical period; finding exactly what determines the timeliness of a paper or an idea for its particular conceptual and social context is a problem of key importance for the history of science.

*Because Kuhn is difficult to pin down, the reader may make his own interpretation from this passage in the original Preface: "Or again, if I am right that each scientific revolution alters the historical perspective of the community that experiences it, then that change of perspective should affect the structure of post-revolutionary textbooks and research publications. One such effect—a shift in the distribution of the technical literature cited in the footnotes to research reports—ought to be studied as a possible index to the occurrence of revolutions."

Highly cited papers are almost never isolated but tend to aggregate in small coherent groups. The tendency toward aggregation could be evidence for simultaneous discovery within an area, provided the papers had the same publication dates. Robert Merton has stressed the frequency with which "multiples" occur in science (16), and it appears that such a multiple will emerge in citation data as a cluster of highly cited documents. Successive as well as simultaneous discoveries will be related and grouped. Thus, the cluster will consist, typically, of a core of discovery papers (either simultaneous or successive) surrounded by several follow-up or derivative works, which have been built on the original discoveries. Any or all of these may be of theoretical, methodological or experimental import.

Co-citation and Bibliographic Coupling

Clearly, the way relations between the highly cited papers are derived is critically important for interpreting the significance of the groupings. Frequency of co-citation, which has been used to determine the relations among highly cited papers, is simply a count of the number of documents citing both of the highly cited documents in a specified time period (in this case, one year). Co-citation, therefore, reflects the association between highly cited papers as perceived by the current population of specialists who have themselves published papers. If, then, highly cited papers can be placed in some kind of correspondence (not necessarily one-to-one) with cognitive components (theories, experiments, methods, etc.), co-citation becomes a measure of cognitive association. In that case, the changes in such patterns of association from year to year can tell us something about the history of ideas.

At this point we might well pause and consider why some alternative methodologies were not employed. Considerable work in information science relates to the use of language associations, or term-term associations for defining related words and for clustering documents (17). To the extent that words can be associated with ideas, the results would be similar to those obtained by using co-citation. This could be done say, by counting the frequency of co-occurrence of words in titles of papers. But difficulties arise when the SCI is used: With a large multidisciplinary data base, words like *plasma* or *complex* which have different meanings in different fields, cause word associations to give rise to many false linkages between fields when such homographs are encountered. This weakness of natural language is not inherent in the reference citations in the form of document surrogates. The latter are virtually unique: No two of them can have the same author, journal, volume, page, and year. Hence, they are ideally suited for automatic manipulation.

Co-citation (18) should also be distinguished from another familiar coupling procedure, bibliographic coupling (19), to which it is related by a kind of mirror symmetry. In co-citation, earlier documents become linked because they are later cited together; in bibliographic coupling, later documents become linked because they cite the same earlier documents. The difference is that bibliographic coupling is an association intrinsic to the documents, the authors themselves having established it by citing one or more of the same works. Co-citation, in contrast, is conceived as a linkage extrinsic to the documents and one that is valid only so long as the community of specialists chooses to co-cite them. Thus, co-citation depends on the collective choices of a population of scientists who have published in the source year. Therefore, discovery papers, either simultaneous or successive, are highly co-cited—as well as being highly cited—and form a cohesive cluster. Further, the reason the groupings so derived correspond closely with what have been regarded as "specialties" is immediately visible, for only the "specialists" capable of understanding and utilizing the discoveries will cite and co-cite the discovery papers in their work. This identification of clusters with specialties is one of the firmest results to emerge from clustering studies using co-citation. Stated in different terms, discoveries or innovations are almost always specific to a problem area that is the domain of a specialty group. Notable exceptions exist: Certain methodologies (e.g., protein determinations) are applied in several specialties. Documents of this kind tend to span several specialties as a kind of methodological structure superimposed on a more fine-grained and intensely interactive specialty structure (20), like the role of function words in language analysis.

Clearly, bibliographic coupling and co-citation are closely related and derivable from a general treatment of pathways in a directed citation graph. Neither should be assumed to be implicitly superior as a measure of association or relatedness of documents. Each measure has its application. For example, since clusters obtained by using bibliographic coupling would presumably be no different from those obtained using co-citation, the only difference is operational—grouping together the citing papers in the former case and the cited papers in the latter. But once a cluster of citing papers is formed by bibliographic coupling, a list of the cited items the citing papers had in common can be generated; and likewise, once a cluster of cited papers is obtained by co-citation, the citing items can be retrieved. Since the procedures lead to the same end-result—namely, a clustering or classification of the literature—the choice of procedures is determined by the nature of the phenomenon to be investigated and the interpretation sought.

Social and Cognitive Structure of Science

From experiments on clustering using co-citation, the specialty appears to be a natural unit of structure and organization in science. Studies of the social structure of specialties, which have been pursued for some time now by sociologists, therefore appear most relevant and appropriate. These include studies of informal communications (21) and contacts among scientists and the so-called "invisible colleges" (22). To clarify the connection between citation studies and the social structure of science, the types of relations implied by the citation linkages must be considered. The three types of linkages are (a) direct citation, the citing of one document (scientist) by another, which is analogous to a sociometric choice—that is, highly cited papers have been "chosen" frequently and the authors of these works assume the role of leaders or "stars"; (b) co-citation, the equivalent of citing the "stars" together, which may or may not reflect the existence of informal contacts among the stars; and (c) bibliographic coupling, the choosing of one or more of the same stars another person has chosen. None of the bibliometric linkages require that social contacts lie behind them, but the existence of strong patterns of coupled documents (clusters) suggests that underlying social factors are at work. Just as a coherent body of knowledge about a fairly narrowly defined subject would be inconceivable without an underlying network of informal communications among specialists, a cluster of documents probably reflects an underlying social network. The extent to which these structures (documental and social) are congruent is not known.

The document networks derived through citation analysis are believed to reflect *both* the cognitive structure and the social structure of specialties. However, since this hypothesis will require much further elaboration and testing, it will only be stated here. A proper test would involve comparison of informal communication patterns with the bibliographic relations established by citations. Studies of this kind may also help unravel the relations between cognitive and social factors in the development of specialties and to determine the extent to which one is dependent on the other.

The importance of obtaining clusters that correspond to specialties can be illustrated by contrasting results here with alternative outcomes, which could have occurred but did not. First, it was possible that none of the highly cited papers in the *SCI* were co-cited. In that case, no clusters would have formed, and each highly cited paper would be an island unto itself. Second, all the highly cited papers could have been so strongly interrelated that only one gigantic cluster emerged, no matter how the clustering parameters were manipulated—the unity of science, forged presumably by interdisciplinary research, proving to be so powerful as to defy a breaking-up of knowledge into smaller subdivisions. But in fact, the outcome fell between these two extremes.

An important question is whether this outcome followed from the clustering algorithm applied. Practitioners of cluster analysis know that the outcome of a hierarchical clustering algorithm is as it was described in the previous paragraph—at one extreme a set of independent entities; at the other, a single cluster consisting of all the entities. The "natural" structure, if one exists, will depend on how the entities (in this case, cited documents) behave between the extremes—that is, whether they maintain a set of stable groupings over changes in level or strength of connection. Using a mountain range analogy, at the highest altitude the peaks are visibly separate; as the altitude is lowered, the peaks begin to merge into the mass; and finally, at the lowest level, all are united in a single mass. In terms of this analogy, science appears to consist of many sharply divided peaks which, for the most part, remain separate over a wide range in altitude. At very low levels, however, large merges occur, and eventually little, if any, of science remains unconnected at ground level. The point at which this merging begins is when co-citation reaches about 10% between papers (e.g., when one-tenth of all citations to the two documents are co-citations). At levels or altitudes only slightly higher than this (e.g., 15%), significant fragmentation remains.

The mountain range can also be more than an analogy, as shown by the following example. A cluster of six papers in particle physics (strong interactions) is depicted in Figure 1 as a contour diagram, with Feynman's 1969 paper on the parton model as the highest peak. This configuration or "landscape" was derived in a completely objective manner. First, each document was interpreted as a normal probability hill equal in volume to its citation frequency. Second, the distance between two hills was calculated by allowing the hills to interpenetrate until the volume of interpenetration was equal to the co-citation frequency. After all the distances were thus determined, multidimensional scaling was used to find that configuration of points in two dimensions that best fit the given distances. Finally, the contour lines were drawn so that along a given line the density of citing authors was constant.

The emergence of small groups of highly cited documents as the dominant pattern deserves special consideration, especially in terms of the clustering algorithm employed. The procedure is usually described as a single-link algorithm (23), because only one link of sufficient strength is required for membership in a cluster. This is, perhaps, the simplest of all clustering procedures, and it was dictated by the size of the data base being clustered—for example, typically about 10,000 highly cited documents. The single-link algorithm also has the weakest possible criterion for cluster membership, and it has been criticized for giving rise to "chaining"—the stringing together in a cluster of objects that bear little similarity to one another. Nevertheless, the application of single-link clustering to co-citation data does not exhibit this tendency, except at very low levels of co-citation: the average cluster size is only about five documents. Therefore, the small-cluster outcome appears to be intrinsic to the data rather than the result of the clustering methodology.

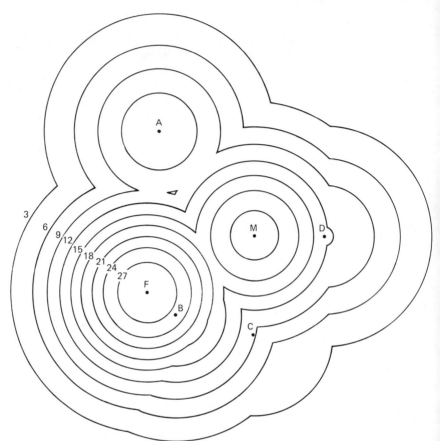

Figure 1.   Contour map of six papers in particle physics. Each paper (designated by a letter code) is represented by a normal probability density function in the plane, with the volume of the hill set equal to the paper's citation frequency. Distances between the hills are determined by allowing the hills to overlap until the volume of overlap equals the co-citation frequency. The positions of the papers in the plane are determined by metric scaling (M-D-SCAL) on the set of 15 input distance. Along any contour line, the density of citing papers is constant and equal to the number on the line. *Key:* A = Amati, 1962; B = Benecke, 1969; C = Caneschi, 1969; D = DeTar, 1971; F = Feynman, 1969; M = Mueller, 1970.

The fact that clusters do eventually merge by chaining when the co-citation level is very low allows a further manipulation of the data, which provides a picture of how the specialties relate to one another. If, for example, a set of clusters is derived at a co-citation level of 11 (i.e., all documents co-cited 11 or more times appear in the same cluster), the clusters can be related by summing co-citation linkages between documents in different clusters of a strength of 10 or less—which would be, to use the mountain range analogy, low ridges connecting the sharp peaks. With the intercluster links (called "cluster co-citation"), "maps of science" can be constructed in terms of specialties. It is possible then to examine the way physics clusters relate to chemistry clusters and how the latter in turn relate to clusters in the biomedical sciences. Studies of this kind would not be possible if the *SCI* were not a multidisciplinary data base. This overall mosaic of specialties has important implications for studying the nature of interdisciplinary activity, since linkages between specialties of diverse subject matter indicate an exchange or a sharing of interests or methodology.

The map of biomedical clusters for 1972, shown in Figure 2, was derived by the application of two thresholds: No cluster with fewer than five documents is included, and must be linked with another by a co-citation strength of 100 or more. A linkage between two clusters means that a number of authors are citing documents in both clusters and thereby creating inter-cluster co-citations. Such citations should reflect the degree of interdependence of one specialty on another or the extent of interdisciplinary effort.

The map shows four major regions corresponding to major research areas in present-day biomedical research: Chromosomes and RNA virus work (viral genetics), at the upper left; immunology, upper right; biological membranes, lower right; and cyclic AMP, lower left. A 1972 specialty now known to have been a key growth point is "microtubule protein" (in the lower left-hand corner), which developed linkages with all major areas on the map in 1973.

The goal implied by the term "mapping science" is most nearly realized by the use of cluster co-citation in dealing with links between specialties rather than links between documents. Many of the document-level techniques (e.g., graphing, multidimensional scaling) can also be applied to the specialty level. The final step in achieving the goal of mapping science is to relate large disciplinary units (biomedicine, physics, chemistry) to one another. The specialty clusters exhibit some tendency toward hierarchical or nested structure—that is, for clusters in chemistry to merge at lower co-citation levels and to form a macrocluster. This tendency may be taken as evidence for a larger structure by discipline, in which the specialties are ordered.

The present system takes advantage of that assumption to arrive at a large-scale map of science. The documents are first clustered at a very low level (e.g., 7), which has the effect of grouping some, but not all, of the specialty clusters into disciplinary clusters. Then cluster co-citation is determined for the disciplinary clusters. A "map of science" that includes some of the largest clusters at Level 7 is shown in Figure 3. Nonmetric multidimensional scaling of cluster co-citation frequencies was used to determine the configuration.

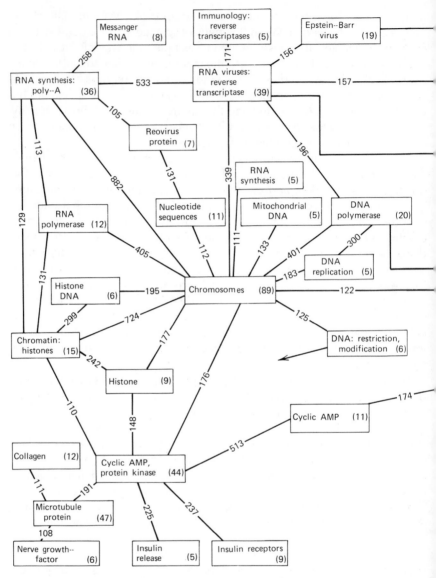

Figure 2. This is a map of major biomedical clusters in 1972. Each box is a cluster of highly cited documents (name of cluster and number of documents given in each box) obtained from the 1972 *Science Citation Index* at thresholds of 10 citations per document and 11 co-citations per docu-

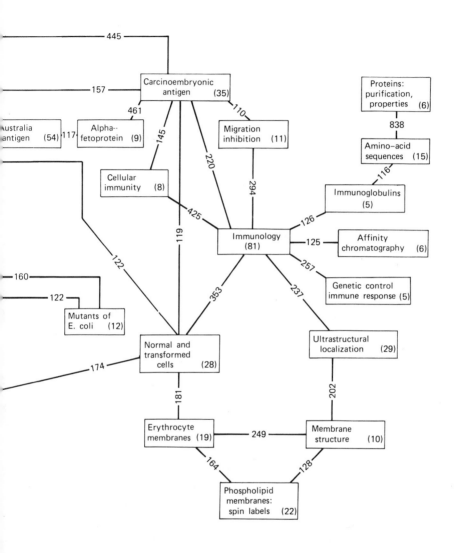

ment pair. In constructing this diagram two further thresholds were used: Clusters were connected by lines only if there were at least 100 co-citations between documents in the two clusters (see number on each line), and only if the clusters contained at least five highly cited documents. Reprinted with permission of *Mosaic* magazine. Originally appeared in S. Aaronson, "The Footnotes of Science," *Mosaic*, **6**, 22–27 (1975).

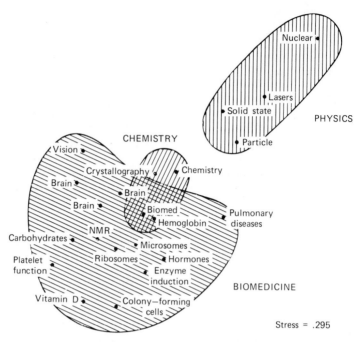

Stress = .295

Figure 3. Major disciplinary clusters obtained at Level 7 from the 1973 *Science Citation Index* and containing 30 or more highly cited papers. A matrix of cluster co-citations is constructed and the matrix is used as input to Kruskal's multidimensional scaling (M-D-SCAL), using its nonmetric option. Stress is a measure of goodness-of-fit of the input similarities to the final configuration. The configuration of specialities shows three regions: a physics region, a biomedical region, and a chemistry region that is between physics and biomedicine and partially embedded in biomedicine.

The notion of mapping implies dealing with objects or entities that have a location in a space of some number of dimensions in which the distance between objects is meaningful and well-defined. Indeed, the language used in talking about science is filled with spatial metaphors. Physics, chemistry, and biology are "fields," related logically, socially, or in terms of shared subject matters; one field is said to have a "close bearing" on another. Information scientists have long posed the problem of finding a measure of distance between classification headings, and a library or a classification scheme is set up (theoretically) to make related subjects physically close to one another. Mapping science is an attempt to arrive at a physical representation of fields and disciplines—and, at a lower level, of individual papers and scientists—in which the relative locations of entities is depicted.

But do the maps of science derived from citation data have reality in this strict spatial sense? Measures of association derived from citation data do not necessarily imply the existence of a metric space. Spatial representations can be obtained merely from ordinal data, as shown by multidimensional scaling. The ontological status of maps of science or other cognitive maps will perhaps remain speculative until more has been learned about the structure of the brain itself. Whatever their physical reality, maps of science are certainly useful as heuristic tools. The same might be said of the "mental maps" being constructed by geographers (24).

MODELS OF SCIENTIFIC CHANGE

Up to this point, clusters and maps of clusters have been examined in terms of a single source year. But it is also possible, since *SCI* is a multiyear file, to examine shifts in clusters over time and to investigate the nature of change in specialties. The problem of change is extremely complex, and we can hope only to point to certain directions in which research must proceed. Historians and philosophers of science have long regarded development and change in science as a central problem, and a number of theoretical statements have emerged that have far-ranging implications, most notably those of Kuhn (13) and Toulmin (25). In fact, Toulmin's recent book contains a set of diagrams that resemble quite closely "historical maps" derived using citation relationships.

Moreover, Toulmin's distinction between intellectual "disciplines" and "professions" parallels our distinction between clusters of cited documents and corresponding clusters of citing authors. The model Toulmin proposes is an adaptation of evolutionary theory and may be directly testable by using citation data. The test might involve identifying his "variants" of an idea among the papers citing a particular seminal work and then seeing which, if any, of these works becomes highly cited itself. In this interpretation, high-citation frequency is evidence for what Toulmin calls "selection" of an idea, and variation occurs when authors cite a previously "selected" idea. Toulmin employs a sort of replacement or gradual-improvement model of science, while Kuhn employs a model in which sudden shifts of perspective alternate with periods of relative stability. Kuhn's model can also be formulated in citation terms and tested.

The manner in which clusters of cited documents change over time should be examined. Periods of stability or gradual turnover should occur when the document clusters continue and maintain a constant configuration. There should also be sudden changes (perhaps in the course of a single year) in the configuration and the document set, with very little overlap between succeeding years. Even a third model is conceivable (perhaps to be called the Popper model [26])* in which a continuing "revolution" in all specialties is evidenced by a large and continuous turnover of highly cited documents in the cluster. We do not attempt a systematic test of the models here; but we do offer some evidence to suggest that the empirical situation is highly complex, and examples can be found to suit all the above theories.

The technique for studying change with citation data and clustering methodology is to identify corresponding clusters in successive annual files. This can be done easily, because at least some of the highly cited papers in each cluster persist from year to year. At the same time, the entrance of new highly cited documents and the exit of old ones can be seen. The core of continuing documents also changes its configuration because of weakening or strengthening of selective linkages. These changes may be depicted and analyzed by the topological approach (graphic theoretic methods) or the spatial approach (multidimensional scaling techniques). The topological approach focuses on the cluster as a formal graph (nodes and edges) where all linkages above a minimum threshold are considered equal. The spatial approach employs an ordination technique (e.g., multidimensional scaling) to assign positions in N-dimensional space to each of the documents (27). The study of change then involves tracking the documents' motion through the space over time. Of the two methods, the second (spatial) approach appears more powerful: It makes use of more of the information and it appeals to a body of highly sophisticated statistical techniques. Graphing, however, can be useful in gaining a qualitative impression of how the structure is changing. These patterns of change are complicated by mergers, sometimes large-scale, or splits in clusters from one year to the next. Such changes probably reflect boundary shifts between specialties—for example, a new subspecialty breaking off from its parent specialty or two previously separate specialties joining forces, perhaps to attack a common problem.

The sociometric and historical literature on specialty development (28) emphasizes the rapidity with which changes can occur in the early phase of development, when growth can be exponential. Recent research suggests that some biomedical specialties can emerge in as little as six months after the publication of the discovery papers (29). Our clustering studies generally confirm this potential for extremely rapid growth, although not all specialties follow the pattern. Studies of specialties also reveal leveling-off periods and periods of decline. These results suggest that specialties go through various

---

*Popper apparently believes that science is, or at least should be, continuously revolutionary: "In my view the 'normal' scientist, as Kuhn describes him, is a person one ought to be sorry for" (p. 52).

phases of development from birth to death—a kind of life cycle. Thus far, cluster data have not revealed such a deterministic pattern but the following composite picture has emerged from a number of cases.

From one to three highly cited and highly co-cited papers appear as a tightly knit cluster in the first year. These papers contain simultaneous or successive discoveries quickly recognized as breakthroughs. Occasionally, an old paper included in this initial "discovery" group may have the formal role of precursor. Usually, at least one of the papers is very recent (perhaps only one year removed from the source year). Often the discovery group is weakly tied to an older cluster of methodological importance. One might hypothesize that the old cluster provides a legitimizing context for the new ideas or perhaps a source of manpower for the new specialty. After one year the nucleus of discovery papers expands dramatically, and sometimes explosively, to include perhaps severalfold the number of highly cited papers—not discovery papers but early working-out papers that have exploited the new ideas. Since they are closely dependent on the discovery papers, they are all highly co-cited with the discovery papers. Concurrent with the rapid growth in the new cluster, the old methodological cluster disappears or declines in importance.

The third or fourth years display increasing stability, although that stability can be short-lived. This middle period may involve, for example, the appearance of review papers in the cluster or movements into applied science or technology. Some of the original working-out papers disappear and are replaced by more recent up-to-date papers. The distribution of cited papers by publication date settles down to the average for all of science, with the mode at about two or three years before the source year. In most cases the original discovery papers persist as highly cited papers in the cluster and provide a kind of framework for later developments. Stability is often followed by decline, which can be manifested in a disintegration of the cluster into smaller fragments. Any dramatic novelty in the specialty would initiate a new sequence of events, similar to the one just described, in which the role of the methodological cluster would be played by the original cluster. Apparently, specialties must face either eventual demise or substantial transformation to incorporate new material.

Much of our discussion must be regarded as hypothesis based on examination of numerous examples; we do not wish to convey the impression that specialty development can be fitted into a neat predictive theory. A four-year (1970–1973) study of some of the 31 continuing specialty clusters shown in Figure 2 yields the percent-change results listed in Table 1. The overall mean rate of document continuation is about 55% for these clusters, but the variation in the percentage can be large. The change from one year to the next in the same specialty varies from gradual to dramatic. In the

Table 1. Percent Change in Sample of 31 Continuing Clusters, 1970–1973
c = continuing, d = dropping, n = new documents

Specialty	Direction of Change	1970–1971 (%)	1971–1972 (%)	1972–1973 (%)
Nuclear levels	c	58	45	25
	d	21	55	17
	n	21	0	58
Adenosine	c	67	25	67
triphosphatase	d	0	50	22
	n	33	25	11
Australia	c	55	54	57
antigen	d	4	26	30
	n	41	20	13
Proton-proton	c	50	7	44
elastic scattering	d	50	21	25
	n	0	72	31
Ultrastructure of	c	50	43	60
secretory cells	d	12	57	0
	n	38	0	40
Nuclear magnetic	c	37	55	23
resonance	d	13	9	54
	n	50	36	23
Polysaccharides	c	46	44	36
	d	46	34	7
	n	8	22	57
Crystallization of	c	100	100	100
polymers	d	0	0	0
	n	0	0	0
Affinity	c	60	67	72
chromatography	d	20	0	14
	n	20	33	14
Leukocytes: chronic	c	40	63	33
granulomatous disease	d	13	5	53
	n	47	32	14
Collagen	c	80	40	27
	d	20	0	40
	n	0	60	33
Erythrocyte membranes	c	9	15	58
	d	64	5	42
	n	27	80	0
Delayed hypersensitivity	c	77	46	50
	d	15	27	29
	n	8	27	21

Table 1. *(continued)*

Specialty	Direction of Change	1970–1971 (%)	1971–1972 (%)	1972–1973 (%)
Fission of deformed	c	45	63	36
nuclei	d	22	25	7
	n	33	12	57
Malignant hyperpyrexia	c	29	45	70
and hyperthermia	d	14	22	0
	n	57	33	30
Transfer RNA	c	52	29	67
	d	0	71	33
	n	48	0	0
Crystallography	c	33	43	40
	d	11	26	20
	n	56	31	40
Subacute sclerosing	c	60	80	66
panencephalitis	d	20	0	17
	n	20	20	17
Marek's disease	c	33	42	12
	d	27	50	63
	n	40	8	25
Tumor-specific immunity	c	100	100	100
	d	0	0	0
	n	0	0	0
Solid state:	c	75	44	64
disordered systems	d	0	0	18
	n	25	56	18
Hepatic porphyria	c	43	22	33
	d	57	11	56
	n	0	67	11
Immunoglobulin-A	c	50	38	75
	d	39	46	12
	n	11	16	13
Spectrophotometric	c	80	100	100
studies of complexes	d	20	0	0
	n	0	0	0
Myocardial contractility	c	31	20	25
	d	38	33	58
	n	31	47	17
Virus-specific	c	50	80	100
proteins	d	17	20	0
	n	33	0	0

Table 1. *(continued)*

Specialty	Direction of Change	1970–1971 (%)	1971–1972 (%)	1972–1973 (%)
Plasma hormones	c	29	50	27
	d	28	33	9
	n	43	17	64
Magnetic properties	c	100	100	100
of alloys	d	0	0	0
	n	0	0	0
Lesch-Nyhan syndrome	c	60	21	26
	d	40	0	67
	n	0	79	7
Multidimensional	c	100	100	100
scaling	d	0	0	0
	n	0	0	0
Pseudopotentials	c	40	50	34
	d	53	37	8
	n	7	13	58
Mean percent	c	56	53	56
	d	21	21	22
	n	23	26	22

four-year period, about one-third of the 31 specialties experienced major shifts in the set of cited documents—that is, all but one or two of the cited documents in the cluster dropped out, and an almost entirely new set appeared. If dramatic shifts of this kind can be correlated with the occurrence of revolution in specialties, we might hypothesize that a specialty will undergo, on the average, one revolution every 12 years. In this regard, the percentage of continuing documents in a cluster is a good indicator of whether a revolution is occurring.

A common feature in the 31 cases studied was the way in which change occurred: Documents moved in and out of clusters in groups rather than singly, and entire clumps of documents would disappear in one year and be replaced by new clumps the next year. We can hypothesize that these changes represent shifts in the leadership of the specialty from one school or group to another.

An example of a specialty in this sample of 31 cases that has undergone a major shift over the period 1970–1973 is research on the protein collagen, the major component of all connective tissue. Comparing the 1970 and 1973 networks (see Figure 4) reveals that there are no papers in common: The specialty has undergone a complete shift in the cited-document set. The shift is clearly evident in the 1972 diagram, with the appearance of a sub-cluster of five 1971 papers centering around the works by Bellamy and Layman. Here we are dealing with a multiple discovery, which occurred in 1971 and became evident in citations the following year, of a new substance called "procollagen," the biosynthetic precursor of ordinary collagen. The papers by Layman and Bellamy are cited by other 1971 papers as those first announcing the finding. The new subcluster is also attached to the old collagen cluster through Piez's 1963 paper, which is primarily of methodological importance. In 1973 vestiges of the old collagen cluster disappear, and the cluster consolidates around the new 1971 work.

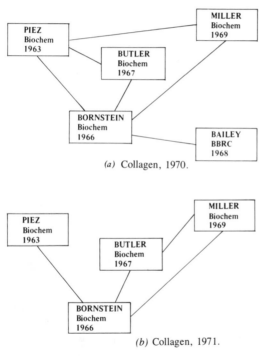

(a) Collagen, 1970.

(b) Collagen, 1971.

Figure 4. Development of a specialty cluster, 1970–1973 (citations to journal articles). The figure shows the evolution of the collagen cluster over the four-year period 1970–1973. Boxes contain the names of first authors of the highly cited papers and years of publication. Lines connect papers co-cited at least 11 times in the corresponding source year. (a) Collagen, 1970. (b) Collagen, 1971. (c) Collagen, 1972. (d) Collagen, 1973.

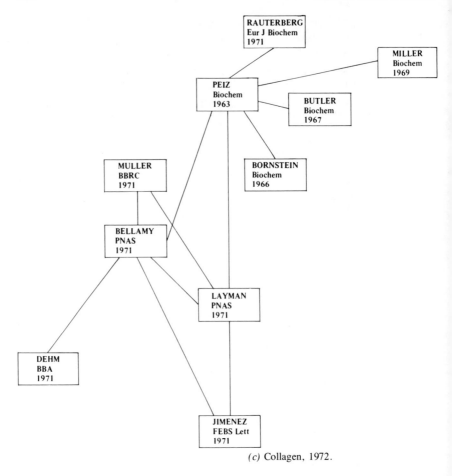

(c) Collagen, 1972.

To validate this picture of the development of the collagen specialty, the citing authors, who were assumed to be specialists currently working in this area, were asked to respond to a questionnaire. The specialists surveyed were not shown any of our results but were only asked to answer a series of questions such as: What are the most important scientific advances or developments in collagen research in the past five years? and What papers were the first to describe these advances? All respondents so far have named the discovery of procollagen as one of the most important advances in the past five years. Of the nine papers said to contain the important discoveries, five are in the cluster, one is in a neighboring cluster (on genetically different collagens), one is not in any cluster, and two are 1973 papers that could not

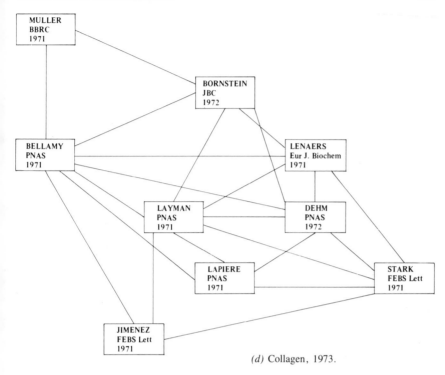

(d) Collagen, 1973.

be expected to appear in the cluster until 1974. The specialists were also asked whether collagen research had undergone a conceptual shift in the past five years. So far, all of the specialists have responded affirmatively, giving similar reasons for the shift. Although it would be difficult to characterize the change in this specialty as a "revolution" in the Kuhnian sense, a major redirection of research has clearly taken place and the specialty has been revitalized by the infusion of new ideas.

These discussions of revolutions in science and life cycles for specialties may seem to be contradictory: How can a specialty be subject both to periodic upheavals and to gradual and orderly progression from birth to death? The reconciliation of these views is consistent with the data at hand and constitutes an hypothesis, deserving further attention, on the development of specialities. In brief, we suggest that revolutions are most likely to occur at the end-point of specialty development—for example, when a state of decline or stability has been reached. What emerges after the major shift in concepts is an essentially new specialty having certain vestigial links with the past, as symbolized by older papers that persist through the revolution.

From Basic to Applied Research

The use of co-citation methodology for the specialty of amorphous semiconductors (30) exemplifies the identification of research trends from basic to applied. After deriving the cluster of cited documents, the citing papers on the subject were used to identify persons and organizations responsible for the research and to detect a movement toward practical application of those devices which had been mainly of basic-research interest.

The co-citation patterns for the period 1968–1972 were investigated. It was observed in 1973 that a new group of papers tied to the previous group began to appear, and that the new group was related to applications of earlier research. This was determined through examination of the titles in the citing and cited articles. Further examination showed that, although all but one of the previous group worked at universities, all the new authors worked in industrial organizations (see Kochen this volume). Evidently, there was movement toward application in the 1972–1973 period, as compared to basic research in the 1968–1971 period.

The examples given, of course, only illustrate some of the capabilities of the methodology and indications for future research; they do not prove or disprove any of the theories of change discussed. What exactly would constitute a confirming or refuting instance is not clear. But it is clear that changes in scientific theories must first be reinterpreted in citation terms— that is, how do citation patterns change in response to social or conceptual change in specialties? This may involve redefining or narrowing some conceptions. It also involves finding appropriate "correspondence rules" between the theoretical constructs and observable citation data. For example, does the appearance of a new cluster of highly cited papers correspond to the cognitive and social event of the emergence of a new specialty? One method of validation is to survey the opinion of the specialists, as was done in the case of collagen. Another would be to examine what science writers regard as newsworthy developments in science and compare their selections with specialty clusters. This method has been pursued in an informal way for news items appearing in *Physics Today* (see Holton, this volume) with resulting agreement between their selection and what appear to be our "hottest" physics clusters. Another method would be to determine the correspondence between informal patterns of communication and citation networks. The development of such correspondence rules is essential if citation or other bibliographic data are to be used in sociological or historical research.

However, it would be wrong to require agreement of citation data with more traditional techniques of investigation. The problem of validating citation data is more complex than that: It involves a new tool or instrument (a citation index), which provides a new perspective to be interpreted. We must learn to do that and then to relate the new information to what is already known. In other words, the interpretations of citation data, not the body of data, are what need validation.

Another frequent question is "Can we use citation data in science forecasting?" For example, can any property of the current specialty structures discussed be used to fortell its future configuration? We do not yet know whether research on the life cycles of specialties will turn up any consistent patterns of change that would be of predictive value. The powerful potential for change contained in the unanticipated discovery is quite apparent. But sudden and dramatic shifts occur in clusters from one year to the next (e.g., in the collagen specialty); and to predict where these will occur on the scale of relationships between specialties—where single events may have less impact—is beyond current capability. We could look for trends in interdisciplinary linkages, such as immunology moving to link with cyclic AMP or biological membrane work moving to link with viral genetics. On the largest scale we might look for corrections forming between disciplines—for example, physics forging links with biomedicine.

Clusters as Science Indicators

We should also ask in what sense are the clusters to be considered a form of science indicators? A cluster of highly cited papers is perhaps an indicator of consensus (31)—at least in the sense that a number of researchers, by focusing their attention on a narrowly defined problem, implicitly agree that it is a worth-while object for their attentions.

The cluster is an indicator of this focusing of attention by the community (see Cole et al., this volume). It points to the problems scientists regard as important and of immediate priority. Hence, clusters are specialty indicators: They provide the information that a certain number of scientists are directing their attention to research on the Epstein-Barr virus or plate tectonics, and they indicate whether that activity is related to work in any other specialties. Furthermore, something can be discerned about the current rate of change in the specialty—whether it is undergoing a revolution or moving through a period of stability. Presumably, this is the kind of information needed to ascertain how specialties are progressing—that is, whether some need revitalization or could use the stimulus of support—according to external measures of priority having their origin in the larger society (see Ezrahi, this volume).

At present, a fully computerized system exists at ISI for clustering annual cumulations of the *SCI* in an off-line mode. Another project currently under way is the development of a fully on-line system (now in the horizon stage) for performing many of these functions. In its final form, it will allow on-line access to the data base in both linear (teletypewriter) and two-dimensional (graphical display) modes (see Kochen, this volume).

The system will automatically structure large fields into clusters, so that at each hierarchical level a manageable amount of information is presented. The researcher could then select substructures to analyze, and the system would automatically proceed to the next level and once again display a manageable amount of information. At the lowest level the system would display the direct citation relations between the various elements of the field or specialty being studied. For example, the researcher might start with a general query on high-energy physics. The graphic system would display the half-dozen or so major clusters in that field, and the user could then select one of the subclusters on particle interactions. The next level might consist of subclusters on strong interactions, weak interactions, and other areas. Finally, by the selection of one of the subclusters, the system will respond with a network of papers in the user's selected area.

Perhaps not so basic in research orientation as the mapping of science but still important in relation to developing indicators are studies—those already carried out and those planned for the future—that use citation data for determining the achievement or impact on science of individuals and organizations. In these studies data from the *SCI* are used to provide quantitative measures of impact. The citation data obtained for the study sample are then used to establish the relative standing of individuals or institutions. The studies are not intended to rate or grade individuals or institutions without regard to other factors, since the major purpose is to provide an indication of impact and not an absolute measure. As more and more of these studies are undertaken, extensive data are being collected that will be used in developing techniques for establishing confidence limits for the results. Methods for integrating the variables associated with the citation data obtained are also being studied.

The questions being addressed in these efforts are: Is a citation count alone a sufficient indicator of impact? Is the average number of citations per paper or per department a better indicator than simple citation counting? What is the relation between an individual's age or the date of his Ph.D. and citation patterns? What normalizations must be established for different age-groups?

Studies of the characteristics of papers cited with high, average, and low frequency will help determine more exactly the relation between the content of a paper and citation data (see Kochen and Cole et al., both papers this volume). Finally, studies of citation patterns of different fields of science should help to normalize for this variable. Certain facts concerning citation characteristics of different fields are already known—for example, the chemical-physics articles, on the average, cite about 20 papers compared to fewer than 10 for mathematics papers. However, there are more chemical-physics papers than mathematics papers to cite—a point often forgotten when people worry about the problems as Janke did (32).

This paper has centered on the application of bibliometrics—in particular, citation metrics. The literature of science, as a by-product of scientific work and sometimes as the culmination of that work, has great potential for the study of science and for developing indicators of the condition of science.

The research directions described should provide indicators for measuring the degree of scientific activity, the quality of research, and scientific achievements. These measures may also have value in identifying the scientist's and the public's options in regard to support for mission-oriented research as opposed to basic research. With such means available, we should be able to deepen our understanding of the problem of setting priorities in relation to societal goals (see Ziman, this volume).

This issue of balanced support of research is also important. Indicators that can be derived from citation analysis could help to identify underdeveloped or currently neglected areas of science. Measuring the degree of research activity in such areas as mental illness and drug addicton, for example, should generate good information to help us decide whether, in terms of our sense of priorities, we have allocated resources correctly.

Citation data can also be used as a measure of national and international science activity. The computer file can be expanded to include the addresses of both citing and cited authors, thus permitting measurements of dependence and independence by individuals and countries. Such analysis would be much more useful then the simple tabulation of the contributions of different countries to various fields presented in *SI-72*. What needs to be measured is international exchange of scientific ideas.

Finally, as the term *indicator* implies, we must be concerned with the evolution of systems over time and the sampling and measuring of systems at successive points in time. Citation data probably will afford little material for the advanced futurologist, but short-term extrapolation may be feasible. The life cycles of specialties must be studied—from their emergence as small clusters of highly cited discovery papers through the explosive initial phase of growth to the stabilization of patterns and the eventual decline or revitalization. Such studies have ramifications beyond an increase in general knowledge of how specialties are born and mature. Perhaps the indicators derived from such study will help us to foresee the need for new journals and books (33). Studies of the new terminology associated with the specialty will aid in controlling for the literature, thus permitting better anticipation of the words and phases to use in retrieving the new literature.

The ability to study change in science—provided by citation data—is an important contribution that could be even greater if an *SCI* were available for the period 1900–1960. With such a compilation, we would have a continuum for the entire twentieth century, and the sociological and historical studies that are so important to the basic question with which we are struggling could be greatly advanced.

REFERENCES

1. E. Garfield, "Citation Indexing for Studying Science," *Nature*, **227**, 669–671 (1970).
2. W. O. Hagstrom, "Inputs, Outputs, and the Prestige of American University Science Departments," *Sociological Education*, **44**, 375–397 (1971).
3. A. Pritchard, "Statistical Bibliography or Bibliometrics?" *Journal of Documentation*, **25**, 358–359 (1969).
4. J. R. Cole and S. Cole, *Social Stratification in Science*, The University of Chicago Press, Chicago, 1973.
5. D. J. de Solla Price, "Networks of Scientific Papers," *Science*, **149**, 510–515 (1965).
6. E. Garfield, I. H. Sher, and R. J. Torpie, *The Use of Citation Data for Writing the History of Science*, Institute for Scientific Information, Philadelphia, 1964, 86 pp.
7. *Science Citation Index 1973: Guide and Journal Lists*, Institute for Scientific Information, Philadelphia.
8. E. Garfield, "Citation Analysis as a Tool in Journal Evaluation," *Science*, **178**, 471–479 (1972).
9. H. Small, "Characteristics of Frequency Cited Papers in Chemistry," Final Report NSF Contract C-795, September 1974.
10. A. Einstein, "Zur Elektrodynamik Bewegter Korper," *Annalen der Physik* **17**, 891–921 (1905).
11. M. Polanyi, *Personal Knowledge*, Harper & Row, New York, 1958.
12. "Exit polywater," *Scientific American* **229**, 66 (1973).
13. T. S. Kuhn, *The Structure of Scientific Revolutions*, The University of Chicago Press, Chicago, 1962, p. xi.
14. O. H. Lowry, N. J. Rosenbrough, A. L. Farr, and R. J. Randall, "Protein Measurement with the Folin Phenol Reagent," *J. Biol. Chem.* **193,** 265 (1951).
15. N. Bohr, "On the Constitution of Atoms and Molecules, Part I, *Philos. Mag.* **26**, 1–25 (1913).
16. R. K. Merton, "Singletons and Multiples in Scientific Discovery," *Proc. Amer. Philos. Soc.* **105** (5), 470–486 (1961). Reprinted in *The Sociology of Science: Theoretical and Empirical Investigations*, The University of Chicago Press, Chicago, 1973, pp. 343–370.
17. K. Sparck-Jones and D. M. Jackson, "The Use of Automatically Obtained Keywork Classifications for Information Retrieval," *Information Storage and Retrieval* **5**, 175–201 (1970).
18. H. Small, "Co-citation in the Scientific Literature: A New Measure of the Relationship between Two Documents," *Journal of the American Society of Information Science* **24**, 265–269 (1973); I. V. Marshakova, "System of Document Connections Based on References," *Nauchno-Teknicheskaia Informatsiia*, **2**, 3–8 (1973).

19. M. M. Kessler, "Bibliographic Coupling between Scientific Papers," *American Documentation*, **14**, 10–25 (1963).

20. H. Small and B. C. Griffith, "The Structure of Scientific Literatures I: Identifying and Graphing Specialties," *Science Studies*, **4**, 17–40 (1974).

21. B. C. Griffith and N. C. Mullins, "Coherent Social Groups in Scientific Change," *Science*, **177**, 959–964 (1972).

22. D. Crane, *Invisible Colleges: Diffusion of Knowledge in Scientific Communities,* The University of Chicago Press, Chicago, 1972.

23. P. H. A. Sneath and R. R. Sokal, *Numerical Taxonomy: The Principles and Practice of Numerical Classification,* Freeman, San Franciso, 1973, pp. 188–308.

24. P. Gould and R. White, *Mental Maps,* Penguin Books, Harmondsworth, England, 1974. See also Garfield, Sher and Torpie, *op. cit.*

25. S. Toulmin, *Human Understanding: The Collective Use and Evolution of Concepts,* Princeton University Press, Princeton, 1972, pp. 201, 203, 205.

26. K. Popper, "Normal Science and Its Dangers," in *Critism and the Growth of Knowledge,* I. Lakatos and A. Musgrave, eds., Cambridge University Press, Cambridge, 1970, pp. 51–58.

27. B. C. Griffith, H. Small, J. A. Stonehill, and S. Dey, "The Structure of Scientific Literatures II. Toward a Macro- and Microstructure for Science," *Science Studies*, **4**, 339–365 (1974).

28. W. Goffman, "A Mathematical Method for Analyzing the Growth of a Scientific Discipline," *Journal of the Association for Computing Machinery*, **18**, 173–185 (1971); G. Magyar, "Bibliometric Analysis of a New Research Subfield," *Journal of Documentation*, **30**, 32–50 (1974).

29. B. C. Griffith, "On the Nature of Social Science and its Literature," unpublished paper.

30. A. E. Cawkell, "Search Strategy, Construction and Use of Citation Networks, with a Socio-Scientific Example: Amorphous Semi-conductors," and S. R. Ovshinsky, *Journal of the American Society for Information Science*, **25**, 123–130 (1974); "Mapping Science by Co-citation Networks," unpublished report.

31. J. M. Ziman, *Public Knowledge: The Social Dimension of Science,* Cambridge University Press, Cambridge, 1968.

32. N. C. Janke, *Science*, **182**, 4118, December 21, 1973; and reply by E. Garfield.

33. E. Garfield, "Citation Analysis of Pathology Journals Reveals Need for a Journal of Applied Virology," *Current Contents*, No. 3, 5–8, January 17, 1973.

Measuring the Cognitive State of Scientific Disciplines

9

Stephen Cole
Jonathan R. Cole
Lorraine Dietrich

The purpose of developing science indicators, according to the Introduction to *Science Indicators 1972 (SI-72)*, is "to measure and monitor U.S. science—to identify strengths and weaknesses of the enterprise and to chart its changing state" (1). In attempting this goal, the National Science Board has presented an impressive report containing large amounts of quantitative data—number of scientists, number of papers published, number of citations to these papers, and amount of money spent on science—and the results of surveys reporting attitudes of scientists and laymen. These indicators are quantitative: they tell us how much science is being done. They are also global in that they deal with large units of analysis such as countries or with research areas such as physics and geophysics.

We thank Richard Alba, Hanan Selvin, and Judith Tanur for methodological advice. We thank the following people for their comments on an earlier draft of this manuscript: Harvey Brooks, James S. Coleman, Manfred Kochen, Robert K. Merton, Burt Singer, and Harriet Zuckerman. This research was supported by a grant from the National Science Foundation to the Columbia Program in the Sociology of Science, NSF SOC-72 05326.

209

Problems occur in interpreting the indicators presented in *SI-72*. Are we to infer that two countries in which the same number of scientific papers is produced, or two fields in which the same number of scientific papers is published, are in the same state of health? Two fields appearing to be similar on a set of quantitative indicators may be qualitatively in quite different stages of development and exhibit very different cognitive structures. One field may be divided into small groups of researchers turning out masses of trivial and disconnected papers; the other may be a cohesive group working on highly significant problems at the frontiers of knowledge. Clearly, we need indicators of the qualitative as well as the quantitative aspects of science. The research reported in this paper is a part of an ongoing effort to develop indicators of the cognitive structure of scientific disciplines.

Recently, the focus of attention of many sociologists of science has shifted from studies of the social structure to studies of the cognitive structure of science. This work has often proceeded without any clear definition of what is meant by cognitive structure. Probably the absence of an attempt to specify the components of cognitive structure can be explained by the absence of any clear idea, as yet, of what the most relevant components are. However, though we do not know all that should be included within the definition of this blanket concept, we have an idea of some components of cognitive structure that should be studied. This paper focuses attention on one component—the level of cognitive consensus among scientists.

Many other lines of inquiry into the cognitive structure of science are not considered at all in this paper. Consider only a few questions that might be raised about the cognitive structure of different scientific research areas. First are questions concerning the epistemological grounds for establishing certified knowledge in the various sciences. How do scientists decide what will be called a scientific fact? How do they decide whether a particular observation supports or contradicts a theory? How do scientists come to accept certain methods or scientific instruments as valid means of attaining knowledge? How does knowledge selectively accumulate? And at what point can it be said to be certified and thus an acceptable point of departure for new work? What are the procedures and requirements for "confirming" or "falsifying" theories?

Second, inquiry might be focused on the internal structure of scientific theories in various research areas. How do research areas differ in the extent to which their theories conform to a logico-deductive system? Are these theories formed, modified, or discarded by inductive or deductive processes? What logical form do the theories take—are they correlational, causal, functional, and so on? Do the theories have a large or small descriptive component? Are they explanatory? Do they predict events or behavior? What are the levels of abstractness, generality, and formalization of theories in various research areas?

A third line of inquiry into cognitive structures would examine the relation between the theoretical and empirical content of scientific work. Under what conditions, both intellectual and social, will we find a close fit between the work of theorists and experimentalists? How good is the fit between concepts and indicators in various fields? How often do debates about the plausibility of theories involve arguments over the validity of indicators or measurement procedures?

A fourth focal point would deal with the social psychology of scientists. What do scientists believe to be the relation between empirical data and theory? To what extent do they believe that theories can be falsified or that all scientific knowledge is relative? What conditions their judgments about the validity of facts, the value of experiments and theories, and the work of colleagues? In short, how do they perceive the process of discovery? Problems in perception and cognition might be studied by psychologists; problems in the structure of cognitive systems, by structural anthropologists.

Finally, the line of inquiry perhaps most amenable to sociological analysis is the examination of social responses to scientific ideas. How is intellectual consensus on problematics, acceptable methods, plausible theories, and the work of fellow scientists formed? How is it preserved? How does it differ among scientific disciplines, and what determines fluctuations in the extent of consensus over time?

In this paper we report the results of an interrelated set of research projects aimed at defining and measuring various aspects of intellectual consensus in scientific disciplines. But before we present our findings, we must address this question: What is the theoretical significance of consensus in understanding how science develops and changes?

The work of Michael Polanyi in 1958, Thomas Kuhn in 1962, John Ziman in 1968, and Imre Lakatos in 1970 has suggested that consensus is a necessary, if not a sufficient, condition for scientific progress (2). This idea is perhaps most clearly stated in Kuhn's *The Structure of Scientific Revolutions.* Kuhn sees science as going through two different phases, which he calls "normal science" and "revolutionary science." Normal science is characterized by a high degree of consensus on a paradigm. Although Kuhn uses the term "paradigm" ambiguously, it frequently refers to a shared theoretical orientation acting as a guide to research (3).

Kuhn distinguishes between fields having a paradigm and those in a pre—paradigmatic phase. A pre—paradigmatic field has no generally accepted theory and it is split into several competing schools. Kuhn suggests that a science can make progress only after a paradigm is developed. Without that consensus, scientists cannot build on a corpus of completed work that is accepted as the given state of knowledge. Referring to the state of physical optics before the development of the Newtonian paradigm, Kuhn says, "Being able to take no common body of belief for granted, each writer on physical optics felt forced to build his field anew from its foundations" (4). But in a field with an accepted paradigm that is ot the case: "When the individual scientist can take a paradigm for granted, he need no longer, in his major works, attempt to build his field anew, starting from first principles and justifying the use of each concept introduced" (5).

Only consensus on what is currently held to be true, on acceptable methodologies, and on acceptable scientific instruments and technologies permits "normal science," which results in the expansion of knowledge. Without such consensus, scientists would spend their energies in continual debate over fundamentals. Such debates often become philosophical and social instead of remaining empirical and theoretical.

In sociological terms Kuhn is saying that consensus is a functional requirement, although not a sufficient condition, for scientific progress. During periods of normal science, consensus is maintained by adherence to the paradigm, sometimes even in the face of negative empirical evidence. Scientists accept almost without question (or, as Karl Popper [6] calls it, "criticism") the dominant scientific theories in their research area. Empirical evidence that cannot be explained by the theory, or even evidence that suggests the theory is wrong, will not necessarily lead to its rejection.

Kuhn gives many examples of theories maintained in the face of significant counterevidence or anomalies. For 60 years after Newton put forth his laws of dynamics, the observed motion of the moon's perigee diverged sharply from what his theory had predicted. But this divergence caused no one to suggest that the theory should be abandoned (7).

In a work published before Kuhn's book, Polanyi shows how consensus frequently is maintained by ignoring anomalies.* For example, experimental evidence casting doubt on the validity of Einstein's theory of relativity was presented by D. C. Miller in his presidential address to the American Physical Society in 1925. The evidence was largely ignored by the physics community which did not question the theory.

> By that time they [physicists] had so well closed their minds to any suggestion which threatened the new rationality achieved by Einstein's world-picture, that it was almost impossible for them to think again in different terms. Little attention was paid to the experiments, the evidence being set aside in the hope that it would one day turn out to be wrong (8).

Polanyi also describes how chemists continued to accept a theory after the discovery of facts that cast serious doubt on its validity:

---

*Lakatos also gives many examples of theories being preserved in the face of negative empirical evidence (2).

The theory of electrolytic dissociation proposed in 1887 by Arrhenius assumed a chemical equilibrium between the dissociated and the undissociated forms of an electrolyte in solution. From the very start, the measurements showed that this was true only for weak electrolytes like acetic acid, but not for the very prominent group of strong electrolytes, like common salt or sulphuric acid. For more than thirty years the discrepancies were carefully measured and tabulated in textbooks, yet no one thought of calling in question the theory which they so flagrantly contradicted. Scientists were satisfied with speaking of the 'anomalies of strong electrolytes', without doubting for a moment that their behavior was in fact governed by the law that they failed to obey (9).

Some philosophers of science and scientists, admitting that theories are frequently maintained in the face of negative evidence, would find such behavior opprobrious. But Polanyi argues that because consensus is necessary for scientific progress, scientists must frequently ignore anomalies in order to maintain consensus—and that this behavior facilitates rather than impedes scientific progress.

It is the normal practice of scientists to ignore evidence which appears incompatible with the accepted system of scientific knowledge, in the hope that it will eventually prove false or irrelevant. This *wise* [emphasis added] neglect of such evidence prevents scientific laboratories from being plunged forever into a turmoil of incoherent and futile efforts to verify false allegations (10).

In a well-known paper, Polanyi maintains that it is better to reject a theory that is "right" than to be too ready to accept new theories at the expense of consensus (11). He gives a telling example from his own autobiography. As a young chemist he proposed a theory of adsorption that ran counter to currently held ideas and was resisted—though many years later it was accepted as "correct." Polanyi maintains that this was proper. Most new and contradictory ideas turn out to be of little value. If scientists were to accept every unorthodox theory, method, or technique, the established consensus would be destroyed and the intellectual structure of science would become chaotic. Scientists would be faced with a multitude of conflicting and unorganized theories, and would lack research guidelines and standards.

Although there is widespread agreement among historians and sociologists of science about the importance of intellectual consensus, little is actually known about how consensus is formed. One hypothesis is that by creating "stars" with intellectual authority, the stratification system of science operates to preserve consensus (12).

One of our central concerns is how the cognitive structures of research areas differ. Kuhn suggests that fields differ in the extent to which their paradigms are developed and that the social sciences may be examples of fields that have yet to acquire paradigms (13). Zuckerman and Merton have suggested that fields differ in their degree of "codification" (14). Although they did not define the concept in detail, codification seems to refer to the extent to which a field's paradigm or theoretical orientation is systematically developed: "Codification refers to the consolidation of empirical knowledge into succinct and interdependent theoretical formulations." On this view, fields such as physics and chemistry are the most codified, and social sciences such as psychology and sociology, significantly less codified. Zuckerman and Merton hypothesized that the highly codified fields would have higher levels of consensus than less codified ones.

> The comprehensive and more precise theoretical structures of the more codified field not only allow empirical particulars to be derived but also provide more clearly defined criteria for assessing the importance of new problems, new data, and newly proposed solutions. All of this should make for greater consensus among investigators at work in highly codified fields on the significance of new knowledge and the continuing relevance of old (15).

The physicist John Ziman, in his book *Public Knowledge*, suggests that consensus differentiates science from other systems of knowledge: "What distinguishes Science from its sister 'faculties'—Law, Philosophy, Technology, etc.? The argument is that Science is unique in striving for, and insisting on, a consensus" (16).

Wide agreement that consensus is a relevant dimension of cognitive structure leads us to begin our research by attempting to measure field differences in consensus.

FIELD DIFFERENCES IN CONSENSUS

We wanted to measure the extent of consensus among scientists in evaluating the work of colleagues in their own fields. Here, right at the start, we encountered our first difficult problem. What is the appropriate unit of analysis in which to measure consensus? Is it the traditional academic discipline such as physics, chemistry, biology, or sociology? Is it the specialty such as high-energy physics, analytic chemistry, genetics, or social stratification? Or is it a research area that may cut across specialty as well as traditional disciplinary lines? Because the traditional disciplines are currently treated as meaningful social entities, we decided to begin by using them as the units of analysis. But to anticipate our conclusions, we found these fields to be so broad and diverse in their intellectual problems that they prove not to be units of analysis appropriate for studying many problems of cognitive structure.

The first set of data was obtained from a questionnaire sent to random samples of academic scientists in five fields (physics, chemistry, biochemistry, psychology, and sociology).* The questionnaire listed the names of 60 scientists in the appropriate field, each recently promoted to full professor at an American Ph.D.-granting institution. For each name, the raters were asked to respond to the following:

> Please indicate the relative importance of the work of the following scientist: has made very important contributions, has made above average contributions, has made average contributions, work has been relatively unimportant, unfamiliar with work but have heard of this scientist, have never heard of this scientist.

Responses obtained from such a questionnaire item have several evident limitations. One is that the question asks respondents to evaluate the significance of a colleague's work without taking into account the extent to which they are actually familiar with that work; some of the scientists answering the question may have only a hazy superficial idea of a rated scientist's work, while others may have detailed familiarity with it. We do not know whether the degree and extent of this type of knowledge varies from field to field. Another potentially serious problem with the use of this question to measure consensus is that of intersubjectivity. How do we know that the responding scientists use similar criteria in evaluating importance or even if they mean the same thing by the word *importance*? Because of such limitations in the data obtained from this question, we must conclude that it provides only a rough measure of consensus at best. This calls for caution in treating the results as anything more than guidelines for future research.

Acknowledging the limitations of the indicator, we computed the standard deviation of the ratings received by each of the 60 scientists. The smaller this standard deviation, the greater the agreement in evaluation. We then computed the mean standard deviation for each of the five fields. As Table 1 shows, differences among fields followed the expected pattern, with physics displaying the most agreement on evaluation and sociology the least. The differences, however, were quite small and only that between physics and sociology was statistically significant at the .05 level. The difference between any other pair of fields—for example, sociology and chemistry—was statistically insignificant. Despite the inadequacies in the indicator, the results of this preliminary investigation suggest the hypothesis that there may not be major differences among the fields in the degree of consensus in evaluating the contributions of samples of scientists. Of course,

---

*For more complete discussion of this study, see S. Cole; and Cole and Cole (17).

Table 1. Consensus on Evaluating Scientists by Field (60 Scientists in Each Field)

Field	Number of Raters	Mean Standard Deviation of Ratings
Physics	96	.63
Chemistry	111	.69
Biochemistry	107	.71
Psychology	182	.74
Sociology	145	.76

an adequate test of this hypothesis would require more sophisticated indicators of consensus.

Another set of data enabled us to measure this form of consensus in a second way. On the same questionnaires, we asked the raters to list the five scientists who had contributed most to their discipline in the past two decades. Presumably, scientists in highly codified fields would show greater agreement in this choice than those in less codified fields. We tabulated the results in three different ways: the percentage of total mentions going to the five most frequently named scientists, the percentage going to the 10 most frequently mentioned scientists, and the number of different scientists mentioned divided by the total number of mentions. The higher the first two proportions and the lower the last, the greater the approximation to consensus. The results, reported in Table 2, again suggest that there may be no systematic differences in this type of consensus between fields of varying levels of codification. In fact, if we use the percentage of mentions going to discrete names as a measure, sociology has the highest degree of agreement,

Table 2. Consensus on Rating Contributions of Scientists in Past Two Decades, by Field (60 Scientists in Each Field)

Field	Percent Mentions Received by Five Most Mentioned Names (%)	Percent Mentions Received by Ten Most Mentioned Names (%)	Percent of Mentions Going to Discrete Names (%)	Number of Mentions
Physics	47	65	23	375
Chemistry	34	47	39	377
Biochemistry	41	56	21	473
Psychology	32	44	36	435
Sociology	36	52	19	661

and chemistry the lowest. We should note again that the indicator of consensus used here may measure only impressionistic images of scientists' work rather than concrete knowledge of it. This may reflect the images created in the stratification system of a discipline which creates "stars" and reinforces authority. The results are therefore only suggestive.

A third possible indicator of consensus is the distribution of citations in a scientific journal. In fields having a distinct consensus, we would expect to find a heavy concentration of references to a relatively small number of papers and authors. Using the Gini coefficient as the measure of concentration,* we first compared the distribution of references in the *Physical Review* for 1965 and in the 1970 volumes of the *American Sociological Review* and *Social Forces* combined (18). The Gini coefficients were virtually identical: .24 for physics and .23 for sociology.

We have completed an analysis of the distribution of citations in 108 leading scientific journals. We computed two different Gini coefficients for each journal, using the cited article as the unit of analysis for one and the cited author for the other. Problems arise in using both of these measures —in the first case because contributions are often reported in a series of articles, and bibliographic practices, especially in the social sciences, tend to be sloppy†, and in the second case because many scientists have worked on many different research problems. Because of the problems in using each of the two units of analysis, we must consider both sets of data in drawing any conclusions from this analysis.

Examining the Gini coefficients, we indeed find significant differences in the concentration of citations in the various journals. Data on the range and mean found for journals in the various fields are presented in Table 3. Whether the article or the author is used as the unit of analysis, relatively small differences are found in the mean Gini coefficient for the various fields.

---

*The formula for the Gini concentration ratio is

$$\text{Gini Index} = 2 \sum_{i=1}^{K} (X_i - Y_i)(X_i - X_{i-1})^k$$

where X and Y are the respective cumulative percentages and K is the number of units. For the Gini coefficients we have computed, in one case X is the cumulative percentage of authors and Y is the cumulative percentage of citations to these authors; and in the other, X is the cumulative percentage of articles and Y is the cumulative percentage of citations to these articles.

†Citing authors frequently misspell the name of the cited author or use an incorrect date, an incorrect journal reference, an incorrect page number, or an incorrect first author of a collaborative work. These errors result in citations to the same paper appearing as citations to several papers.

Table 3. Concentrations of Citations, by Article and by Author, to Research Articles in Selected Fields [a]

Field	Number of Journals	By Cited Article Gini Coefficients		By Cited Author Gini Coefficients	
		Mean	Range	Mean	Range
Mathematics	6	.09	.06–.13	.38	.33–.43
Physics	10	.18	.06–.35	.48	.37–.62
Chemistry	12	.15	.06–.27	.46	.30–.56
Biochemistry	10	.21	.05–.34	.44	.32–.51
Geology	7	.10	.04–.23	.40	.29–.53
Psychology	8	.16	.05–.29	.42	.32–.54
Sociology	7	.09	.05–.11	.34	.30–.39

[a] Omits Specialties and Fields with Fewer Than Five Journals Available.

When the article is the unit of analysis, the two social sciences, psychology and sociology, have about the same mean Gini coefficients as geology and mathematics. Examination of the complete range of data shows that, for example, the psychology journal with the highest Gini coefficient, *Journal of the Experimental Analysis of Behavior,* had a higher coefficient than any journal in chemistry, geology, or mathematics. Clearly, considerable differences in the coefficients occur for different journals in the same field. For instance, in chemistry, with the article as the unit of analysis, *Analytical Chemistry* has a Gini coefficient of .06, whereas the *Journal of Chemical Physics* has a coefficient of .27. When the author is used as the unit of analysis, the Gini coefficients for chemistry journals vary from a low of .30 to a high of .56. Similar variances are found in other fields that we intuitively believe to have a high level of codification; also, significant differences in coefficients are obtained for journals in fields that we believe to have a lower level of codification. A good example is psychology, where the scores vary between .05 and .29 with the article used as the unit of analysis, and between .32 and .54 with the author as the unit. When the other two social sciences were taken as one group and all fields as another, an analysis of variance showed that the differences were not statistically significant with the article as the unit of analysis. When the author was used instead, the difference between the two groups was significant at the .01 level, but the variable "field" explained only 10% of the total variance on the Gini coefficients.*

*We thank James S. Coleman and Judith Tanur for their suggestions on summaries and statistical analyses of these data.

Some evidence exists to support the hypothesis that consensus of the type measured by a Gini coefficient is related to codification. A field like sociology generally has lower scores than a field like physics. Also, specialties within a field that we intuitively believe to have lower levels of codification generally have lower Gini coefficients. Thus in psychology, a clinical journal has a lower score than an experimental journal. In physics, optics and accoustics have lower scores than other specialties, and within the sections of the *Physical Review* itself, particle physics has a higher score than the other major specialties. In chemistry, analytic chemistry, which we intuitively believe to be less codified than other specialties, does indeed have a relatively low Gini coefficient. The *Journal of Geophysical Research*, which might just as easily be classified as a physics journal, has a considerably higher coefficient than any of the other geology journals.

The analysis of the Gini coefficients suggests that specifying the boundaries of a scientific area will be a critical problem in studying the cognitive structure of science. Should we compare broad areas such as the traditional scientific fields (physics, chemistry, psychology, etc.), research published in a particular journal, research in a specialty or set of specialties—or should the unit of analysis be defined in some other way? The insignificant differences in extent of agreement found in the various fields suggest that these fields taken as a whole may no longer be meaningful intellectual entities. In contemporary science, intellectual variation within a field is often as great as across fields.

To what extent are the intellectual concerns of scientists working in the same field in fact related? The biological sciences are perhaps the most difficult to deal with. Recent developments in biology have had the effect of splitting the biological sciences into a multiplicity of distinct fields; work done in one area frequently has no relation to that done in other areas. The organization of the biological sciences is rarely the same from one university to another. Similar problems exist in many other fields: A solid-state physicist often has more intellectual interests in common with a physical chemist or a biophysicist than with a high-energy physicist; an experimental social psychologist in a sociology department more often than not has greater intellectual interests in common with some members of the psychology department than with the sociologists in his own department. The point is that meaningful dissimilarities of interest may be as great within a single field as between fields.

However, the finding of similar degrees of agreement about significant contributors in fields such as physics and sociology does not mean that these levels of agreement in the two fields have similar sources. For example, if an entire field is treated as the unit of analysis, we may find lower levels of agreement on some matters in physics than in sociology. Physics is a more highly specialized and differentiated field: high-energy physicists and solid-state physicists employ distinctly different theories and methods. Thus, if both groups of physicists are considered part of the same unit for analytic purposes, the level of agreement may be reduced. Because sociology is a less clearly differentiated discipline than physics, researchers in many different areas of sociology may utilize the same theories and methods. Thus the ideas of Parsons, Goffman or Merton, and techniques such as elaboration or regression analysis are used in virtually all the specialties. This uniform employment of diverse theoretical ideas and methods tends toward consensus across research areas, if only because sociologists in one research area can read and comprehend as well as utilize work produced in most others. If, however, we were to take the research area as the unit of analysis, we *might* find different results. We might find more agreement within a research area such as high-energy physics than in any research area in sociology.

The Gini coefficients of citation concentrations in scientific journals suggest that variations in agreement there may be due less to codification than to the scope of the research published in the various journals. The more specialized and restricted in subject matter or orientation the journal is, the higher the Gini coefficient. For example, with the article as the unit of analysis it is instructive to compare the coefficients for the *Physical Review Letters* and for the four sections of *Physical Review*. These journals publish essentially the same research, done by the same scientists. But the *Letters*, which covers research results from all specialties, shows a lower level of agreement (Gini coefficient = .16) than the sections of the *Review*, which divides up research by broad areas within the field. The Gini coefficients for the different sections of the *Physical Review* range from .35 to .21. The field of psychology provides another example: The .54 coefficient obtained for the *Journal of the Experimental Analysis of Behavior* is one of the highest among the 108 journals. According to colleagues in psychology, this is a Skinnerian journal which publishes only research done in a limited area of psychology.

The size of the Gini coefficient for literature in a particular journal is probably a function both of the degree to which the journal publishes research on a limited range of topics and of the degree of agreement in the research areas included. All scientific fields probably have a relatively high level of agreement. However, the differences found in these indicators of consensus do tend to coincide with our intuitive judgments of the degree of codification of scientific fields.

These preliminary investigations of consensus lead us to raise two basic questions: First, are the traditional scientific fields more meaningful as sociological entities than as intellectual entities? The first set of data suggests that this is the case. If we want to study the cognitive structure of science, we should probably use the research area as the basic unit of analysis. Second, does the level of intellectual agreement result from degrees of codification? For our data, differences in the extent of agreement on the different matters we have measured are not as great as one would expect. These results could be artifacts of the rough indicators of consensus we have employed, or they could be adequate reflections of the cognitive structure of the fields we have studied. Until more sophisticated indicators and techniques of operationalization are developed, we tentatively hypothesize that variations in agreement may be greater within a specific research area over time than between areas. Let us emphasize that we are not yet willing to question the idea that different scientific fields and specialties are differentially codified. We are only raising the question of the extent to which the various forms of cognitive agreement vary from field to field.

Of course, thus far we have attempted to measure only two of the many aspects of consensus—the degree to which scientists agree in evaluating the significance of colleagues' contributions, and the extent to which citations are concentrated on a small number of articles and authors. Other dimensions of consensus should be measured, and they may show significant differences between more and less codified fields. To what extent do scientists agree on the evaluation of the significance and worth of a specific idea? For example, most solid-state physicists might agree that the theory of superconductivity of Bardeen, Cooper, and Schrieffer has been a valuable contribution to the specialty, whereas sociologists studying deviance might disagree on the value of the Merton theory of opportunity structure and anomie or the Lemert-Becker labeling theory. Despite the difficulty of comparing this type of consensus across research areas, these problems warrant further research.

MEASURING CODIFICATION

Codification is said to affect the level of consensus in a field. To date the only systematic indicator of codification proposed is an indirect one: the age of citations in scientific literature. In highly codified fields, we should find a faster rate of what Merton has described as "obliteration by incorporation" than in fields with lower levels of codification. It is not that the research is no longer important but that it has been incorporated into the general body of knowledge and is, therefore, no longer explicitly cited.

Zuckerman and Merton have suggested that the age of references in journal literature may be an appropriate indicator of codification: "The journals in fields we intuitively identify as more highly codified—physics, biophysics and chemistry—show a larger share of references to recent work;

they exhibit a greater 'immediacy,' as Derek Price calls it" (19). In the work of Kuhn, and of Zuckerman and Merton, the suggestion is at least implicit that rapid incorporation of old work makes the discovery of new ideas more probable, since workers in these fields need not continually return to first principles or develop their own logical framework. Rapid incorporation and a corresponding high immediacy of citations is an indicator of the extent to which a science is growing in a cumulative fashion. The extent to which recent work is utilized in current research may thus be seen as an indicator of the presence of conditions necessary for rapid scientific advance.

The preliminary study by Zuckerman and Merton of the age distribution of journal references provided some data that support the use of this measure as an indicator of codification. They report data (some collected by Price and others by them) which seem to show a correlation between the degree of codification (intuitively determined) and the proportion of references to work published in the most recent five-year period. However, two problems are encountered with the data presented in their paper. The less significant one is that the reported data are based on a small number of journals unsystematically selected, and the reported statistics are based on small samples of articles within each journal. The more significant problem is that no control is provided for the age distribution of the entire body of literature in each field.*

Consider the second point first: Suppose that in a particular field, 50% of all the articles ever published have been published in the past five years. If, in this same field, 50% of all references in current literature are to work published in the past five years, there is no immediacy effect. For, if citations were drawn from the literature at random, one would expect, based on the age of the literature, that roughly 50% of them would refer to work published in the previous five years. That is, recent work would have no higher probability of being cited than older work, and the age of references would be simply a function of the growth rate of the research area. To determine the immediacy effect, the distribution of references in the current literature must be compared with the age distribution of that literature.

To see whether controlling for the growth of the literature would yield results different from those reported by Zuckerman and Merton, we examined the age distribution of all the references in the same 108 journals for which we had computed Gini coefficients (see Table 3). To estimate the age distribution of the literature in each field, we used the number of abstracts listed in the appropriate abstracting journal. (These figures yielded only rough estimates.) To do this we were forced to make the assumption that

*Zuckerman and Merton cite both Price (20) and MacRae (21) on the necessity of making such controls.

Table 4. Immediacy Effects, Selected[a] Scientific Fields (Data for 108 Scientific Journals)

Field	Number of Journals	Work Published in Previous Five Years		Work Published in Previous 10 Years	
		References (Average %)	Immediacy Effect	References (Average %)	Immediacy Effect
Mathematics	6	45	14	68	12
Physics	10	57	19	81	21
Chemistry	12	55	25	78	30
Biochemistry	10	62	30	85	31
Geology	7	48	21	71	24
Psychology	8	47	24	75	38
Sociology	7	40	b	67	b

Sources. Growth of literature—mathematics, *Mathematics Review,* physics; *Science Abstracts;* chemistry, *Chemistry Abstracts;* biochemistry, *Biological Abstracts;* geology, *Bibliography of North American Geology;* psychology, *Psychological Abstracts.*
[a] Fields with fewer than five journals available were omitted.
[b] No reliable method of calculating the size of the literature within the various time periods was found and therefore no immediacy effects were computed.

work published before 1920, which constitutes only a small fraction of all scientific work, is essentially irrelevant today. Clearly, we are not implying that pre-1920s scientific work is irrelevant in the sense that it has not been crucially necessary for scientific advance but only in the sense that most of it has been fully incorporated and is, therefore, rarely explicitly cited. (In fact, very few citations to work published before 1920 appeared in any of the journals.) Using the abstracting journals, we were able to get at least rough indicators of the age distribution of the relevant literature in each of the selected scientific fields. This distribution was compared with the distribution of references in current literature.*

For Table 4, the immediacy effects for the five-year references were computed as follows: Using an abstracting journal, we added up all journal articles listed between 1920 and 1970 and determined what portion of them were published between 1966 and 1970. This number was subtracted from the proportion of references to work published in that five-year period. For example, in psychology, 23% of the 1920–1970 literature was published in

---

*For biochemistry, the distribution of references in the *Journal of Biological Chemistry* was compared with the age distribution of abstracts in *Biological Abstracts.* If *Chemical Abstracts* had been used, the results would have been the same, because chemistry and biology have grown at about the same rate.

the last five years. Since 47% of the current citations refer to work that is five years old or less, the immediacy effect is 24. The 10-year immediacy effect was computed by subtracting the proportion of literature published in the last 10 years from the proportion of references to work published in those years.

A few preliminary observations can be made about the problem of analyzing immediacy effects. The most difficult problem in computing them is to obtain even roughly accurate data on the growth rates of research areas. We were forced to use data on the growth rates of literature in an entire field, and even those data are barely adequate. Since we have concluded that the research area should be the unit of analysis, we should ideally know the growth rate of the literature in each of the research areas.

Although the data on the size of the literature are clearly inadequate and must be refined by further research, several patterns about the immediacy effect are notable (see Table 4).

1. The immediacy effects of psychology, which is presumably less codified than physics, chemistry, or biology, are, nevertheless, essentially similar to the immediacy effects for those fields. When we examine the immediacy effect for the last 10 years, the psychology journals show the highest figure. Given these data, we are not surprised that an analysis of variance showed no statistically significant difference in the immediacy effects between psychology journals and journals in the natural sciences.

2. Journals publishing research letters and results of experimental research tend to have high immediacy effects—for example, *Physical Review Letters, Journal of Experimental Psychology,* and *Analytical Chemistry.*

3. Some journals that have unusually low Gini coefficients—for example, *Analytical Chemistry,*—have very high immediacy effects. And, although we cannot compute the immediacy effect, engineering journals have a high proportion of references to recent work. Engineering, an applied field, presumably is not as highly codified as some of the physical and natural sciences.

Although these data require much more analysis, the immediacy effect may be primarily a function of the degree to which the literature in a journal consists of reports of experimental results rather than the degree of codification. In general, we have found that theoreticians and theoretical articles receive more citations than do experimentalists and experimental papers. The citation analysis leads then to another question: If citations do capture the duration of actual utility of scientific ideas and if obliteration through incorporation does not significantly distort the indicator, why are theoretical rather than experimental discoveries utilized for longer durations? Although a satisfactory answer must await additional inquiry, this is probably because experiments are superseded relatively quickly by technically more efficient ones. Also many authors cite data compilations and review articles that summarize experimental results rather than all the separate articles.

What do these results suggest? First, the lack of differences could, in part, result from measurement error—we may have measured the immediacy effect inadequately. Second, the immediacy effect may not be a good indicator of the conditions that produce rapid scientific advance. Finally, these results suggest the hypothesis that rates of advance may differ more widely within the various research areas in scientific disciplines than between them. Thus, at any given time in the history of the sciences, there may be several research areas in economics in which more significant advance occurs than in particular research areas in chemistry or physics. In the same historical period, the correlative may also be true—some research areas in chemistry and in physics are developing more rapidly than some in economics.

The data in Table 4 indicate that scientific fields have a slight tendency to favor recent work in their citation patterns. Is this tendency a peculiar and distinguishing characteristic of science, or does the same pattern appear in the journal literature of the humanities? Derek Price's data showed substantially lower proportions of citations to recent work in humanities journals than in either the natural- or social-science journals (20). But Price considered neither the growth rate of the relevant journal literature nor the differences between citations to "data sources" and citations to other work in the area. For instance, if a literary critic is writing an article on the seventeenth-century sermons of John Donne, some references will be to the works of John Donne (the data source) and others to the work of other literary critics or other sources. These two types of references must be distinguished before the immediacy effect for a journal in the humanities can be computed.

We have completed a small pilot study of journal citations in English literature journals.* A random sample of 25 articles was drawn from the 1971 volume of the *Proceedings of the Modern Languages Association* (*PMLA*)and another 25 articles from the 1971 volume of *Studies in English Literature*. The references were then separated into what we termed "data citations" and "influence citations" to provide the distinction mentioned above. The mean age of all references made in the 50 articles from the two journals was 43 years. However, and unsurprisingly, the mean age of the

*The data for this study were collected and analyzed by Jim Risse, a Stony Brook graduate student.

Table 5.   Immediacy Effect in Two English Literature Journals (463 Influence Citations; Data Citations Excluded)

Age (in Years Before 1970)		Studies in English Literature		PMLA	
Of Cited Work	Percentage of all Literature[a] (%)	Influence Citations (%)	Immediacy Effect	Influence Citations (%)	Immediacy Effect
5	29	13	−16	21	−8
10	50	43	− 7	50	0
15	66	69	3	65	−1
20	76	83	7	79	3

[a] References listed in MLA *International Bibliography* 1927-1970. Linguistics excluded. Sections included were: American, English (language and literature), medieval and neo-Latin, general and miscellaneous.

300 data citations was 83 years, and the mean age of the 463 influence citations was 18 years.

Nevertheless, even when we separate influence citations from data citations and control for the growth of the literature,* we find a negative immediacy effect for both journals. Thus, although approximately 29% of the literature has been published in the last five years, only 13% of the influence citations in *Studies in English Literature* and 21% of the influence citations in *PMLA* referred to work published in that period. The negative immediacy effects were, respectively, −16 and −8. The data for the two journals are summarized in Table 5.

Clearly, the citation practices of scholars in English literature differ from those of scientists. The former are likely to draw equally on work produced in the past and work produced in the past few years. Although we must study the humanities disciplines in greater detail, the immediacy effect may enable us to distinguish between a literature that is scientific and one that is not, even if it may not allow us to distinguish between highly codified and less codified scientific fields.

MAPPING THE FINE STRUCTURE OF RESEARCH AREAS

The first part of this paper has dealt with preliminary attempts to compare the extent of agreement among research areas. The indicators used were global in the sense that they treated the research area as the unit of analysis.

*Determined from the number of references listed in the Modern Languages Association *International Bibliography* from 1927 to 1970 (see Table 5 for selection details).

In this section, we discuss our effort to measure the level and describe the type of consensus *within* four research areas—two areas in sociology with which we are familiar, and two areas in physics with which we are not.

We are particularly interested in whether scientific work produced in a research area can be subdivided into distinct groups. Furthermore, if subdivisions are found in the fine structure, along what lines are these divisions drawn? They could take various forms. At one extreme would be a research area in which all the participating scientists do similar work employing the same theoretical perspective and the same research techniques. We would say that the internal cognitive structure of such a research area is characterized by a high degree of agreement, approaching consensus. At the other extreme would be a research area divided into a multiplicity of small groups of researchers doing different types of research that employ differing theoretical perspectives and differing research techniques, with no interconnections between the separate groups. Such a research area should be characterized by a very low degree of agreement.

This type of problem allows us to apply the sociological perspective to an analysis of the internal cognitive structure of science. In the past, internal analyses of cognitive structures have been conducted almost exclusively by historians and philosophers of science. Indeed, a significant difference remains between the kind of internal analysis conducted by members of these sister disciplines and the kind we are conducting. We do not examine in any detail the actual content of an individual theory or experiment. Neither do we dissect theoretical orientations in terms of level of abstraction or of the extent to which they conform to principles of a logico-deductive scheme. Instead, we are examining the ways in which a field is subdivided cognitively and how the different subdivisions hold together. Examination of levels of agreement among sociologists or physicists about a theory or theoretical orientation is not identical with examining the theory itself. Although both types of analysis are necessary for understanding the cognitive structure of research areas, here we concentrate on the aspect of cognitive structure which sociology may be best suited to interpret.*

We consider two questions. Can we develop quantitative indicators that will map the subgroups existing in research areas and that will chart the changing structures of these areas over time? How do the fine structures of the sociology areas compare with each other and with the two areas in physics? In a sense, we can report here only the part of the answers which emerges from our quantitative analysis. We would like, of course, to go on to compare the results obtained through the use of our techniques with mappings made by historians of science and by experts in the research areas.

The techniques currently being developed to analyze change in fine structures include adaptations of numerical taxonomy, multidimensional scaling,

*This part of the analysis has benefited from a discussion with Harriet Zuckerman.

and computer programs used by biologists studying evolutionary patterns.* We report the results obtained from the application of two elementary techniques: factor analysis and the analysis of co-citation matrices. Data are presented for the sociology of deviant behavior and the sociology of science and for high-energy and nuclear physics.

First, we describe how the data were collected and the risks entailed in generalizing from them. All the analytic techniques we are currently employing use citations in journal articles as the primary data. The assumption is that a citation is an indicator of intellectual influence, but that such use of citations involves several limitations (22).These can be grouped into two kinds: Overcitation, in which works that have had little or no real influence on the author are included in the citations, and undercitation, in which works that have had a significant influence on the author fail to be included in the citations. Overcitation creates primarily technical problems which can be handled fairly easily. The excessive "noise" in the data—created by inclusion with the truly significant citations of a mass of trivial or irrelevant citations—is reduced by discarding from the data set all citations to authors who are not cited a minimum number of times during the period under consideration. This procedure is based on the empirically supported assumption that on the average truly influential ideas will be more frequently cited than relatively insignificant ideas (23).

The second drawback, undercitation, creates primarily conceptual problems. With the data source limited to citations, the problem of undercitation cannot be handled technically; but it is essential to consider how the type of data used may limit or even negate the conclusions drawn. First, consider the difference between undercitation and the resistance to new ideas discussed by Barber (24) and Cole and Cole (25). An idea that ultimately becomes important but is currently ignored is being resisted or experiencing delayed recognition. Since a resisted idea has not yet had its influence, we need not be concerned that citation analysis will cause us to overlook resisted ideas. The term undercitation refers to ideas that have, in fact, already had an influence but are either not cited at all or are less cited than their significance might merit.

Well-known work is sometimes cited implicitly in the text rather than explicitly in a reference. Thus, physicists might feel that a precise reference to one of Einstein's papers is unnecessary (26). Similarly in sociology, not all articles that utilize Durkheim's concept of anomie make explicit reference to Durkheim in the references (22).

Another type of undercitation occurs through "obliteration by incorporation" (27). In the course of selective accumulation of knowledge in science,

*Lorraine Dietrich is currently working on this problem as part of her doctoral dissertation.

certain important ideas are extended, improved or in some way altered. Frequently, explicit reference is made only to the descendants rather than to the parent. Sometimes an author is unaware that his own ideas are heavily dependent on a precursor. If we look only at explicit citations, we ignore the significant influence of work that has been obliterated through incorporation. In our conclusions we must remember that the technique we have adopted leads us to ignore significant influences not formally cited. The extent to which this limitation of the technique distorts or invalidates results is a serious question that deserves further investigation.

The research reported in the first part of this paper led to the tentative conclusion that a research area or specialty is one appropriate unit of analysis in studying the cognitive structure of science. If we want to analyze citations in a research area, the first problem is to identify a population or sample of papers in that area. We could define all the articles published in a particular specialty journal as a sample of the literature in that area. Thus, articles published in *Physical Review D* would be a sample of work being done in high-energy physics, and articles published in *Physical Review C* would represent a sample of research currently being done in nuclear physics. Likewise, articles published in a journal like the *Sociology of Education* would be taken as a sample of the work being done in that specialty.

Difficulties arise, however, when we want to study research areas that do not have specialized journals or when we want to include a broader sample of work than a specialty journal publishes. In these instances we can derive a sample by having a few experts in the area classify articles as either inside or outside the area. This is the method we used with the two sociology research areas. For example, in studying the growth of ideas in the area of deviance between 1950 and 1973, our sample consisted of all articles dealing with deviance published in the following four major journals of sociology: *American Sociological Review, American Journal of Sociology, Social Forces,* and *Social Problems.* This procedure yielded a list of 533 articles about topics we classified as being within the area of deviance. A similar procedure for the same 24-year period was followed in drawing the sample of 195 articles in the sociology of science, where the journals were: *American Sociological Review, American Journal of Sociology, Social Forces, Social Problems, Sociology of Education, British Journal of Sociology, Minerva, American Behavioral Scientist,* and *Science* (28).

If the intellectual development of a research area is to be studied over a period of time, it becomes virtually impossible to include all the literature in that area and related areas; sampling therefore becomes compulsory. The procedure for the two sociology research areas was to have the sample consist of all articles in the research area that were published in several leading journals—thus automatically excluding from the analysis the great bulk of relevant literature, for example, books, U.S. journals except the few most prestigious, all foreign journals, many small specialty journals, and all applied journals. If the literature from these various sources were included, we would probably find patterns that differ to some extent from those observed in the leading journals. The extent to which citation patterns differ in various sectors of the relevant literature must be studied.

After the sample of articles in a research area was chosen, all pairs of citations (a citing author and a cited author) were extracted. For the analysis presented in this paper, in all four specialties we excluded all self-citations and all multiple citations by the same author to another author—thus preventing distortion of the results by a few authors citing a particular person frequently. Therefore, a report that an author received ten citations *means that he was cited by ten different authors*. Furthermore, with citing papers having several authors, we used only the first author in constructing pairs of citing and cited scientists. Thus, if Smith and Jones wrote a paper in which they cited Brown, that counted as only one citation for Brown. This prevents disproportionate weight being given to citations by authors of quantitative papers, which tend to be more frequently coauthored. After extracting all pairs of citations, we wanted to construct a matrix with the cited authors in the rows and the citing authors in the columns. But if all cited authors were included, the matrix would have been too large to manipulate. We therefore included only the most frequently cited authors in the matrix, using arbitrary cutoff points.

An alternative procedure was used in collecting data for the two physics specialties. First, we gathered data for only one year—1971. Second, instead of classifying articles in a set of journals, we began with lists of the most-cited scientists in each specialty. For nuclear physics, we took the 73 most frequently cited authors in the 1971 volume of *Physical Review C*, which publishes research in nuclear physics and closely related areas. For particle physics, we took the 83 most-cited authors in *Physical Review D*, which publishes research in particle physics and closely related areas.* We then recorded the names of all scientists in the 1971 *Science Citation Index* who cited these heavily cited physicists in articles published in any section of *Physical Review* or *Physical Review Letters*. There were 711 citing authors in nuclear physics and 937 citing authors in particle physics. Of course, virtually all citations to nuclear physicists appeared in *Physical Review C* or the *Letters;* nearly all citations to particle physicists appeared in *Physical Review D* or the *Letters*. We analyzed about 2700 citations in nuclear physics and

---

*We made these calculations from the tapes provided to us by the Institute for Scientific Information. The number of physicists included for each specialty was arbitrary: For nuclear physics, all authors with 17 or more citations; for particle physics, all authors with 30 or more citations.

about 6500 in particle physics. This procedure of sampling yields, in the end, the same matrix as that obtained by following the procedure adopted for the two sociology specialities.

Mapping Fine Structure by Factor Analysis

The first technique employed to study the fine structure of the four research areas is factor analysis. The results obtained for all four science specialties are examined, but for the purpose of describing our methodology, we focus only on sociology of deviance. The sampled deviance literature included 533 articles—51 published between 1950 and 1954, 82 between 1955 and 1959, 128 between 1960 and 1964, 152 between 165 and 1969, and 120 between 1970 and 1973. (No doubt, other sociologists abstracting the four sociology journals would have excluded some articles we included and included some that we excluded.)

For each period, we created a separate matrix of citations to the most-cited authors for that period. Each cited author was defined as a variable, and each citing author was defined as a case. We asked whether each citing author cited the most frequently cited author, the next most frequently cited author, and so on. A positive answer was scored 1 and a negative answer 0. A correlation matrix was then computed that showed the extent to which one sociologist was likely to be cited by the same people as another. A correlation of 1.0 between authors 1 and 2 (or between variables 1 and 2) would mean that every author who cited author 1 also cited author 2 and that every author who did not cite author 1 also failed to cite author 2. This correlation matrix was the basis for the factor analysis.*

A major problem in factor analysis is to decide how many factors to extract. We did not use the arbitrary procedures wherein the analyst continues to extract factors until the proportion of additional variance explained by each new factor declines below some arbitrary point. Instead, we first rotated 10 factors and then reduced the number until we had the smallest number of factors that made substantive sense. This process, of course, required substantive knowledge of the field.

The results obtained from the factor analysis of the four research areas are presented in Table 6. This large and complex table requires explication. It is

---

*Using a matrix of product-moment correlation coefficients in conducting the factor analysis may not be a satisfactory procedure: Given the unequal totals of citations received by different authors, all of the correlation coefficients are artificially reduced in size. Thus, if A has a total of 6 citations and B has a total of 40, even if all 6 people who cite A also cite B, the correlation between A and B will be positive, but low. If Yule's Q were used as the measure of association, the correlation would be 1.0. We are currently experimenting with different measures of association.

Table 6. Factor Analysis of the Fine Structure of Four Scientific Research Areas

Period	Factor[a]	Group Members
	A. Deviance Literature in Four Sociological Journals: Rotated Factors in Five Periods	
1. 1950–1954	I. Criminology, differential association	Clinard; Dunham; Lindesmith; Merton; Sellin; Sutherland; Tappan; Thomas; Znaniecki
	II. Empirical study of delinquents	Bronner and McKay; Glueck, Glueck; Harris; Healy; Jenkins; Shaw; Thomas; Wirth; Znaniecki
	III. Social class, urbanism, mental health	Faris; Hollingshead; Park; Redlich; Sorokin; Warner; Wirth
2. 1955–1959	I. Correctional institutions	Glueck, Glueck; Korn; Mannheim; McCorkle; Ohlin; Parsons; Redl; Weinberg; Wineman
	II. Mental health	Dunham; Faris; Hollingshead; Malzberg; Redlich
	III. Anomie	Durkheim; Henry; Lazarsfeld; Merton; Mills; Parsons; Porterfield; Short
	IV. Differential association, delinquency	Cavan; Clinard; Cressey; Dunham; Lindesmith; McKay; Reckless; Redl; Reiss; Shaw; Sutherland; Wineman

232

3. 1960–1964

I. Juvenile delinquency

Cloward; Cohen; Dinitz; Glueck, Glueck; Matza; Miller; Nye; Ohlin; Porterfield; Reckless, Reiss; Rhodes; Short; Yablonsky

II. Correctional institutions

Becker; Clemmer; Cressey; Goffman; Messinger; Schrag; Sutherland; Sykes; Wheeler

III. Mental illness

Clausen; J. Cumming; E. Cumming; Gurin; Hollingshead; Kohn; Redlich

IV. Anomie

Bell, Meier; Durkheim; Merton; Srole; Williams

4. 1965–1969

I. Juvenile delinquency

Clark; Cloward; Cohen; Glueck, Glueck; Lander; Matza; McKay; Miller; Nye; Ohlin; Reckless; Reiss; Rhodes; Sellin; Shaw; Short; Strodtbeck; Wenninger

II. Correctional institutions

Clemmer; Cloward; Cressey; Korn; McCorkle; Messinger; Schrag; Sykes; Vinter; Wheeler

III. Mental illness

Dunham; Faris; Hollingshead; Kleiner; Kohn; Langner; Redlich; Srole

IV. Anomie

Durkheim; Gibbs; Gold; Henry; Martin; Porterfield; Reckless; Short

V. Symbolic interaction, labeling

Becker; Clinard; Cressey; Erikson; Kitsuse; Lemert; Lindesmith; Matza; Sutherland

233

Table 6. (continued)

Period	Factor[a]	Group Members
5. 1970–1973	I. Symbolic interaction, labeling	Becker; Bittner; Douglas; Erikson; Gibbs; Gove; Kitsuse; Lemert; Lofland; Parsons; Scheff; Schur; Simmons; Tannenbaum
	II. Juvenile delinquency	Akers; Cloward; Cohen; Cressey; Empey; Hirschi; Matza; Nye; Ohlin; Short; Strodtbeck; Sutherland
	III. Mental illness	Hollingshead; Langner; Michael; Pasamanick; Redlich
B. Sociology of Science Literature in Nine Journals: Rotated Factors in Two Periods		
1. 1950–1964	I. Historical perspective	Clark; Fox; Frank; Gilfillan; Gillispie; Hall; Hans; Kuhn; Ogburn; Planck; Stern; Whitehead; Znaniecki
	II. Sociological perspective	Avery; Barber; Kidd; Kornhauser; Marcson; Parsons; Shepard; Thompson; Whitney
	III. Academic systems	Caplow; Manis; McGee; Meltzer; Wilson
2. 1965–1973	I. General	Andrews; Barber; Bayer; Ben-David;

234

factor
(loadings ≥ .40)

Berelson; Caplow; Cartter; Cole, J.;
Cole, S.; Dennis; Folger; Garfield;
Gaston; Hagstrom; Hargens; Kaplan;
Keniston; Kessler; Lazarsfeld; McGee;
Meltzer; Merton; Orlans; Parsons;
Pelz; Derek Price; Storer;
Thielens; Wilson; Zuckerman

II. Others
(loadings < .40)

Abelson; Anderson; Blau; Cardwell;
Clark; Coleman; Davis; Glaser; Gordon;
Gouldner; Greenberg; Harmon; Hirsch;
Hofstadter; Kornhauser; Kuhn; Lakatos;
Lipset; Marcson; Menzel; Mills; Polanyi;
Reif; Roe; Rosse; Rossi; Shepard; Shils;
Toulmin; Watson; Weber; Weinberg; Wolfle; Ziman

C. Nuclear-Physics Literature: One Period

I. Nuclear spectroscopy (s, d) shell

Blaugrund; Endt; Glaudemans; Litherland;
Warburton, Wildenthal; Wilkinson

II. Nuclear-reaction theory

Bassel; Bayman, Drisko; Perey; Satchler

III. Nucleon-nucleon problems, Effective n-n interaction in nuclear matter

Becker; Brandow; Hamada; Kuo;
Moszkowski; Shakin

IV. Theory of collective nuclei

Bohr; Kisslinger, Kumar, Mottelson

1971

235

Table 6. (continued)

Period	Factor[a]	Group Members
	D. Particle-Physics Literature: One Period	
1971	I. Weak interactions and fundamental-scattering theory	Adler; Brandt; Callan; Coleman; Dashen; Gell-Mann; Glashow; Gross; Jackiw; Johnson; Mathur; Pagels; Schnitzer; Weinberg; Wilson
	II. Dynamics and exchange processes (Regg theory)	Amati; Bali; Barger; Chew; Freund; Gribov; Lovelace; Mueller; Veneziano
	III. Scaling theory, deep and elastic phenomena	Abarbanel; Bjorken; Bloom; Chang; Cheng; Chou; Feynman; Jackiw

[a] The reader must understand that the labels given to the various factors involve some degree of imprecision. The label is an attempt to convey the dominant orientation of group members. However, there may be some members of groups who do not fit the label precisely. For example, Paul Lazarsfeld, who has not focused on the sociology of science appears in the general factor for the sociology of science literature for 1965–1973, which is of course composed primarily of specialists in the field.

subdivided into four sections, one for each research area. Within the subsections (A–D), we present the names of scientists who appear in each factor, which is labeled to characterize the substantive work of its member scientists. For the two sociology research areas, the factor analysis has been done separately for several time periods, but because we had data at only one time for the physics research areas, factor analysis there is restricted to the one period. To simplify the presentation, we omitted the individual factor loadings and the number of citations received by each scientist.* Before presenting the results, we should note that the factor-analytic technique excludes from all factors individual scientists whose work had not been *differentially* cited. Thus, the most highly cited author in the sociology of science, Robert K. Merton, does not appear in any of the 1950–1964 sociology of science factors because his work was cited extensively by others referring to authors in all three clusters. With reference to the data presented, let us consider how this type of analysis can increase our understanding of the fine structure of research areas.

By identifying the sociologists receiving the largest number of citations and examining the factors that emerged, we can identify dominant theoretical orientations at different times in the history of the two research areas in sociology.

A schematic overview of the intellectual structure of the research area of deviance between 1950 and 1973 appears in Figure 1. In the early period, differential association appears to have been the dominant theoretical orientation; in the 1960s, anomie theory became dominant, although differential association remained important. And finally, in the past few years the symbolic-interactionist approach has become increasingly important.

In general, subject matter seems to have been a more significant differentiator than theoretical orientation. Throughout the period a distinctive group studied mental illness, with a special emphasis on the influence of social class and urban living on mental health. Hollingshead appeared in the mental-illness group in all five periods. In three of the five periods groups of sociologists studied prisons and other correctional institutions. This group contained many sociologists who might be considered "traditional" criminologists, but it was held together more by a common research site than by common theoretical interests. In all five periods an eclectic group of sociologists studied juvenile delinquency. In fact, juvenile delinquency was the major research topic in the 24-year period.

Unquestionably, there were distinctive theories of delinquency—the theory of "Social Structure and Anomie" (SS&A); Cohen's theory, which depended heavily on SS&A; Cloward-Ohlin's theory, which attempted to

---

*For more extensive, and substantive analysis of the data on deviance, see S. Cole (22).

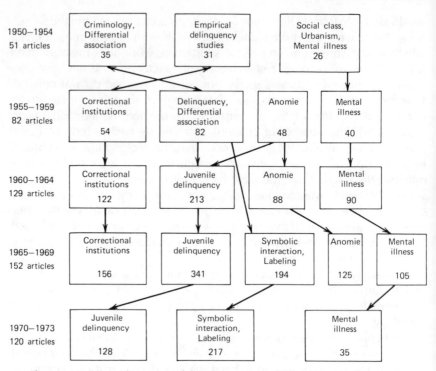

Figure 1. Intellectual structure of deviance research, 1950–1973. (Numbers = citations received; arrows = hypothesized direct influences.)

integrate SS&A with differential association; the cultural theory of Miller; and the normative-drift theory of Matza. But sociologists who were doing research on juvenile delinquency drew ideas from all these theories. Instead of being organized into a group of researchers primarily interested in a theory and attempting to apply and test it on different research sites, the area was organized into groups of researchers with similar substantive interests. For example, throughout the period the number of sociologists appearing in the anomie factor was small—Durkheim, Merton, Srole, and a few others. Sociologists doing research that employed anomie theory were more likely to be cited by others studying juvenile delinquency than by colleagues using anomie theory to study other subjects. It is typical that, in the fourth period, Cloward, who employed anomie theory in both his work on juvenile delinquency and his work on prisons, appears in two different factors. The symbolic-interactionist group, in the last period, is the closest example we have of organization centered on a theoretical orientation. Although these sociologists have written on several different substantive areas, they all employ a similar theoretical orientation.

Sociology of Science. In the sociology of science we find that factor analysis reveals a fine structure differing considerably from that of deviance. Using a set of varied indicators, Cole and Zuckerman recently described the process of institutionalization of the sociology of science (28). Between 1950 and today this research area has moved from what Robert Merton in 1952 described as an "unfulfilled promise . . . cultivated by social, physical and biological scientists. . . ." (29) which then showed no signs of impending institutionalization, to a research area populated by sociologists who had phenomena of science as their principal research interest. All the various indicators reported by Cole and Zuckerman converge to suggest a preinstitutionalized phase from 1950 to 1959, a phase of initial growth and consolidation between 1960 and 1964, and a take-off phase after 1965. For purposes of this analysis, we divide the papers published in this area into pre- and post-1965 groups. Factor analyses of the research area for these two periods indicate the fundamental change in cognitive structure.

In the first period, a substantial portion of the most frequently cited authors were not professional sociologists; another substantial percentage had only a minor or peripheral interest in the sociology of science or none at all. Among the sociologists were Parsons, Lazarsfeld, Wilson, Caplow, Weber, McGee, and Mills. Among the nonsociologists were Kuhn, Roe, Dewey, Conant, Lasswell, Kluckhohn, and Goudsmit.

This *prima facie* classification is supported by the factor analysis for the first period. No dominant theoretical orientation can be identified. The effect of rotating only two factors was essentially to separate sociologists from nonsociologists. Thus, the first factor contained a variety of scientists and scholars such as Kuhn, Whitehead, and Planck; and the second factor contained specialized sociologists such as Barber, Kornhauser, Caplow, Marcson, and McGee. When we rotated three factors, the first two remained basically unchanged; the third factor contained those sociologists who examined academic organizations but who were cited in the sociology of science literature. As we have noted, Robert Merton, who was the most highly cited individual in the period, does not appear in any of the three rotated factors because his work was cited by authors who cited individuals appearing in all the factors.

In the second period, substantive examination of the factor loadings finds only one factor—a general factor of sociologists of science (see Table 6, section B). This means that the leading members of the specialty could not be divided into groups cited by different people, for virtually all leading members of the specialty are cited by the same authors. Thus, authors who cite Warren Hagstrom are usually the same authors who cite the work of Diana Crane or Joseph Ben-David. Members of the specialty differ more in their frequency of citation than in the types of scholars who utilize their work.

When we tried to rotate two factors for the post-1965 period, we were able to subdivide the group of sociologists of science into an empirical-research group and a theoretical group; but this subdivision produced substantial overlap of the two factors. When we rotated more than two factors, we found that the stability of the factors was weak and that many sociologists appeared in more than one. Essentially, rotating the factors produced no stable distinctive groups. We therefore divided these cited authors into two groups based on their loadings on the general factor—one group of sociologists with loadings of .40 or higher, the other with less than .40.

Examination of the fine structure of the sociology of science in this later period suggests the predominance of the theoretical orientation initially developed in the work of Merton and elaborated in the work of followers and students whose theoretical orientation is consistent with or complementary to Merton's. Although at the present time we see only one dominant factor in this research area, the possibility that other distinct orientations will emerge is by no means precluded. This was true for deviance research, and it may turn out to be true for the sociology of science. For example, a group of young British sociologists of science have recently been critical of what they call the "North American school." They see their own work as a revolutionary departure from Merton and other influential American scholars in the sociology of science. If the work of this group does turn out to be distinctive and widely utilized, that should be reflected in the emergence of multiple factors within the sociology of science literature in the near future.

Nuclear and Particle Physics.   Because of our firsthand substantive knowledge of sociological research areas, we began our fine structure analysis there. Without such knowledge of an area, no set of analytic techniques or indicators, regardless of its sophistication, will provide the required insight into cognitive structure. The indicators we are developing are meant to be used as analytic tools by historians, philosophers, and sociologists of science to supplement the more traditional techniques of analysis.

Since our primary goal is to compare the cognitive structures of many research areas in both the natural and social sciences, we have applied the factor-analysis technique to the nuclear-physics and particle-physics data sets.* This analysis is in its first phase; results reported here are preliminary to a more complete analysis.

---

*Robert McGrath of the Stony Brook physics department and Mark Sakitt of the Brookhaven National Laboratory physics department helped to identify the factors in nuclear and particle physics, respectively.

We were able to identify at least four factors in nuclear physics and three in particles (see Table 6, section C). We are interested in differences in the cognitive structures of nuclear and particle physics as well as differences between the physics and sociology specialties.

A qualitative analysis of the content of the factors indicates that nuclear physics has clearer subdivisions than high-energy physics. In nuclear physics the same substantive factors appear when we rotate different numbers of factors; in short, the factors have a great deal of stability. For example, when we rotated three factors instead of four, factors 1, 2, and 3 remained the same (with one exception—i.e., one physicist in factor 2 was different), and factor 4 dropped out. This means, of course, that the same groups of individual scientists continue to reappear, with no additions or deletions. Factor 3 always remains the same, no matter how many factors we rotate. This is not so for high-energy physics where, as we rotate different numbers of factors, the composition of the various factors changes. Also in this research area the same scientists frequently appear in more than one factor. (This same pattern was observed in the latter period for the sociology of science.) That virtually never occurs in nuclear physics. Further analysis might reveal that there is a general factor in particle physics.

The differences between the physics specialties also became evident when, with the aid of some physicist colleagues, we attempted to identify the substantive content of the factors. That was fairly easy to do for nuclear physics, but considerably more difficult for particle physics. It was especially difficult to give a name to the first factor in particle physics. This factor contains many physicists who have worked on a good number of fundamental theoretical problems*. Particle physics seems to be highly integrated theoretically. Perhaps that corresponds to what is meant by codification.

Cognitive consensus.   To what extent did cognitive agreement exist among scientists in each of these research areas about the identity of important contributors to their research area? We continued to use the Gini coefficient as a measure of consensus. The coefficients for deviance—.23, .29, .40, .48, and .30 in each of the five periods—indicate first a growth in this type of cognitive agreement from 1950 to the late 1960s and then an erosion of consensus. (All citations in the articles in our samples in each period were used in computing the Gini coefficients for both deviance and sociology of science.)

---

*With papers, rather than authors, as the unit of analysis, we might have been more successful in differentiating the many problems worked on by the physicists in factor 1.

The pattern for the sociology of science indicates a growth toward consensus throughout the entire period. If for these purposes we divide the specialty into five-year segments, we obtain the following Gini coefficients: .22 for 1950–1954; .23 for 1955–1969, .33 for 1960–1964, .42 for 1965–1959, and .47 for 1970–1973. By the most recent period, the sociology of science had achieved a level of cognitive agreement equalling that among researchers in deviance during its period of greatest cognitive agreement.

For the two physics areas—using the author as the unit of analysis—we found a Gini coefficient of .62 for particle physics and .50 for nuclear physics.

The difficulty in using citation distributions as a measure of consensus is that there may be a few scientists who receive most of the citations, but each one may lead a different school of thought, with each being highly cited by different researchers. Factor analysis, however, offers us a way of potentially measuring the number of distinct groups within a research area. We are currently studying the utility of using the amount of variance explained by a set of factors as a rough indicator of the way in which a research area is organized. Before we can apply this technique, we must solve some technical problems, including determining what type of input matrix should be used for the factor analysis of these types of data.

Two variables might turn out to be significant in interpreting the results of a factor analysis.* The first is the presence or absence of a "general" factor. A general factor is one on which most variables have a high loading; it indicates that there is some variance common to the variables under examination. The second variable would be the total amount of variance explained by a set number of factors. The more distinct and the less overlapping the factors are, the more variance will be explained by any given number of factors. By considering both variables, a typology can be constructed that may prove to be useful in interpreting the results of factor analyses of citation networks.

		General Factor	
		Present	Absent
Total amount of variance explained by a set number of factors	Low	I	II
	High	III	IV

*This section has benefited from discussions with Edward Nadel and Judith Tanur.

Consider each of these four types of cognitive structure. Factor-analysis results that identify the cognitive structure of a research area like Type I would suggest an extremely cohesive cognitive structure. The presence of the general factor would indicate that there is one thread linking the work of the scientists in the area. The small amount of variance explained by a set number of factors would mean the absence of distinctive intellectual groups in the research area. A research area dominated by one paradigm in which all the scientists worked on similar and interrelated problems would look like this.

A research area with a Type II cognitive structure would be less cognitively cohesive than one with a Type I structure. This type would have distinctive groups of scientists, perhaps working on different problems or holding slightly differing theoretical or methodological orientations. However, as indicated by the small amount of variance explained, these groups would have a good deal of overlap. Scientists would be drawing upon ideas from several different groups.

A research area with a Type III cognitive structure would have one dominant intellectual approach that influenced most of the work being done in the area. There would also be several other significantly weaker orientations which would overlap neither with the dominant orientation nor among themselves. A research area might look like this soon after a certain level of institutionalization had been achieved or after an exciting new theory or approach had emerged as a dominant orientation. In both cases scientists following a group of weakly related older orientations may continue to contribute to the field.

A research area classified as Type IV would have many distinctive groups of about the same significance. Research areas having many different topics with little overlap in citation practices, areas marked by weak cognitive consensus, and areas in a pre-paradigmatic phase or having several competing paradigms should all produce this pattern.

Analysis of Co-citation Matrices

The use of co-citation matrices provides another technique for analyzing the cognitive structure of a research area. A co-citation may be defined as the citation of two different documents in the same paper. Small and Griffith (30) have outlined a method for identifying clusters of papers in research areas (For a detailed discussion of co-citation, see the Garfield, Malin, and Small paper, this volume). They took all documents cited in the *Science Citation Index* 10 times or more during the first quarter of 1972 as their sample (1832 documents). Then they made up lists of all documents that were cited in common (co-cited) by another document. The number of co-citations received by a pair of documents is a measure of the extent to which they are connected. For instance, if a paper by Murray Gell-Mann and a paper by Richard Feynman are cited in the same paper, they would have one co-citation. A pair of articles receiving a large number of co-citations is defined as being more closely connected than a pair receiving few or no co-citations. Small and Griffith then used a clustering program that linked all documents with a co-citation strength of 3 or more. Thus, an article would be considered part of the cluster if it were cited at least three times in common with at least one other document in the cluster.

The procedure we have followed in analyzing co-citation matrices differs significantly from that employed by Small and Griffith. First, from the group of sociologists who appeared in each factor analysis, we constructed a symmetric matrix in which the cell entries were the number of co-citations. Thus, in the first period, if Merton and Burgess were cited in common three times, a 3 would be entered in the appropriate cell. We are using level of co-citation as an indicator of intellectual similarity. The basic underlying assumption in adopting this approach is that the work of a pair of scientists who are frequently co-cited has more in common than the work of another, rarely co-cited pair.

Second, we wanted to order the co-citation matrix so that scientists who were most "alike" would appear next to each other in the matrix.* To do this, we wrote a program based on the technique developed by Beum and Brundage (31). Through use of this technique, the columns and rows of the matrix are simultaneously rearranged to maximize the value of the cell entries surrounding the major diagonal. This maximization is accomplished by assigning weights to the cell entries according to their location in the rows of the matrix, computing the average weights of the columns, and rearranging the columns in descending order by average weight. The rows are rearranged in the same order as the columns.† Authors who appear next to each other at the upper left end of the diagonal can be seen to have similar co-citation patterns, as do those who appear next to each other at the lower right end of the diagonal.

---

*We thank David P. Phillips for his aid in conducting the analysis reported in this section.

†The reordering is accomplished by assigning weights to matrix elements according to the inverse of their row location. In an n by n matrix, a weight of n is assigned to elements in row 1, a weight of n-1 to elements in row 2, and so on down to a weight of 1 in row n. For each column, the appropriate weight is multiplied by the cell entry. These products are summed and then divded by the sum of the unweighted column entries, yielding an average weight for each column. These average weights are then ranked from highest to lowest, and the columns are rearranged in that order. The rows of the matrix are also rearranged to correspond to the new column order. This procedure is repeated until, according to Beum and Brundage, further

Table 7. Names As Ordered in Co-Citation Matrix for Deviance Literature in Four Sociological Journals, 1970-1973 (Factors Identified in Table 6, Section A)

Sociologist	Factor Number	Sociologist	Factor Number
1. Skolnick, J. H.		32. Merton, R. K.	
2. Simmons, J. L.	1	33. Blau, P. M.	
3. Douglas, J. D.	1	34. Clinard, M. B.	
4. Scott, R. A.		35. Cohen, A. K.	2
5. Tannenbaum, F.	1	36. Rushing, W. A.	
6. Becker, H. S.	1	37. Matza, D.	2
7. Lemert, E. M.	1	38. Sutherland, E. H.	2
8. Kitsuse, J. I.	1	39. Duncan, O. D.	
9. Bittner, E.	1	40. Hollingshead, A.B.	3
10. Lofland, J.	1	41. Redlich, F. C.	3
11. Szasz, T. S.		42. Pasamanick, B.	3
12. Schur, E. M.	1	43. Sellin, T.	
13. Scheff, T. J.	1	44. Dinitz, S.	
14. Erikson, K. T.	1	45. Cressey, D. R.	2
15. Wheeler, S.		46. Reckless, W. C.	
16. Goffman, E.	1	47. Blalock, H. M.	
17. Gove, W.	1	48. Michael, S. T.	3
18. Davis, K.		49. Langner, T. S.	3
19. Gibbs, J. P.	1	50. Srole, L.	
20. Parsons, T.	1	51. Hyman, H.	
21. Messinger, S. L.		52. Wolfgang, M. E.	
22. Briar, S.		53. Cloward, R. A.	2
23. Piliavin, I.		54. Ohlin, L. E.	2
24. Schuessler, K.		55. Bell, W.	
25. Durkheim, E.		56. Hirschi, T.	2
26. Black, D. J.		57. Strodtbeck, F. L.	2
27. Toby, J.		58. Short, J. F.	2
28. Reiss, A. J.		59. Nye, F. I.	2
29. Akers, R. L.	2	60. Empey, L. T.	2
30. Chambliss, W. J.		61. Clark, J. P.	
31. Ball, J. C.		62. Rosenberg, M.	

iteration results in no change in the order of the columns or in an alternating order. At first we intended to allow the program to iterate until it reached an order that did not change. However, when some of the matrices did not reach a final order after more than 300 iterations, we decided to compare the Spearman rank order correlation coefficient between successive iterations. The iterative procedure was halted when the Spearman exceeded 0.98. For one of the matrices, we compared the order produced using this criterion with the order produced after 150 iterations and found only slight differences. This implies that for matrices such as the ones we are analyzing, there is no one best order. Many may be roughly equivalent and equally "good." Accordingly, we used the Spearman criterion in ordering all the matrices.

Instead of presenting the results of this analysis for each specialty and for each period, we have selected as an illustration the last period in the deviance specialty. The names in the order in which they appeared in the matrix are listed in Table 7. For each sociologist who received a high enough loading (>.40) to appear in any of the factors, the number of the factor is shown beside the name.

What information does analysis of the co-citation matrix give us? This entirely different method essentially corroborates the results of factor analysis. Most of the sociologists appearing in the same factor also cluster together in the co-citation matrix. In the matrix, names that appeared in more than one factor are located between the sociologists who appeared in the relevant two factors. In addition, the matrix allows us to locate scientists who do not appear in any factor and to gauge the relationship between the various factors themselves. Those factors that appear adjacent to each other in the matrix are most similar.*

Co-citation matrices may actually be of greater utility in analyzing the structure of the entire research area than in analyzing subdivisions within it. After ordering the matrix, we can utilize aspects of information theory to tell us how much information would be obtained by dividing the matrix into various numbers of groups.† To do this, we use the information-theory statistic a_L, which is a noninterval measure of interdependence used to measure the predictability, or information, gained by dividing one group into a series of subgroups. The value of a_L varies from 0 to +1 and tells us the extent to which treating particular sets of individuals as subgroups enables us to

---

*There is one important potential drawback to the use of this technique. When there are many zero entries in the matrix (i.e., when there are clear nonoverlapping subgroups), adjacent authors are apt to be similar. However, when there are few zero entries, the fact that the location of a particular author depends on his pattern of co-citation with many other authors may bring about a misleading contiguity. For example, an individual who has co-citations with authors clustered at both ends of the diagonal will be "pulled" toward the middle of it—where his and others' names will be adjacent, although they may not share many co-citations.

†The data from the co-citation matrices can be analyzed in at least one other way. We can compute the level of connectedness in each matrix. Our procedure differs from the one used by Small and Griffith (30). They used the same level of co-citation for all of their clusters. The problem with using so uniform a procedure is that, clearly, if the average number of citations received by the scientists in the matrix is large, a high level of co-citation and connectedness will occur by chance alone. We have not yet completed this part of the analysis. However, we plan to compute "connectedness" in the following way: For each pair of cited authors in the matrix, we shall compute the number of co-citations that they would have in common, assuming independence of chance. Thus, if the total of citers is 100 and Scientist A has 10 citations and Scientist B has 20, the number of citations they would have in common, assuming independence, would be (10/100) (20/100) 100=2. Therefore, if Scientists A and B have three or more co-citations, they are co-cited more frequently than expected by chance.

predict more about their patterns of co-citation than we can predict by treating all of them as composing one group. The value of a_L is always highest when each individual is treated as a separate group, reaching 1 in the case of a completely structured situation—for example, an author is co-cited only with someone who was co-cited only with him, thus constituting a unique group. The value of a_L would reach 0 if every author was co-cited with only one other author and the same author was co-cited in each case. Our ability to predict co-citation patterns would not be increased by looking at authors in terms of subgroups because there would be no substructure in the research area. In other words, the value of a_L will be close to zero for relatively unstructured groups and will increase as the substructure of a group becomes more definite.*

For our purposes, the most important statistic is obtained by treating every author in the matrix as a separate group.† The more alike the authors are, the less information will be obtained by treating them that way. The maximum

---

*For an i by j matrix, we compute a_L as follows:

$$a_L = \frac{H_j - H_j|i}{H_j}$$

where L is the number of possible values of i. H_j measures the dispersion of j and is computed by

$$H_j = -\Sigma p_j \, \log_2 p_j$$

where p_i is the proportion of all observations that fall into category

$$H_{j|i} = -\sum_{i=1}^{L} p_i \, \Sigma_j p_{j|i}^- \, \log_2 p_{j|i}$$

where $p_{j|i}$ is the proportion of all cases in category i that fall into category j.

For a discussion of the utilization of information theory to analyze sociometric matrices, see Phillips and Conviser (32).

---

†Information theory may also be used to determine the best cutting points in the matrix, if we want to divide the matrix into any number of groups. For example, in the two-group situation we can compute the point at which to divide the people in the matrix into two groups so that those within each group are most similar and the two different groups are most distinct. We used information theory to divide several of the deviance co-citation matrices. In those periods (e.g., the first and second) where there were fairly distinctive groups, the information-theory program divided the matrices in a way similar to the factor analyses. In those periods (e.g., the fourth) where groups were not so distinct, however, the divisions made by the information-theory program made considerably less substantive sense than those made by the factor analysis.

Table 8.   Maximum Possible Information Obtained from All Co-citation Matrices

Research Area	Maximum Possible Information
Sociology of deviance	
1950–1954	.21
1955–1959	.13
1960–1964	.10
1965–1969	.07
1970–1973	.11
Sociology of science	
1950–1964	.22
1965–1973	.05
Nuclear physics	
1971	.17
Particle physics	
1971	.07

amount of information obtainable for each of the five periods in the deviance literature was .21, .13, .10, .07, and .11, respectively. These figures parallel those obtained from the analysis of the Gini coefficients. By using both indicators, we obtain a curvilinear pattern. When the information scores obtained for all the analyzed matrices are summarized (see Table 8), the scores closely fit our intuitive sense of the extent to which the various research areas have been cohesive disciplines at various times. The scores are lowest for the middle periods in the deviance studies, for the last period in sociology of science, and for particle physics; they are higher for the first and last periods in deviance, for the first period in sociology of science, and for nuclear physics. The factor analysis results suggesting differences between the two physics specialties are corroborated by an examination of co-citation matrices for the two specialties. The maximum information obtained for nuclear physics is .17 and for particle physics, .07.

CONCLUSIONS

This paper reports the results of several empirical studies constituting part of a larger research program that is only in its first stages. We have just begun to experiment with various research strategies. Strong conclusions on the basis of the data presented here would be premature, but several questions can be raised that require further research.

One of our primary concerns has been the measurement of field differences in approximations to consensus. On the basis of the work of Kuhn, Zuckerman and Merton, and others, we expected to find such differences. We have examined the following types of consensus: degrees of agreement in (a) evaluating colleagues' work, (b) designating the most significant contributions, (c) rating usefulness of work, and (d) comparison of importance of recent and past work. We have also examined consensus within the fine structure of research areas. All these studies suggest that the differences among fields are rather complex.

The following hypotheses emerge from this research:

1. The scientific field is not the most appropriate unit of analysis for studying consensus. The research area, which is frequently a narrower area than a specialty, appears to be more appropriate.
2. For the types of consensus measured, there are no substantial differences between entire disciplines and between the research areas within fields. For example, all research areas in chemistry do not necessarily have higher levels of agreement than all research areas in psychology.
3. Historical patterns of levels of consensus within a single research area may vary as much, or more than, the pattern of differences found between research areas.

With regard to the types of consensus we have examined, differences between the social sciences and the natural sciences do not seem to be clear-cut. To the extent that these types of consensus result either from adherence to a paradigm or from the level of codification, the validity of the assumptions that the social sciences are in a pre-paradigmatic phase or that research areas in the social sciences invariably have lower levels of codification than those in the natural sciences may be questioned. But before those hypotheses can be rejected, we must, of course, further refine the measures of consensus that we employed, and we must develop measures of consensus on matters that have not been tapped at all—for example, the extent to which consensus exists about the important problems within a research area.

Although we cannot now measure certain significant differences between the cognitive structures of the social and natural sciences we remain persuaded that differences exist. At this point, we have not yet identified the range of these differences. This plainly calls for continuing to develop new indicators of consensus and of codification until we find precise indicators that show the differences that we intuit, or until we are forced to change our intuitive assumptions.

Consensus in science is not, of course, a sufficient condition for the advancement of scientific knowledge, although it is widely agreed that it links up with development. The cognitive development of science would scarcely be indicated by a "hollow" consensus founded on ideological agreement or on cognitive ideas that come to a dead end. Thus the development of indicators of cognitive consensus must be coupled with the identification and measurement of other conditions that attend the advance of scientific knowledge.

The measures of consensus examined in this paper direct us to the formation of science indicators in a relatively new, largely uncharted domain, namely, the quantitative historical study of scientific growth and decline at the micro-level of analysis. As we noted at the outset, most science indicators have been global "input-output" measures. Plainly, the use of science indicators to assess the "health of science" more adequately will require enlarged attention to the development of indicators of the cognitive state of major research areas in the sciences.

References

1. *Science Indicators 1972: Report of the National Science Board, 1973,* National Science Foundation, Government Printing Office, Washington, D.C., 1973.
2. Michael Polanyi, *Personal Knowledge,* Routledge and Kegan Paul, London, 1958; Thomas S. Kuhn, *The Structure of Scientific Revolutions,* The University of Chicago Press, Chicago, 1962; John Ziman, *Public Knowledge,* Cambridge University Press, Cambridge, 1968; Imre Lakatos, "Falsification and the Methodology of Scientific Research Programmes," in Imre Lakatos and Alan Musgrave, Eds., *Criticism and the Growth of Knowledge,* Cambridge University Press, Cambridge, 1970.
3. See Margaret Masterman, "The Nature of a Paradigm," in Imre Lakatos and Alan Musgrave, Eds., *op. cit.*
4. Kuhn, *op. cit.,* p. 73
5. *Ibid.,* pp. 19–20.
6. Karl Popper, *Logic of Scientific Discovery,* Basic Books, New York, 1959. (First printed in 1935.)
7. Kuhn, *op. cit.,* p. 32.
8. Polanyi, *Personal Knowledge,* p. 13.
9. *Ibid.,* p. 292
10. *Ibid.,* p. 138.
11. Michael Polanyi, "The Potential Theory of Adsorption," *Science,* **141,** 1010–1013 (1963).
12. Jonathan R. Cole and Stephen Cole, *Social Stratification in Science,* The University of Chicago Press, Chicago, 1973.
13. Kuhn, *op. cit.,* p. 15.
14. Harriet Zuckerman and Robert K. Merton, "Age, Aging, and Age Structure in Science," in Merton, *The Sociology of Science,* The University of Chicago Press, Chicago, 1973.
15. *Ibid.,* p. 507.
16. Ziman, *op. cit.,* p. 13.

17. Stephen Cole, "Scientific Reward Systems: A Comparative Analysis." Paper presented at the Annual Meeting of the American Sociological Association, Denver, 1971; Jonathan R. Cole and Stephen Cole, "The Reward System of the Social Sciences," in Charles Frankel, Ed., *Controversies and Decision: The Social Sciences and Public Policy*, Russell Sage Foundation, New York, 1976.

18. Stephen Cole, "Age and Scientific Behavior: A Comparative Analysis." Paper presented at the Annual Meeting of the American Sociological Association, New Orleans, 1972.

19. Zuckerman and Merton, *op. cit.*, p. 508.

20. Derek J. de Solla Price, "Citation Measures of Hard Science, Soft Science, Technology and Non-Science," in Carnot E. Nelson and Donald K. Pollock, Eds., *Communication Among Scientists and Engineers*, Heath Lexington, Lexington, Mass., 1970.

21. Duncan MacRae, Jr., "Growth and Decay Curves in Scientific Citations," *American Sociological Review*, **34**, 631–635 (1969).

22. Stephen Cole, "The Growth of Scientific Knowledge: Theories of Deviance as a Case Study," in Lewis A. Coser, Ed., *The Idea of Social Structure: Papers in Honor of Robert K. Merton*, Harcourt Brace Jovanovich, New York, 1975.

23. Cole and Cole, *Social Stratification in Science*, Ch. 2.

24. Bernard Barber, "Resistance by Scientists to Scientific Discoveries," *Science*, **134**, 596–602 (September, 1961).

25. Cole and Cole, *Social Stratification in Science*, Ch. 7.

26. Eugene Garfield, "Uncitedness III—The Importance of Not Being Cited," *Current Contents*, **8**, 5–6 (February 21, 1973).

27. Robert K. Merton, *On Theoretical Sociology*, Free Press, New York, 1967, pp. 26–27; Zuckerman and Merton, *op. cit.*, p. 508; Robert K. Merton, *On The Shoulders of Giants: a Shandean Postscript*, Harcourt Brace Jovanovich, New York, 1965, pp. 218–219; Eugene Garfield, "The Obliteration Phenomenon in Science—and the Advantages of 'Being Obliterated'," *Current Contents*, **51/52**, 5–7 (December 22, 1975).

28. Jonathan R. Cole and Harriet A. Zuckerman, "The Emergence of A Scientific Specialty: A Sociological Study of the Sociology of Science," in Lewis A. Coser, Ed., *The Idea of Social Structure: Papers in Honor of Robert K. Merton*, Harcourt Brace Jovanovich, New York, 1975.

29. Robert K. Merton, "Foreword", Bernard Barber, *Science and the Social Order*, Free Press, New York, 1952.

30. Henry Small and Belver C. Griffith, "The Structure of Scientific Literatures I: Identifying and Graphing Specialties," *Science Studies*, **4**, 17–40 (1974).

31. Corlin O. Beum, Jr. and Everett G. Brundage, "A Method for Analyzing the Sociomatrix," *Sociometry*, **12**, 141–145 (1950).

32. David P. Phillips and Richard H. Conviser, "Measuring the Structure and Boundary Properties of Groups: Some Uses of Information Theory," *Sociometry*, **35**, 235–254 (1972).

Difficulties in Indicator Construction: Notes and Queries

10

Hans Zeisel

INDICATOR AND INDICES

As a rule, we use the terms indicator and index interchangeably* whenever we gauge the dimensions of an object (indicand) that is not itself directly accessible because the physical task would be difficult or because it lies in the future. Change in the barometric pressure indicates impending weather change; a rise in body temperature indicates infection; the Intelligence Quotient (I.Q.) indicates the level of what we call intelligence; the Consumer Price Index (CPI) indicates changes in the price of the cost of living. (See Kruskal, this volume.)

It may be useful to distinguish between indicators and indices and reserve the latter term for custom-made measures, such as the I.Q. and the CPI, which aim at measuring complex objects and have by themselves no separate significance; and to reserve the term indicators for measures that have such a separate significance and are in addition indicators of another object to which they stand in some numerical relationship: barometric pres-

---

*"A system of science indicators that would fulfill these several purposes must include indices of both intrinsic and extrinsic aspects of the enterprise." (1).

253

sure to the weather; infant mortality to national health; arrest rates to police efficiency.

INDEX AND INDICAND

The relationship between an index and its indicand may be clearly defined as in the Consumer Price Index, or it may be defined vaguely so that the index becomes the operative definition of the indicand as is the case with the Intelligence Quotient. For the CPI the index-indicand relationship is secured through statistical sampling of the costs of the "market basket." Since no two households buy the same market basket, the construction of the index has been simplified in two directions. It has been limited to sampling the cost of living of the wage-earning population and it allows the various categories of the budget to be represented by typical representatives: by the price of bread, of chicken, of cleaning a dress, of gasoline, and so on. The prices of these samples of expenditures are recorded in a sample of stores, and the emerging weighted averages form the CPI, computed separately for the regions and the major cities of the country. The validity of the CPI is derived from its being a probability sample of its indicand.

The validity of the I.Q., the index of intelligence, is of a different nature. Its numerical value is derived from the test score of answers to a number of standardized questions, expressed as deviation from the mean for the testee's age group. The value of the I.Q. derives from its being sufficiently related to present and future performances and achievements in which "intelligence" comes into play. Since we have no direct access to "intelligence" as we have to the "cost of living," the I.Q., the index, becomes the operative definition of the indicand.

Ideally we should begin our search for indicators and our construction of indices after we have clarified the indicand, the object we want to measure. In practice we often begin search or construction without such clarification and postulate, or try to establish their relationship later.

SCIENCE INDICATORS

Efforts go in two directions. The first attempts to define the dimensions of the scientific enterprise and to develop the measurements appropriate for those dimensions. The second looks for measurements, already available as by-products of the administrative process, that promise a connection with dimensions of the scientific enterprise.

We have notions about the scientific establishment and its endeavors--where and how it operates, how it ensures its continuity, how much it costs, what is produces, and what the quality of the product is. Both the conceptual and social outlines of what is meant by "science" are vague, but they can be sharpened by thinking about how their dimensions might be gauged.

To determine the dimensions of the science enterprise that deserve watching and measuring (see Ziman, this volume) the following questions might be asked.

How many people are engaged in scientific development?

What facilities are at their disposal?

How successful are the scientists, that is, what do they contribute to scientific advance?

What quantity and quality of teaching will be available to future scientists?

How is the effort distributed between the sciences? Is there a proper balance? (This question implies some thought as to which sciences ought to be included in the "science" to be measured.)

How much of the scientific endeavor is dedicated to basic research, how much to applied research (granting the difficulty of making the distinction)?

With respect to both types of research, what are the major substantive directions—such as energy research?

How do long-run and short-run goals determine emphasis (e.g., the present search for energy sources)?

Beginning at the other end we might ask which of the readily available measurements have a bearing on the dimensions of the scientific enterprise. In terms of input, we have such measures as science expenditures by government and other public bodies; science expenditures and staff enrollment in schools and universities; Research and Development budgets in private industry, and so on.

The following measures relate to the output: number of scientific publications, citations of scientific publications, awards and honors, patents awarded and income therefrom, and productivity increase in specified industries.

The goal is to coordinate the available indicators with the relevant dimensions. One of the more useful approaches towards that goal is the analytic dissection of the indicators.

Commonsense inferences about the relation between indicator and indicand are easily developed. But the need for validation remains. The task necessarily involves also measuring the indicand. But then, if there is a way of measuring the end result (the indicand), what need is there any longer for the indicator, admittedly now a substitute? The answer is twofold. First, it may not be possible to measure the totality of the indicand but only some of its facets, as when we relate achievements to I.Q. scores. Second, it may be more economical to produce the indicator continually if, for instance, it is a normal by-product of the routine administrative process. Therefore, one might leave it at establishing the relationship, at least once, on a sampling basis. Depending on its scope, such effort would constitute initial validation by indicating the general magnitude of the relationship, a point important for the acceptance of the indicator.

DISSECTING THE INDICATOR

Labels attached to indicators often promise more than they can keep. Questions such as these require answering (see Kruskal):

What kind of data went into its making?
Where did they come from?
Are the sources complete?
Are they accurate?
How do the data combine?
Are these combinations meaningful?

The so-called arrest rate of crimes, often used as an indicator of police efficiency, offers a good example of what transpires when an indicator is dissected.

The police figures for New York City, for instance, showed the following figures for 1971: 510,000 felonies reported to the police and 102,000 felony arrests made by the police, an arrest rate of 20%. The arrest rates computed separately for the major crime categories are given in Table 1.

The arrest rate for the category "Narcotics and other felonies" is unusually high. This stems from there being two types of crime—those where the victim complains and the police either find or do not find the perpetrator, and crimes such as trade in narcotics, gambling, carrying of illegal weapons where it is most often the police arrest that creates the complaint. Clearly, the more cases of the latter sort are included in the overall arrest rate, the more it will grow. Hence, the "effectiveness" measure can be manipulated upward by making and including more arrests for "Narcotics and other felonies." Limiting the arrest rate to the victim-crimes will therefore produce a more meaningful measure of police effectiveness. When this is done, the arrest rate for New York City, as Table 1 shows, drops from 20 to 13%.

In the field of science, the dissection of indicators would open up similar questions. One might ask, for instance, what makes for more or fewer citations of a scholarly work? Enduring quality? Topicality? Controversial character? Or simply having been written by a member of a set of scholars who prefer to cite each other?

Table 1.   Arrest Rates by Type of Crime, New York City, 1971

	As Percentage of All Crimes (%)	Arrest Rate (%)
Violent (homicide, assault, rape, robbery)	(22)	27
Property (burglary, auto theft, other larceny)	(68)	8
Weighted average of these	(90)	13
Narcotics and other felonies	(10)	88
Average, all crimes	(100)	20

As for the number of scientific papers published, might it not be affected by the number and spaciousness of the journals in the field?

URGENCY AND PURPOSE

Validation of indicators is a never-ending process. How much of it needs to be done will depend on the uses the indicators are put to. If the purpose is only to serve the historian of science (with apologies for the "only"), a presumption of validity might go a long way even if the evidence for it is modest. But if science indicators are to serve policy making,* as most economic indicators do, validity assumes heightened importance. In addition such indicators must be developed speedily to allow timely action and be manipulation proof—that is, safe from the influence of those who might profit from changes in the indicator.

If science indicators are to be used to stop or reverse undesirable trends, time lags may become dangerous. It takes time before indicator data are collected and published; it takes time before they are studied, and it takes time before they lead to actions, presumably administrative decisions of some sort. And then it takes time before these decisions are carried out and take effect. Unlike welfare payments which achieve their purpose the minute they are made, actions in support of science need time before they mature. Conceivably, the sum of these delays could make certain developments irreversible, if, for instance, scientists had migrated in the meantime to foreign countries.

One should, therefore, look for early-warning indicators in science as in other domains (see Holton, this volume). The watchdogs of our economy look at orders in the steel industry as an early indicator of investment trends, and at the levels of wholesale prices as early indicators of retail prices.

*For the full development of complementary properties in the political contexts of indicator use, see Ezrahi, this volume.

Similarly, the number of young people starting a career in science might be an early indicator for science.

WHICH "SCIENCES"?

Science indicators such as those in *SI-72* have been limited to the natural sciences, perhaps because, traditionally, they have monopolized the term *science* or, perhaps because they have been thought to have an applied branch and hence a direct economic payoff. Across the borderline are what we call the humanities. The criteria for the dichotomy have never been made quite clear. The division is similar to the German distinction between *Natur-* and *Geisteswissenchaften*. (See Introduction and Thackray in this volume.)

Some disciplines, such as anthropology, psychology, and sociology, have been allowed to shift, partly or totally, from the humanities to the sciences. There are probably two reasons for this shift. First, these sciences have been adopting more and more of the canons of inquiry that dominate the natural sciences. And second, there is little doubt of the increasing practical values which these sciences in their applied forms transmit to society.

The very fact that the borderline is movable suggests that we give thought to which sciences should be included in the field for which we seek indicators. Several advantages may obtain if science indicators cover a broad field of scientific endeavor rather than just the "sciences," in the old meaning of the term.

The first reason has already been mentioned, the advantage of avoiding a fluid borderline that will shift again. The second reason for including the humanities in our concern is precisely that they do not yield a calculable economic payoff, but only a spiritual gain that is difficult to measure and may therefore get lost in the accounting process. We should make particularly sure that the indicators for these nonpaying humanities be included once science indicators become institutionalized.

The final reason for making sure that the humanities are included rests on the proposition that the most important ingredient of science, the scientific spirit, nourishes all the sciences. To continue along the spectrum, by including the arts in the system of indicators might be the next step.

NOTE

1. *Science Indicators 1972: Report of the National Science Board, 1973*, National Science Board, National Science Foundation, Government Printing Office, Washington, D.C., 1973, "Introduction," p. xii.

PART **III**

CONTEXTS

From Parameters to Portents—and Back

11

John Ziman

INTRODUCTION: "DISCIPLINED ECLECTICISM"

Robert Merton has advocated "disciplined eclecticism" as the mental set to be adopted in approaching a topic as complex and inchoate as science indicators (see Introduction and Thackray, this volume). The present paper seeks to exemplify this attitude by suggesting the variety of ingredients that might go into the recipe for such indicators, baked to firmness by persuasive argument.

On the one hand, it is important not to hobble the imagination by attending only to a few conventional statistics that happen to have proved useful in policy making or in academic discourse; on the other hand, the utility of any particular indicator depends ultimately on the accuracy of the observations on which it is based, on the validity of the unstated assumptions by which it is accompanied, and on the logical consistency of the further processes by which it is reduced to operational form (see Holton and Ezrahi papers, this volume). Except to the bigot, eclecticism and mental discipline are not contradictory qualities; but intellectuals are often reluctant to mount attacks on a complicated subject from many sides, in the knowledge that these cannot be coordinated by a simple unifying theory. Whatever they may be or may in due course become, science indicators can only be treated pluralistically, as no more than a miscellany of measures of the secondary characteristics of the multifarious, multidimensional human activity that is loosely referred to as "science."

The basic premise of this paper is that the processes by which a social indicator may be produced and interpreted can be broken into a sequence of characteristic steps which are worth separate attention. *Logic* is too strong a word for the relations between the successive stages of arguments that seldom claim more than plausibility or rhetorical force. By assigning to each category an alliterative label, we avoid the temptations and ultimate dangers of sharp definition: The schematization is a metaphorical map, not a blueprint. But in trying to construct such a scheme, we learn a little about some aspects of the central problem, and remind ourselves of many potential errors. Formal citations are eschewed; a whole library of reference material is implied in the topics hinted at here.

PARAMETERS

To begin with, we need some parameters: What features of science can be given numerical expression? Most primitively, we can count the elements in distinct categorial sets. *Men* (including, or differentiated from, women, according to context) can be numbered by head in sets characterized by an immense variety of labels, such as academic qualifications, age, employment, and institutional membership. In many cases, social and legal principles permit us to count aggregates of men—that is, corporate *Institutions* such as universities—as elements in larger sets. Since instruments, materials, and other *Hardware* play an important part in scientific research, these too can be counted or measured in reasonably precise terms—for example, how many nuclear reactors are available in India for neutron-diffraction experiments or for making plutonium? On the output side, the conventions of the communications system of science make it quite easy to count such software items as *Publications, Citations* and *Patents*—arranged by categories such as authorship, national origin, medium of publication, or subject matter. Methodological questions such as the weight to be assigned to books or review articles need not concern us for the moment.

The facts of financial accountancy and the fictions of market-exchange equivalence are so compelling that it is usual to lay great stress on the single quasi-continuous parameter, *Money*. It is assumed, for example, that the miscellany of hardware can be reduced to this common measure along with buildings, administrative services, postage stamps, and paper napkins for back-of-envelope calculations at the lunch table. Such technical niceties as discounting instrumental obsolescence or assessing the exchange value of the ruble against the dollar introduce much less error than other factors we shall encounter. For example, the pressure of technical sophistication can inflate a scientific budget more grossly than any ordinary economic force.

PERTINENCE

In choosing parameters, we assess their pertinence. An essential requirement is *Consensibility*: There must be reasonable agreement between different observers on the values to be assigned to the variables. Counting heads and dollars is easy enough, but the boundaries of the categories into which these are to be divided are often ill-defined: Witness the controversy concerning the proportion of "Puritans" among seventeenth-century English scientists. For this reason, it is often convenient to use legal categories ("holders of Ph.D. degrees," "budget appropriations for the fiscal year") which are sharply defined according to public criteria even though such sets may not be coextensive with intuitively significant entities.

We take for granted adequate testing of the *Reproducibility* of numerical data, but this does not say much about the much deeper property of functional *Invariance*. In the physical sciences, not every quantity that can be measured is of theoretical significance. We are interested, for example, not in the coordinates of the planets in some arbitrary frame of reference but in quantities like the radii and periodic times of the orbits, which do not depend on the standpoint of the observer and which can be related to one another by simple equations. In the course of determining such invariant parameters, we may be forced to measure according to some arbitrary conventions, but all "frame-dependent" characteristics eventually must be eliminated from the final theory. It is easy to remark, philosophically, that this process of elimination will itself depend on theoretical assumptions; our justification, as always, will be the ultimate consistency of the whole mathematical model with a wide range of apparently unconnected observational data.

In dealing with social and psychological parameters, the lack of well-established procedures for the reduction of data to invariant form is a grave handicap. Statistical techniques such as analysis of variance or factor analysis do not validate the measures they generate. Consider, for example, the attempt to measure the scientific "ability" of a number of individuals by peer assessment. Out of a diversity of opinions, it may not be difficult to produce a statistically significant rank order; but this does not mean that this parameter ought to appear in a functional relation involving other such observable phenomena as "quality of research." In every case, this is a question to be tested on its own merits, and not to be taken for granted. For this reason, the results of Delphi studies, although often extremely interesting in their variety and operating as a stimulus to the imagination, cannot be regarded as satisfactory ingredients of quantitative indicators. At best, such studies provide information about the opinions held by a particular group of people, but tell us nothing reliable concerning the reality about which these opinions are expressed (see Ezrahi, this volume).

A highly desirable characteristic of a parameter is *Precision*. It is important to remember that the standard techniques of mathematical statistics have been devised to deal with quantities like crop yields, which are distributed approximately "normally" about their modal values. But unfortunately, many of the interesting features of science are very unevenly distributed. Thus, the statistical averages, even over large samples, can be entirely misleading. This is known to be the case, for example, with "scientific productivity" (e.g., publications per author), which has an extremely skewed distribution. Because a very small proportion of authors produces a large fraction of the literature, any central statistic, such as the average number of publications per author, becomes quite meaningless. Indeed, if one takes the intuitively quite plausible view that science makes progress mainly through a few exceptional contributions by a few persons of outstanding ability, one must be skeptical about the significance of nearly all statistical parameters which will inevitably be dominated by the pedestrian products of a mass of uninspired research workers doing "normal science" with little prospect of great success. Until this fundamental structural feature of the knowledge industry is well understood, we must be extremely cautious in the arithmetical reduction of data by conventional methods. As Gerald Holton has emphasized in his paper in this volume, it is essential to preserve the "fine structure" of the data, right through to the final interpretive stages of the investigation.

The same warning is called for against the crude assumptions of *Linearity* or *Aggregability* implicit in some parameters. One may naively suppose, for example, that the more people at work on a certain problem, the more rapidly it will be solved. But that ceases to be true when the literature of the subject has become too bulky to be properly surveyed and where the administrative demands of large research groups divert the most competent scientists from actual research, or when the need to produce a large number of apparently original contributions displaces—toward a mass of trivial elaborations on an irrelevant theme—the primary goal of solving a genuine problem. Here again, the aggregation of numerical data is no substitute for firsthand knowledge of the social, psychological, and cognitive realities they are supposed to represent.

Particularly serious errors arise from the aggregation of data that are not strictly *Commensurable*. This is almost inevitable when everything is reduced to money. For the purposes of financial accountancy, it is perfectly legitimate to lump the stipends of the professors with their annual expenditures on technical equipment and services; but that does not mean that trained intellects are worth precisely what they cost to hire. Within narrow limits—that is to say, within a particular discipline, in a country at a particular level of economic development, over a period of a few years—there may well be a conventional schedule of equipment for a research laboratory that ensures a fairly uniform expenditure on material facilities for each active scientist. In those circumstances, a comparison of total research budgets of institutions or countries is a rough measure of their relative research power. But such a comparison would be nonsensical under wider circumstances: What, for example, would be the proper correction factor for the ratio of the professorial stipend paid by Yale University to J. Willard Gibbs in the 1880s to the salary of a computer programmer at that same university today? Or, would it be fair to say that the research potentialities of the U.S. National Accelerator Laboratory are greater by a factor of 10^4 than those of the physics department of Calcutta University in the great days of C. V. Raman and S. N. Bose?

KNOWLEDGE AS OUTPUT

The foregoing objections apply with irresistible force to attempts to measure the output of science. Considered as an "industry" within society at large, science makes characteristic contributions to technical innovation, economic growth, educational development, and other forces or symptoms of historical change. It provides benefits that can be crudely assessed in economic terms. Unfortunately, however, those benefits are often long-term, generalized, and nonspecific: It is extremely difficult to attribute a particular quantifiable benefit to a limited body of science whose input cost can be estimated, or to assess all the economic benefits that have been gained over a long period from a particular piece of research (see Griliches, this volume). These material benefits flow from the generation by science of an intermediate product—*knowledge*—which becomes available for common use, often at negligible cost to the user. If knowledge itself could be parameterized, the theoretical basis for cost-benefit indicators for science would be immensely strengthened.

But clearly knowledge is not an economic category that can be quantified. This is not merely a statement of prejudice against attempts to achieve such an end; it rests upon simple fundamental principles. Any scheme for the quantification of knowledge would constitute a theoretical representation of certain processes in the real world. Within such a scheme, the logical relations between symbols standing for real operations like the creation of knowledge and its communication, aggregation and use, must be, of course, essentially equivalent to relations between the real processes they represent. Thus, the abstract algebra of the symbolic system—the logic of the theoretical model we have constructed—must be isomorphous with the intrinsic relational structure of the reality it is supposed to depict. A model that violates this principle can produce nothing but nonsense.

Whatever it may be, the intrinsic logic of knowledge does not constitute an algebra with typical arithmetical properties. Suppose, for example, that we have decided to assign the value x to a particular item of knowledge, A (e.g., "the fact that Mars is covered with craters is worth 293 *Kans*"). What arithmetical manipulations might we apply to this number? Suppose we have assigned the value y to some other item, B; would it then be correct to say that knowing both A and B is worth $x + y$? Obviously, this would depend entirely on the context. If B happened to be sure knowledge that craters on the moon are caused by the impact of meteors, then the combination is obviously worth far more than if B were some equally valuable but irrelevant fact such as the structure of DNA. The real process of "adding to knowledge" is not adequately represented by adding the corresponding quantities arithmetically.

Again, what would be the counterpart of the arithmetical process of subtraction? Can a piece of knowledge be taken away, once it is known? The acquisition of knowledge is time-dependent, and essentially irreversible, so that the quantity held by any one person must be a monotone increasing function of time. Or did those physical chemists who had proposed theoretical models to explain the mysterious properties of "anomalous water" lose a few units of knowledge when it was finally shown that this substance was not a genuine thermodynamic phase—and did the skeptics gain significantly when their persistent doubts were thus confirmed?

Presumably, the process of publishing a new discovery is akin to multiplication: When the first photographs showing the surface topography of Mars appeared in the newspapers, the knowledge store of every physicist in the world must have increased by 293 *Kans*. Is this large quantity of knowledge—say 293 *Mega Kans*—in some sense equivalent to what might be found in a library of 29,300 volumes, each containing about 10,000 *Kans* of knowledge?

The confusions and paradoxes suggested by these elementary examples cannot be resolved by mathematical devices such as giving to the combination of two pieces of knowledge the value of the arithmetical product xy: taking logarithms, we are back at the equivalent of simple addition. It must be emphasized that the logical structure of "knowledge processes in society" determines the algebraic structure of its symbolic representation; it is scarcely necessary now to invoke standard theorems of mathematical logic to show that this "knowledge calculus," whatever it may be, could not be reduced to any recognizable form of arithmetic. It is the responsibility of the applied mathematician to represent faithfully what he finds in nature, not to maim reality to fit his own petty systems of thought.

It is tempting to try to quantify knowledge by referring to its exchange value on the market. The buying and selling of secrets is one of the oldest of professions and it gets honorable mention in economic treatises. But a secret is not a piece of knowledge; it is an item of information, utterly bound to time, place, and opportunity. The price paid for a secret is a measure of its catalytic power in an unstable system. Its knowledge content may be utterly banal ("The King spoke to the Ambassador this evening") and lose all value in a brief time. The calculus of secrets may well be of some interest (thermodynamics of irreversible processes, catalytic reactions far from equilibrium, catastrophe theory, decision trees, nonlinear dynamics?), but it is quite irrelevant to the theory of scientific knowledge.

More soberly, we might try to follow up a remark by Kenneth Boulding: "It is certainly tempting to think of knowledge as a capital stock of information, knowledge being to information what capital is to income" (1). Information theory is a well-developed branch of applied mathematics of great practical utility. But the concept of information is there defined in a very strict sense; it is limited to communications whose content is confined, *a priori*, to a finite set of possible alternatives. In this technical sense, and in general usage, "information" is distinguished fundamentally from "knowledge" by the former's restriction to a particular context and language of communication. Nobody need doubt that a list of Stock Exchange prices, for example, has a precise information content (i.e., in "bits") and that it is a commodity with an ascertainable exchange value for a habitual purchaser. But its ephemeral life and its coded form show that it is only information: All the stock-market ticker tapes that have ever been telegraphed do not add up to our knowledge of that particular institution—which encompasses theories, generalizations, historical facts, anecdotes, mathematical models, and many incommensurable quantities, qualities, and images.

This is the reason that "publications" cannot be treated as the quantifiable output of science without considerable reservations. Scientific papers are, of course, very unequal in quality, so that this parameter lacks precision—but that is not the main objection. The fundamental point is that a scientific publication usually communicates information: It is bound by publication date and by subject matter to a definite context, and it is written in the narrow vernacular of a particular scientific discipline. Whether it merely reports experimental data or tentatively sets out new theories, the message is nearly always restricted in advance to one of a set of already formulated alternatives. Genuine novelty in the primary scientific literature is so rare that most papers read to the expert like greetings telegrams: From the title, abstract, and footnotes alone, it is apparent what each one is about—so much so that the American Institute of Physics has been publishing a journal of scientific papers consisting only of title, abstract, and footnote references! (see Holton, this volume). It is the accumulation, mutual interaction, and eventual transformation of this mass of information by various intellectual and social processes that turns it into knowledge, which is what we are really seeking.

PATTERNS

Having determined an appropriate set of parameters, how might we construct an indicator? Into what formulas should the numerical data be substituted? These questions immediately bring to the surface the theoretical models latent in the whole enterprise: Significance cannot be assigned to any parameter or indicator without reference to underlying assumptions about science, about society, and about man. Questions of ideological bias, of the protection of powerful interests, of the indefinability of goals, and of the misinterpretation of motive are bound to arise. These questions are dealt with elsewhere in this volume (see Ezrahi; also, Kochen and Holton papers).

What is certain, however, is that none of our theoretical models is sufficiently precise to justify the transformation of the primary data according to complex algebraic formulas to yield simple numerical indicators. In every sociologist and economist there is, one sometimes feels, a physicist trying to struggle free: He reads about Newton's laws of motion, and Maxwell's equations, and Einstein's theory of relativity and wants to put his own data into some devastatingly simple equation like $F = ma$ or $E = mc^2$. In his school textbooks on mechanics, the answers to the problems are numerical quantities like 1427.2 kilograms for the mass of the elephant on the seesaw or 7.3 km/sec for the velocity of the rocket. Surely he must express the answers to sociological or economic questions in similar numbers, such as dollars of benefit per head or Intelligence Quotient points per generation or percentage rates of increase of GNP!

The practical obstacle to this apparently desirable goal has always seemed to be the statistical variance of the data. The precision of the measured parameters is always so low that only mathematical ingenuity and a fair dose of optimism can separate the significant factors from the accompanying "noise" (see Cole et al., this volume). Regression analysis, analysis of variance, and other statistical devices have often revealed extremely interesting and suggestive correlations between the measured variables, with great utility as empirical indicators and predictors; but the fundamental equations never seem to shine through the murk. It is as if one were trying to discover Newton's laws of motion inside a tank of treacle—or while encaged with a swarm of bees.

Physics enjoyed the luxury of starting with single "particles" (e.g., the planets of Newton's cosmogony). Even in the physics of solids and liquids we deal with ordered, or statistically uniform, aggregates. The real question is whether the "laws" of sociology are of this kind at all. The search for a "social physics," or a "psychophysics," may be entirely vain; what happens in society or inside a person's mind may not be describable in the language of continuous variables linked by partial differential equations. Even if such a description can eventually succeed, we are far from it at the moment and gain nothing by attempting to mold our data into such form. Every arithmetical operation we carry out on the numerical values of our parameters is "theory laden," in this sense, and becomes a potential source of systematic misconception.

Consider, for example, the number obtained through dividing the total expenditure on equipment in some large institution by the total number of research workers. This quantity—"equipment per scientist"—looks like a simple indicator of research power, or of technical sophistication, or something. But what does it really mean? In practice, much of the equipment will be shared by a number of scientists, so that each one really has far more facilities at his command than the indicator suggests. However, apparatus may be unevenly distributed within the institution, so that some research workers have very little. But then these might be the theoreticians, who never actually do experiments. Or there may be, in some corner of the building, a genius who is inventing an entirely new device, using very simple equipment—and it may be that the biggest, most expensive piece of apparatus in the institution is actually being used quite fruitlessly by a nincompoop.

The point is not merely that the single ratio contains all the imprecision of its constituent parameters and hides all the detail that would really be of interest; the danger is that by quoting such a number, we unconsciously give our assent to some very questionable propositions—for example, "The more apparatus a scientist has, the better his research," or "Every scientist ought to have his fair share of equipment," or "If we spent as much per scientific head on cancer research as we do on nuclear physics, we should surely find a cure." This particular example is not, in fact, a case where any sensible person is likely to be seriously misled, but it illustrates the general principle that even the simplest arithmetical formula, such as a ratio of two parameters, implies a whole theory of the social organization of science.

What should be constructed from the parametric data is not a number but a *pattern*—a cluster of points on a map, a peak on a graph, a correlation of significant elements on a matrix, a qualitative similarity between two histograms, or the connectivity of a topological structure. The term is meant quite generally: It could refer to the qualitative fact of the change of a parameter with time, or even to the observably different rates of change of two different parameters. Two-dimensional patterns are easy to represent on paper and to grasp at a glance, but this limitation is not essential to the definition. The key point is that purely numerical quantities are replaced by geometrical or topological objects or relations, with an overall significance that must be assessed by visual inspection and intuition rather than by mechanical arithmetical criteria.

By the intellectual standards half-consciously adopted by philosophers from the themata of physics, this is a retrograde step—almost as if we were descending to the level of analysis we thought we had left behind in descriptive biology. But this is, in fact, a much more appropriate language than the differential calculus for describing social phenomena, and the patterns we discern in our observations may reflect genuine structural features of the social world. The fine structure of the data is not lost, and abnormal statistical distributions and functional nonlinearities are no longer tiresome "anomalies" that have to be allowed for or "corrected" mathematically; they show up as features of the pattern, worth serious consideration in their own right. Instead of imposing an alien algebraic structure on the universe of our discourse, we try to draw from it its own natural logical topology. This process is never, of course, "theory free"; but the models we now have in mind are more "organic" and less mechanical than those implicit in arithmetical formulas. That does not make them softer or less compelling; what may be lost in the indefiniteness of broader qualitative categories is more than made up by the uniqueness of a recognizable pattern or by the many-dimensional correlations implied by the discovery of distinct clusters of objects with many related characteristics. The obvious imprecisions and intuitions in the assessment of the significance of a pattern are less misleading than the uncertainties and systematic errors often hidden in a bare number.

The Psychological Dimension

What aspects of science demand quasi-quantitative study: What might an indicator indicate? It is all too easy to concentrate attention on problems of immediate urgency where policy decisions are being called for, and to fail to observe phenomena with a longer time scale. For this reason, it is desirable, now and then, in those brief moments of lucidity between one research project and the next, to take a look around at science as a whole.

In the attempt to give some sort of formal structure to thoughts on this Protean subject, it is convenient to distinguish among the psychological, social, and cognitive dimensions in which science has its being: it is an activity of highly conscious persons, corporately engaged in the creation of a body of knowledge. The parameters we measure provide information about *personal, political,* and *philosophical patterns*—at the simplest level, independently, but much more significantly in regard to their interactions. If a science indicator is to have any validity, it must be properly located within this framework of potential qualities and categories.

Personal patterns are those that refer to the individual scientist, as a human being, as a professional expert, and as a member of a group. Attention must be given, for example, to his (or her) *Upbringing.* This takes the investigation back into the basic educational system from the primary school onward, where the foundations of vocational attachment are laid and many intellectual habits are acquired. The relation between general education in science and a specialist education for science is extremely subtle, especially at college level, and strongly affects the important task of encouraging and selecting those few persons with the talents required for successful research. Questions of educational *Opportunity* cannot be ignored, especially where the claims of social equality come into direct conflict with the assessment of intellectual merit and promise. All these issues come to a head in the graduate school, where the aspiring scientist becomes socialized to the forms of his chosen profession.

On the material plane, the conditions of *Employment* in the scientific profession deserve close study. As employees of a large institution such as a university, a government bureau, or an industrial corporation, modern scientists are by no means captains of their fate, and they are seldom in a strong bargaining position for a high salary or security of tenure. The effects of government policy decisions and of general financial conditions in the business world can be serious for a group whose indispensability in the short term is always questionable. The very fact that successful research scientists, by solving the problems put to them, are always in the act of working themselves out of a job contributes a peculiar uncertainty that makes their careers quite unlike those of physicians, lawyers, or administrative officers.

Since research is often combined with *Teaching,* the patterns of academic life demand attention. What does it mean to be a professor? Crude indicators of the burden of teaching, such as student/staff ratios and student contact hours per week, have begun to be heard in the discourse of university administrators, faculty deans, and department chairmen. Lectures, seminars, tutorials, practical classes, and examinations are not merely educational techniques; they take up a great part of the time and energies of academic scientists. Curriculum reform is as significant for the instructor as for the pupil. Specialization in teaching may be equally harmful for teacher and student, and as much an obstacle to interdisciplinary research as the conventions of the learned societies and "invisible colleges."

The modern scientist seldom carries out his research alone. Patterns of *Collaboration*, often learned in graduate school, characteristically differ with time, culture, and discipline. Large-scale team research, as in space science, calls for qualities of intellect and temperament that might be entirely out of place in a solitary activity such as pure mathematics. Personal attitudes; management structures; the deployment of expertise, training, experimental design, and the like must all be matched to the scale of the investigation to make the most of the talents available. Simple facts about human behavior, known by instruction or learned from experience by those who manage research laboratories, can easily be lost to sight in the application of mechanical indicators of organizational efficiency. For example, the term *wasteful duplication of research projects* invokes an economic metaphor (conjuring up a vision of, say, a house with two kitchens) and thus fails to recognize the dialectical value of competition and criticism in the pursuit of knowledge.

Among basic scientists, material rewards are supposed to be less important than honorific *Recognition* by the scientific community. Significant questions for the "health" of science (if that is what we are trying to get indicated) arise from the study of this phenomenon, with its overtones of social anthropology and its undertones of depth psychology. The patterns of recognition (e.g., the probability of getting elected to a national academy, or winning a Nobel prize) change in quality as the scientific community grows in numbers. Traditional rewards of esteem may no longer be adequate—or they may have become so distorted as to do more harm than good.

From recognition springs *Authority*, whether informal—through the "invisible college" of a discipline—or bureaucratically enforced—as in the hierarchy of a corporate institution. The subtle relation between these two aspects of personal authority in science is by no means well understood. How does intellectual authority come to be recognized? What are its constituents? What does it mean to those who exercise it? What happens when authorities are in conflict? How far should its power extend? These and many similar questions come to mind when one considers the history of science in various countries and its likely future. To give a very simple example: The average age at which scientific authorities relinquish bureaucratic power (whether by compulsory retirement in their 60s, by senility, or by death) may be one of the most significant indicators distinguishing between American and Russian science at present.

These topics do not exhaust the patterns that might be deduced from observation of the parameters in the personal dimension of science. However, the purpose of this brief survey is not to provide a checklist for possible research projects in the sociology of science but to suggest the width and depth of the imagination needed in the search for "evidences," "symptoms," and other indicators concerning this most sophisticated of all human activities. As we shall see the other dimensions of science demand no less eclectic treatment.

The Social Dimension

What we might call the political dimension of science encompasses far more than the issues that agitate the lobbyists. Modern science is so highly bureaucratized that the pattern of institutions has come to seem more important than the lives of persons. A diversity of *Organizations*—universities, government laboratories, industrial corporations, funding agencies, private foundations, hospitals, technical military corps, publishing houses, learned societies, advisory and testing bureaus, among others—contribute in one way or another to the progress and application of science in society. These are differentiated in size, structure, and specialized function, thus providing a rich variety of environments for research and development activity. Once again, overall statistical parameters do not do justice to the range of institutional conditions that actually exist.

The fundamental question about any such institution is the extent of its *Accountability* to other nonscientific political or social powers. Who eventually decides the objectives of research and oversees their attainment? At the highest level of the government apparatus, decisions on science policy may be centralized in a single office, or they may merely be rationalized by compromises among a number of autonomous agencies competing for support. At each downward level we may observe this conflict between centralized and pluralistic tendencies; its outcome has a profound effect on the quality and nature of the science that emerges from the system. Administrative patterns appropriate for one type of goal, such as the development of a new military device, may be quite unsuitable for a purpose like finding a cure for a mysterious disease. Under some circumstances, the best solution is self-government of the institutions by the scientific authorities, using the familiar mechanism of peer review; in other cases strong lay influence, to attain politically chosen ends on behalf of the general public, may be essential.

Institutional accountability may, in practice, be exercised mainly through control of a *Budget* for research. The pattern of such budgets may be the most obvious indicator of actual social priorities—as for example, in the disproportion between military and civilian research. But the means by which financial resources are made available (whether grants or contracts), the balance permitted between expenditure on personnel and on material facilities, and the degree of detailed control of individual items may all be of

the greatest significance. In many countries, for example, cumbersome accountancy procedures aimed against fraudulent misuse of allocated funds
may have the effect of greatly reducing the efficiency of research by delaying
procurement of small items of equipment.

A significant feature of modern science is the need for enormously expensive experimental *Facilities,* such as rocket systems, nuclear reactors, or
research ships. The institutions that grow up to run such facilities, although
nominally subservient to the research workers who use them, actually acquire their own autonomy. In prosperous times, with expanding budgets, all
is well; but the process of closing down such a facility for economic reasons
is not unfamiliar nowadays. Managerial skill is also called for in the organization of cooperation between research teams from independent institutions
and in the allocation of appropriate financial resources to joint ventures.
Where this cooperation is international, as in CERN and ESRO, successful
research is not the invariable outcome of the attempted collaboration!

The political dimension also includes research by independent corporations for commercial profit. The motivation of such research, its management, and its ultimate profitability, are widely discussed but by no means
well understood. Economies of scale, especially in the development of new
industrial processes, are assumed to favor very large corporations; but the
spreading of financial risk over many uncertain ventures may be just as
important. The great variation of research expenditure between different
industries is a matter for careful study: Is it mere tradition that restricts
fundamental research in food processing to a minute fraction of what is
spent in the pharmaceutical industry? The role of government research on
industrial processes and the effects of government research contracts within
"private" industry may be highly significant.

Another category of scientific institutions is that of the *Learned Societies,*
whose membership cuts right across the boundaries of government and
industrial corporations. The extent to which these institutions have become
like trade unions, defending the professional status of specialized technical
experts, needs to be explored. Does a body like the American Chemical
Society represent a coherent group with an authentic voice on sensitive
political issues such as the use of scientific weapons in the Vietnam war?
How far does a National Academy of Sciences regard itself as an auxiliary of
the nation state, as compared with its role as a geographically defined
section of the world scientific community? Such questions of fact, or of
attitude, are closely related to the "health" of modern science, at home and
abroad.

The learned societies are among the most important institutions for the
Communication of scientific information. Through their publishing activities
and meetings, they largely determine the pattern of the communication

system of science, which lies as much in the political as in the cognitive dimension. Journals have to be edited, printed, and distributed; conferences have to be planned and managed; books have to be written and published; and large secondary services must be provided for the dissemination and retrieval of relevant information. The learned societies are organs of the scientific community itself, but commercial and governmental organizations are also involved in these activities and must be made responsive to the needs of their customers. These customers are not all scientists; the communication of science to the layman is an extremely important political aspect of science.

Once again, this brief survey of a major dimension of science cannot pretend to completeness. But it may help to show the immense richness and variety of studies that might be undertaken in answer to the broad question "How is it with science?" Overemphasis on "science policy" issues gives too much weight to "lobby" indicators, which become weapons in power conflicts but are much too crude to reveal the realities of mismanagement, inefficiency, discontent, and other pathologies of the science system.

The Cognitive Dimension

The philosophical patterns of science—its intrinsic cognitive content and structure—are rightly deemed fundamental, even though they cannot be deduced from quantitative parameters. The categories in this dimension are necessarily qualitative; yet that does not render them useless or meaningless. Experienced and knowledgeable observers can come to a satisfactory agreement about a great many facts; most of the natural sciences depend in practice on no greater criteria of "objectivity" than this level of consensibility.

Although the overall cognitive map of science is quite beyond human capacity to master in detail, it is of extreme interest not only for the mundane purposes of library classification but also as a representation of the natural world. We may scrutinize it, for example, for areas of *Ignorance*. Are there phenomena that remain mysterious, despite intense scientific attack? Are there important questions that could probably be answered if they were given adequate attention? It is often asserted, or assumed, that basic research proceeds with automatic efficiency; the decisions of individually autonomous scientists, familiar with the potentialities for successful research in their own fields, ensure that all good scientific problems are tackled and solved at the appropriate moment by those best qualified to solve them. But this notion of the "hidden hand" directing the self-governing republic of science makes no allowance for excessive disciplinary specialization or for such sociopsychological mechanisms as intellectual fashions, which often leave large gaps in our understanding. Great unevenness in the overall development of scientific knowledge, with displacement of research goals toward elaborate but trivial puzzles, would be a serious pathological symptom in this dimension.

Small regions of the scientific realm need to be mapped in detail, not only for philosophical or historical purposes but also for the evidence that can be provided concerning the process of intellectual change. The dialectic between the conservative paradigm and the revolutionary innovation is supposedly the mechanism by which truth eventually triumphs, but a good deal of *Error* is also generated and perpetuated in the process. Science's pattern of errors characterizes and delimits its truth in the same way that the pathologies of defect and disease define the healthy body.

Study of the content of individual scientific papers instructs us not only on human fallibility but also on the *Redundancy* and rapid *Obsolescence* of much scientific work. Citation linkages may be used to define the topology of the cognitive pattern in a particular field; they also tell a great deal about the degree of originality that the author ventures to display in his work, the extent to which it is essentially duplicated by other work, and the rate at which each contribution is superseded by new discoveries. This, in turn, is related to the incoherence and *Fragmentation* of research; information is acquired and accumulates in the archives at a greater rate than it can be codified and reduced to knowledge. The assimilation of this knowledge into general culture can scarcely be neglected as the final stage in this process.

The spectrum of *Relevance* of scientific knowledge stretches from basic research to technology. The difficulty is always to weigh potentiality for the future against immediate or past actuality. To use what we already know, in familiar circumstances, is merely the exercise of expertise; research is always speculative with no sure return. Experience and intuition advise on the likelihood of success in solving problems of practical relevance, but tend to emphasize the use of well-tried techniques rather than the unknown powers to be gained by the elucidation of fundamental principles. A major technical enterprise with a long time-frame, such as the harnessing of nuclear energy, demands a very broad mix of research projects, from "trouble-shooting" in engineering development to the most academic of problems in pure physics. The pattern of this mixture—with short- and long-term research given relative weights, highly speculative and for the moment apparently unprofitable lines cautiously kept open, genuine problems that impede practical advance recognized, and the new discoveries that may quite alter the balance of future technical priorities duly assessed—comprises important cognitive factors in the application of science and technology.

The question of the influence of basic scientific knowledge on technical *Innovation* cannot be answered by crude quantitative studies intended to validate simple mechanistic hypotheses. The depth of the knowledge itself, its analytical or predictive power, the mechanisms by which it percolates into the realm of technique, the nature of education for professional practice, the mobility of individuals between pure and applied fields, the availability of the necessary human and instrumental resources—all of these and many other factors are involved. Yet this is one of the most important questions for the support and planning of science by corporations and by the state.

To which aspects of human life is scientific knowledge mainly applied? The large place taken by destructive applications—that is, military research and development—should not be neglected. A great deal more could be found out about this subject than is deliberately stated in official documents. Military research ranges from the practical design and testing of battle weapons, through an immense variety of general and technical developments such as aeronautical engineering and electronics, into such realms of higher academic science as the biological principles possibly relevant to chemical and biological warfare. A careful study of the actual patterns of use of such knowledge—for example, in connection with atomic energy and space research—might be extremely illuminating; the attempt to discriminate between the destructive and constructive potentialities of scientific knowledge is one of the most serious questions of our day. The cognitive dimension of science includes an ethical categorization; the patterns of *Misapplication* are part of the picture, as are the codes of professional ethics against which "applications and misapplications" are unconsciously measured.

Multidimensional Interactions

But science is never so simple as to be represented solely by a personal, political, or philosophical pattern. Those are, to exploit our geometrical metaphor, merely projections in one dimension or another of a whole solid object with multiple connections and strong interactions. The temporary separation of the coordinates can clarify our thoughts, but it must not become a source of error. Indeed, it is precisely the failure to look beyond the cognitive dimension that has blinkered academic philosophy of science, just as the analyses and prescriptions of conventional science-policy studies take inadequate account of personal factors. A similar danger exists that the sociology of science may fail to make sufficient allowance for the cognitive structure of the "contributions" made by the "tribes" it studies.

If our indicator patterns are to be realistic, they must be multidimensional and interacting strongly. For example, personal patterns, such as publications, must be related with institutional structures, such as graduate schools, through the cognitive network of citations. To demonstrate interaction among the preceding sections, here let us say that *Misapplications* of science, associated with poor political mechanisms of *Accountability*, may be traced back to faulty *Upbringing* or *Employment* practices. The cognitive structure of science is universal—but to what extent is this universality reflected in international institutions for research, in a worldwide information system, and in personal career mobility across national frontiers? Personal recognition and authority are given for contributions to knowledge, but they are highly influential in the political sphere.

The strength and nonlinearity of these interactions are not to be ignored. For example, the political authority granted for an outstanding scientific contribution may become, by the irrationality of human emotions, a barrier against further progress: The erstwhile young radical reigns as a reactionary old tyrant. The applications of science may have proved so valuable that science is planned, subsidized, politically directed, and eventually corrupted out of all usefulness. Or reliance on the peer-review system for the allocation of financial resources may strongly reinforce the swings of intellectual fashion, to the detriment of progress even in basic science. In choosing our parameters, in the observation of patterns, and in the interpretation of indicators, all such effects must be allowed for.

PERSPECTIVES

A numerical indicator or an indicative pattern, standing alone, has little significance. "What proportion of GNP is spent on R&D this year in the United States? 2.78%, you say. Very interesting, but what does it mean? Is that good, or bad? Should it be more, or less? Has it been increasing or decreasing? What is the corresponding figure in the United Kingdom?" and so on. The data must be given perspective: Comparable data must be available for similar systems outside the immediate object of our study.

The natural coordinates of relativity are *Time* and *Space*. As we have seen, the change of a parameter over a period of years may be a much more significant indicative pattern than its measured value here and now. Yet the apparent invariance over several centuries of a characteristic parameter of science, such as the number of pages published in a lifetime by a typically productive scientist, suggests an extraordinary stability of the system, with a basis in very deep and permanent features of human psychology. And when we see change in a parameter that has previously shown great stability—as with the recent flattening of the growth rate of science which had seemed so uniform for several centuries—we know that profound structural changes must be occurring within science and in its relations to society.

But lack of data may make the complete historical perspective difficult to achieve. How deep in time should we seek to delve? This must depend on the phenomena we are studying, which in many cases have their own natural epochs. Little can be gained, for example, from trying to measure the influence of industrial-research laboratories before the mid-nineteenth century: Such an institution scarcely existed before that date. Again, despite several instructive episodes from earlier times, the mobilization of scientists for war began only in the First World War. Many lessons can be learned from occasional glimpses of the rudimentary forms of such a phenomenon in historical contexts where it could not yet come to maturity; but these lie outside the scope of an indicator.

Yet we must look at present circumstances with the full perspective of time, to see the growth of a parameter or a pattern from its very beginnings. That is why so much of what is done in the name of the sociology of science fails to give a convincing representation of the scientific life: Many of the personal and cognitive patterns observed today go back to the seventeenth century without essential alteration, and cannot be explained in the restricted social vocabulary of the contemporary "organization man." The information explosion, or crisis, seems a phenomenon of our own times, yet the abstract journal has been a necessary service for at least a century. Expensive instruments like those of Tycho Brahe or the mammoth sundials of the Maharajah Jai Singh in North India have always been characteristic of astronomy, although it would be a mistake to confound the managerial practices of modern space research with the aristocratic authority of a wealthy amateur like the Earl of Rosse. From the history of science, in general, we discover the characteristic time-scale of those features of the science of today in which we happen to be interested.

The geographical coordinate is equally important. It is all too easy to take a parochial point of view and see the development of science in one's own country as unique—in fact, so thoroughly unique as to appear entirely normal. This is a particularly serious matter for the United States, whose scientific output is so large a fraction of published world science that it scarcely seems necessary or possible to compare it with the output of other countries. Per contra, smaller countries tend to denigrate their own style of science or to imitate American attitudes, as if "bigger" were surely better. But the United States is a country like any other, and its science can scarcely be so excellent in all respects as to have nothing to learn from the way things are done elsewhere. The significance of an indicator as a statistical quantity cannot be assessed without some idea of its contemporary variance.

International comparisons in science are made easy by the universality of science. Standards of scientific achievement are widely shared; ideally, the "invisible colleges" know no frontiers. International journals and conferences set the level, and the best work travels freely in English or in translation. The ancient tradition of the wandering scholar and modern practices of international scientific and technical collaboration give rise to considerable job mobility; the typical European professor has spent several years of his life (not always voluntarily!) in countries other than his own and is quite familiar with general conditions of academic employment, teaching, laboratory facilities, and so on. In spite of the most extraordinary variations in the background culture, the tasks of the scientist or technologist are much the same whether he works in Benares, Baku, Birmingham, or Buenos Aires. For this reason, comparisons of science parameters and patterns among different countries are much more meaningful than comparisons of social or economic indicators.

It is important, nevertheless, to compare like with like. Since the level of *Economic* development is a dominant factor in scientific activity, little can be gained by matching an indicator measured for Indonesia with the corresponding indicator for Sweden. But comparisons between Indonesia and Nigeria, or between Sweden and Australia, might prove extremely instructive. At the same level of national wealth, *Political, Religious,* and other cultural factors come into play; it would be interesting, for example, to make a comparative study of Rumania, Spain, and South Korea—countries of about the same size, with essentially totalitarian governments, industrializing rapidly out of a semi-feudal peasant economy—What are the really significant factors? What are the successes and failures on the way to creating a viable scientific community? At the most advanced level of industrialization, an honest appraisal of the parameters and patterns of science under the would-be socialist policy of the Soviet Union would provide a most valuable background against which to see in perspective the achievements of capitalist democracy in the United States, Western Europe, and Japan.

PROBLEMS

In using the word *indicator* we imply something more than an interesting theoretical parameter: We are suggesting an evaluation—the answer to some problem. The nature of the indicator and our attitude toward it are not determined objectively, as if in a purely "scientific" investigation; instead, they depend on our situation, the power we exercise, the position we might wish to defend, and our present values and hopes for the future.

From the point of view of the general public, for example, a science indicator or indicative pattern would naturally be related to the question of *Use.* What's to be got out of science for the man in the street? Almost always this character of sociological fiction is presumed to have purely material needs such as nonstick frying pans—although he too might get a kick out of knowing that Mars has craters like those on the moon.

The politician or industrial manager who sees it as his job to maximize material economic growth may try to get a yet more refined indicator, relating to *Profit*. In the very short run, and making large allowances for the speculative element, this is a legitimate exercise, capable of concentrating the mind on essentials—although the numbers arrived at by cerebration are seldom confirmed in the event.

In much the same spirit, if we treat science as a source of technical invention and innovation, we might ask about *Efficacy* of research and development. This would be the problem facing the research manager, who must try to deploy his resources of men and money to the best effect to produce an output unquantifiable in strict financial terms but assessable in terms of general value. The familiar issues of "science policy" in the political sphere revolve about the same concept; Alvin Weinberg's "criteria for scientific choice" are prescriptions for efficacy indicators in this sense.

For the professional scientist, however, a deeper concern might be for the *Health* of science: Potential use may well be threatened by present-day trends within the scientific community or in its relations with the laity. The scientist may feel that long-term potentialities are being sacrificed for short-term gains, and he may look for possible pathologies in less tangible indicative patterns. Comparative measures of the quality of the output may not prove very instructive; this is where the internal sociology of science in all its subtlety should find its application. Unfortunately, there is a catch in this: Undue concern about one's health is a symptom of hypochondria, a thoroughly crippling disease!

The personal dimension should not be disregarded: How strong is the *Satisfaction* to be gained from doing research? On the face of it, this seems no more than a subcategory of the health or efficacy indicators, but it is really a dominant factor. Until quite recently, the moving force of science has been individual motivation—the satisfaction of following a vocation rewarded by the social tokens of recognition and esteem. The problem of satisfying these psychological needs is fundamental.

And at the heart of the matter is the question of *Responsibility*. Knowledge is the product of independent minds. Their freedom is essential to the health of science. But that freedom must be conscientious, related to human use. In the patterns of science we need to see indications of this delicate synthesis between these antithetical psychological and social forces and constraints.

PORTENTS

Indicators are thought to be valuable as a guide to action. They inform us about the future. Or rather, they inform us about the past and the present; we extrapolate them, or policies we can derive from them, into the future. In the hard sciences, such an extrapolation is called a prediction: The unconscious model is a prediction of an eclipse—the utmost, the most convincing demonstration of intellectual power, coming straight from the heavens themselves.

In the softer sciences, this model is, alas, no more than a misleading metaphor. Human capabilities for precise numerical prediction are gravely limited. In economics, they have become laughable; in political sociology, no pretense is made. So what could our indicator foretell? At best, what we might call trends or, with a tinge of human emotion (since we really care), portents. Nothing for use, be it understood; but enough, if we are wise, to suggest corrective action.

In a fortunate era, the indicated trend may simply be toward more of a good thing—that is, *Progress*. Except for certain contradictions inherent in our humanity, progress would be highly desirable; but a good thing for one person often turns out to be a bad thing for another. Knowledge, judged to be a universal good of which we cannot have too much, is an ideal parameter by which to register progress. Indeed, short of a Dark Age and a wholesale burning of books, archival knowledge is almost bound to grow; hence, progress is continual and irreversible. More subtle indicators, bearing more instructive portents, are called for.

In economic terms present science may be regarded as an *Investment* for the future: To the eye of the financier, past profitability portends future payoffs. Broadly speaking, this also is fairly sure, although the cost of producing the knowledge might conceivably exceed the benefits. In practice, the element of uncertainty turns the investment into a gamble; but the prizes keep on growing, the average payoff over many ventures makes it a good bet.

Or could it be that the portent of our indicators is toward *Saturation*? The number of scientists, the amount of research, the output of knowledge no longer grow exponentially, as if forever. As must happen, they reach a certain level, and stop there. This is no mere nightmare; it seems to be happening, and it suggests many consequences.

Does the system have built-in mechanisms that will stabilize it at an optimum level? Might the trend of the second differential coefficient—that is, the rate of change of growth—itself be steady, so that the curve is passing through a peak and will then turn down? None of our present indicators is sufficiently precise to distinguish between saturation and *Decadence*: There are symptoms of the latter in the intellectual dimension.

Suppose that we had energetically set about determining indicative patterns for the health of science; then we should in due course begin to be concerned about portents of *Disease*. Could these be a source of decadence? More indicators, please—and a better understanding of how science really works.

Eventually, sooner or later—in periods of decades rather than centuries —the past patterns of change, the extrapolated and subsequently verified portents, will have aggregated into something new, a *Transformation* of science. The academization of research in the German universities in the 1830s and the industrialization, bureaucratization, and militarization of science since the 1940s are evidences of past transformations. Like auguries, which could be read only by the high priest himself, the indicators will not reveal this future plainly to untutored eyes, yet should not altogether deceive the scrutiny of the wise.

FEEDBACK

The problems that affect us, and their portents, determine our choice of parameters and the interpretation of the patterns they indicate. The enthusiast for research looks for *satisfaction*; his indicators will stem from *publication* parameters, through personal patterns of *recognition* for dispelling ignorance. But the politician is concerned with the *efficacy* of research and will measure it in terms of *money* for *institutions* and *facilities*. In a policy conflict each party may look to its chosen indicator for support. Thus the program of short-term technological development proposed by the politician as an investment is countered by the research man's evidence that it would lead to intellectual decadence. (The general political phenomenon that exemplifies this is discussed by Ezrahi in this volume.) The categorical analysis attempted in this paper is not a simple multidimensional tree; the path from the parameter to its portent closes on itself.

For this reason, the most natural graphic representation of the categorical framework of science indicators is circular—the "mandala" of Figure 1 (see page 284). But the subjectivity of the choice of an indicator as a particular lobby's battle does not reduce the whole activity to mere rhetoric. Taken as a whole, science remains a distinctive, organic, highly interactive social mechanism with its own level of objective reality. Specialized indicators, or indicative patterns, can inform us correctly about particular aspects of this mechanism, crudely abstracted and simplified, lacking in predictive power or theoretical precision, but instructive nonetheless. Provided that we keep in mind the whole framework of categories of which these are such limited samples, we need not be deceived. Particular indicators may involve circular arguments, but the whole enterprise is not spherically senseless.

NOTE

1. "Standards for the Performance of Our Economic System (Discussion)," *American Economic Review,* **50**, No. 2 (1960).

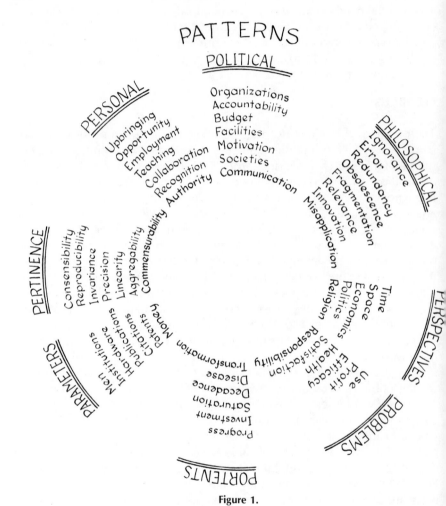

Figure 1.

Political Contexts of Science Indicators

12

Yaron Ezrahi

The idea that information about the "state of science" can be useful for improving public-policy decisions and programs affecting the scientific enterprise has recently become an important part of the growing faith in creating equivalents of the established economic indicators for other areas of social behavior. Bertram Gross, a central advocate and promoter of the idea, describes social indicators as "a continuation of the great information-gathering tradition of Western civilization, particularly of the United States," representing increasing efforts to understand our society (1). Gross believes that such information-gathering, in itself a sign of transition from industrial to postindustrial society, where knowledge plays a larger role in social discourse and action and generally enhances rationality in human affairs, inspired the "indicators movement." According to Gross, "behaving rationally" depends chiefly on "advancing and using sciences and technologies" and "developing and using advanced methods of social-system guidance—with pos-

I am indebted to the editors and to Harvey Brooks and William Kruskal for their very helpful comments on the first draft of this paper. My thoughts on the subject were stimulated also by Edward C. Banfield and members of the Department of History and Sociology of Science at the University of Pennsylvania during the fall semester of 1974.

285

sibilities of 'unitary thought' through general systems analysis'' (2). This outlook on the relation between information-gathering and increasing rationality in human affairs supports the faith in the possibility of effective forecasting and planning for improving public policy.

Similar expectations appear to have led to the goal, stated by the chairman of the National Science Board, of developing indices of scientific activity that "would reveal the strengths and weaknesses of U.S. science and technology . . . and . . . assist in improving the allocation and management of resources for science and technology. . . ." The introduction to *SI-72* proposes that indicators about the state of science "provide an early warning of events and trends which might reduce the capacity of science—and subsequently technology—to meet the needs of the nation. The indicators should assist also in setting priorities for the enterprise, in allocating resources for its functions, and in guiding it toward needed change and new opportunities" (3). Conceding the preliminary quality of data and methodologies in this pioneer effort to provide indicators, the authors assemble statistics on allocation of resources, on training and distribution of personnel, and on research facilities and results. They conclude with a Delphi experiment on problems of science policy and a survey of public attitudes toward science and technology.

As with other types of indicators in the context of public policy, the recent development and expected uses of science indicators appear to be justified by the belief that indicators can furnish "the basis for more informed and enlightened forecasting and action by both public and private agencies" (4). The question about the place and significance of "science indicators" in science policy is therefore a special case of the larger problem of the functions of various categories of information in the context of public policymaking.

The purpose of this paper is to set the questions about the role of science indicators in science policy into this larger context and to analyze the wider problems of developing and utilizing indicators for use in public policy. The focus will be on the various constraints imposed on the uses of indicators to render more rational and enlightened the decisions and actions of public bodies. I examine the role of political-administrative considerations, and their weight in comparison with technical-functional considerations, in the selection and deployment of science and other indicators in the context of public policy.

More specifically, the discussion will be guided by the following questions:

1. How does measurement of various dimensions of science for the manifest purpose of aiding public policy relate to the measurement of science in the context of scholarly research?

2. What effect do publishing measurements of the "state of science" have on the political processes through which the public influences the scientific enterprise? And in what sense—if in any—can information about science be considered a means for improving the rational guidance of science?
3. What is the significance of the changing cultural image of science and the perceived implications of the "state of science" for cherished social values?
4. What are the factors affecting the reception and integration of information about science, as well as other social activities, in the political context? And what is the role of indicators in institutionalizing the process of public policy?

Our preoccupation with constraints on the use of indicators is in no way meant to suggest that the integration of objective and accurate information into the process of public policy is impossible or to imply that such cognitive characteristics of indicators as validity or accuracy are irrelevant. One of the purposes of this paper is to suggest an outline of a theoretical perspective from which the cognitive and technical characteristics of indicators can be considered side by side with their political and organizational uses. Also covered is the possibility of delineating, and in particular cases perhaps even anticipating, the leverage left by noncognitive constraints on the functional uses of indicators as cognitive input into public policy.

MEASURING SCIENCE

The rational collection and interpretation of information for any purpose presupposes systematic selection of aspects of "reality" that are interesting and relevant. A meteorologist and an artillery commander designing tactics for targeting enemy planes may be looking at the same cloud formation, but they are systematically selecting different kinds of information and integrating them into different cognitive and normative sets. Police department and newspaper photographers, both taking pictures of the same street accident, are guided by different interests and different criteria of a "good" picture. In each field of action one of the most strategic kinds of knowledge, the knowledge that underlies professional skills, provides criteria for the systematic selection and organization directing the separation of useful from useless information.

Different purposes then, set boundaries between different kinds of information and substantiate different strategies and principles for its selection and organization. This simple fact is important for our coming to understand that because the making of public policy is a group process which revolves around a multiplicity of often competing, incommensurable, or ambiguous goals, the very grounds for the systematic sorting out of useful from useless information become a problem.

One aspect of that problem may affect the grounds for regarding the manipulation of information as a separate professional activity. In the natural sciences, a sharp division of labor between the information-gathering and the theory-making functions is facilitated by an approximate consensus on the definitions of research purposes and on the conceptual economizers guiding the systematic selection and organization of information. In the social sciences, where the subject matter of research and the comparatively lower level of theoretical agreement generally do not permit comparable consensus on the value and utility of information extracted from phenomena, sharp division of labor between empirical and theoretical tasks is less warranted (see Cole et al., this volume.) In the context of public policy making, the multiple and heterogeneous ends which must be considered impose cognitive limitations on the capacity to rationalize means. Thus, this context is even less congenial for systematically selecting and organizing information and for institutionalizing information-gathering and manipulation as a separate expert activity.

Despite such low levels of agreement, groups or units specializing in the technical activity of developing and communicating information are to be found in both social-science research and in regions of politically controversial public policy. But without the intellectual foundations that substantiate such a division of labor in the natural sciences, the treatment of these information functions as a distinctive professional activity is unwarranted. It may, however, serve the purpose of projecting an image of consensus where dissent prevails, and reinforce a belief in the existence of objective and compelling reality, although reality-definition itself may be a part of a larger scholarly or political controversy.

It may be useful for our discussion to distinguish at this point between *information* and *knowledge* (see Ziman, this volume) as expressions of differing degrees of systematic data selection with reference to any given set of goals. If we use *information* to refer to data not rigorously selected and only loosely related to given goals, and *knowledge* to describe rigorously selected and systematically organized data, the term *scientific knowledge* will presuppose only the specific goals of a particular mode of explaining, predicting, or understanding the world. The term *technological knowledge* would include data systematically selected and organized to enhance specific goals of human action. Thus the same body of data can be seen as knowledge with reference to some goals and information with reference to others. With reference to the same given goals, data may be transformed from the status of information to knowledge and back again, depending on the rigor with which it is systematically selected and organized. With reference to different goals, changes in the bodies of selected and organized data will not represent different bands on a spectrum between information and knowledge, but different kinds of information and knowledge.

Once information and knowledge are conceived as attributes of the relations between bodies of data and goal sets and not intrinsic properties of the data itself, the context within which data are developed and used takes on special significance. From such a perspective, it may be easier to recognize why the status and functions of data are not likely to be the same in a context where the development and uses of data are organized around internally consistent ends as in a context where the goals that guide the process are heterogeneous. Thus a body of data may also constitute knowledge in a goal-consistent context, and yet in a context of heterogeneous goals may vary between information and knowledge—depending on which goal is directing the moment. For example, knowledge in the context of economic theory may be only information when the goal is to control inflation. The same body of selected and organized data may not remain a "constant cognitive package" across diverse normative domains, even in the same general social context. Therefore, where changes in the political and organizational settings within which experts try to develop and deploy indicators would involve—even with adherence to narrowly defined professional standards—changes in their contextual, normative references, they are also likely to involve changes in their status, significance, and effects. Consequently, in political contexts discussions among competing parties about the bodies of data to be accorded the status of relevant information and knowledge are not strictly technical and may even be an extension of the struggle in which each group attempts to establish its own goals as the accepted normative references of collective behavior.

These considerations raise the question whether scholarly works about science as a sociocultural enterprise and its interaction with political and other social forces can be used to produce "improvement" in science policy. As a specific case in point, does this describe *SI-72*? Theoretically speaking, studies of such subjects as the socio-organizational factors affecting the growth of science, the productivity of various scientific communities, and the linkages between science and the economy doubtless can be used to produce aids for "improving" public policy. But juxtaposition of *SI-72* and some of the principal scholarly works on science reveal that the indicators adopted in *SI-72* are neither directly constructed from the scholarly knowledge generated in this field nor designed as aids to scholarly research. No significant systematic effort has been made, so far, to determine whether the studies of, say, Joseph Ben-David on relations between socio-organizational characteristics of academic systems and the growth of science (5), or of Robert Merton on relations between the institutionalization of science and prevailing social values (6) or of Warren Hagstrom on social mechanisms regulating scientific activity (7) can help identify indicators for the state of science and the ways in which that state could be affected by alternative public policies. Without such efforts, these scholarly works remain only loosely relevant to the needs of public policy. The reverse is, of course, also true: Indicators used for policy making are not likely to be useful in scholarly research before their selection and organization are adapted to the intellectual goals that guide scientific inquiry.

Aside from limitations imposed by the complex structure of public policy and the unadvanced state of social science knowledge in this field, the slight contact between the scholarly works on science and the efforts to construct science indicators for use in public policy may result from the pervasiveness of what may be characterized as an informational concept of knowledge. Frequently manifested in the rhetoric of those who advocate indicators as a tool for public policy, that concept is a vulgar version of the positivistic philosophy of science based on a crude separation between facts and theories. It tends to regard measurements of reality as the autonomous basis of theoretical statements and does not consider the role of theories in selecting and defining facts. Adherents to this view believe it legitimate to concentrate upon description and thereby to prepare a foundation for explanation. The accuracy and technical refinement of measurements are therefore treated as valuable independently of the principles by which they were selected or organized.

In no sense, however, can "hard facts" be properly understood apart from "hard concepts." Botanical gardens of indicators cannot constitute a solid base for technically effective and rational intervention in the course of science any more than pre-Darwinian natural-history collections assembling animals, plants, and curious stones could provide a base for technically effective and rational intervention in the course of natural phenomena (8). Yet, in the political context of public policy, the premise that reality is objective and external and therefore compelling has had a special appeal for the experts. Functioning as a support for their social authority, it has helped to depersonalize their claim to be the representatives of the "inescapable logic of factual reality" in human affairs. The same premise has also operated to legitimate the demand that freedom from political and other "extraneous" considerations be accorded to experts in their role as sages and that it be extended to their role as technologists (9).

But particularly in the context of public policy, where differing descriptions of the reality of science have differing implications for the goals and actions of government, such demands cannot be satisfied; construction and deployment of science indicators are simply unlikely to be guided by intellectual considerations alone. Therefore, the two following assertions about science indicators can be compatible: they neither exemplify the properties of scientific knowledge nor are even significantly informed by it; and they can constitute a form of knowledge-for-policy that consists of systematically selected data,* organized to orient and influence decision making.

Indicators and the Rational Guidance of Science

The question now arises about the status of indicators as knowledge for policy making—and principally, about the possible impact of the development and publication of such measurements of the scientific enterprise on political and policy decisions affecting the structure, growth, and direction of science. One can, of course, raise questions about the political and policy effects of developing and publishing measurements—even if they do not constitute a body of valid knowledge that can be thought of as technically applicable to science policy. But the central point remains the expectation that, because they are valid, certain categories of data, or indicators, which refer to "systems" of behavior and institutions, can be used to guide, or even "manage," these systems better and more rationally.

That notion, at least in part, underlies much of the appeal of the idea of using indicators, and is, on the face of it, an eminently plausible expectation. The difficulties begin to emerge, however, when we ask in what sense science (or in connection with social indicators, society) can be thought of as a "system" that can be guided more properly and rationally on the basis of more information. So far, no satisfactory basis seems to emerge for conceiving of complex networks of behavior like science or society as such systems. Neither the process of allocating resources among various social functions, including science, nor the process of allocating resources among various parts of the scientific enterprise itself is organized around sets of internally consistent and universally agreed-upon ends. On the contrary, resource allocation characteristically becomes largely an extension of the fundamental processes of goal setting and trading off for competing preferences and interests, both at the level of the larger society and within the scientific community itself. As Aaron Wildavsky has shown, the national budgetary process is no paradise for rational models of decision making (10). Increased understanding of such difficulties has recently undermined even the earlier optimistic belief of economists—whose example of

---

*It is necessary to keep in mind that terms like *data, measures,* and *indicators* in the following discussion may imply either information or knowledge, or both.

economic indicators inspired the whole indicators movement—that economic knowledge can furnish a basis for the rational guidance of the economy. Former White House economic adviser, George Schultz, has been quoted as saying that "economic policy decision becomes the product of political forces which [can] override the strictures of economic analysis" (11).

This state of affairs means neither that the operation of political forces produces or reflects a kind of irrationality that can be corrected by substituting knowledge for ignorance nor necessarily that economic knowledge has yet to reach a sufficient level of development for appropriate impact. Political constraints on the application of knowledge in various social spheres often constitute only one manifestation of a condition of profound conflict in the social values and ends of the various social groups. Kenneth J. Arrow has already pointed out that since "well known attempts to form social judgments by aggregating individual expressed preferences always lead to the possibility of paradox, . . . there cannot be a completely consistent meaning to collective rationality" (12). The Johns Hopkins University economist Carl F. Christ provided an example of such conflicts in social values when he observed that "the evidence strongly suggests that it is not possible for the American economy, structured as it has been since World War II, to achieve low unemployment and low price rises simultaneously without wage or price controls," and also that "no major industrial democratic country has been able to impose effective wage or price controls in the face of substantially inflationary pressure without creating black markets, inefficiency, misallocation and widespread frustration" (13). Such practical constraints on the simultaneous satisfaction of the politically established value references of public policy mean, of course, that decisions and actions cannot be oriented, economized, and coordinated according to a clear hierarchy of internally consistent ends. This further complicates the problem created by the absence of a solid theoretical basis for understanding the operation of those complex social systems of behavior and institutions which public policy aims at influencing.

Obviously, the significance of additional information for substantiating a more rational approach to social problems is diminished if these problems involve such fundamental value conflicts as are often found between satisfying individual and group interests; regulating resource allocation by the market, on the one hand, and by application of standards of justice on the other; rewarding merit in scholastic achievement and yet promoting equal educational opportunity; enhancing efficiency in government operations and encouraging wider public participation in decision making; or using measurements to aid in grounding a politically effective justification for government science policy, and to aid representatives of the scientific community in rendering that policy more rational or adaptable to the functional needs of science. In areas of such conflicting values a neutral basis for the systematic functional selection and organization of data cannot be found. More information may often serve only to clarify further the differences among the parties and thus reduce their incentives to reach an accommodation.

Given the diffuse and diverse links between scientific knowledge and skills and the various spheres of social action, a clear-cut ordering of scientific fields according to social merit would have had to presuppose an effective consensus on public issues of a kind unlikely to occur in modern democratic societies, especially in peacetime. In the absence of such consensus, the de facto ordering of scientific fields according to social merit tends to result from the interplay of pressures and counterpressures of interest groups whose interaction largely shapes the national budget.

Similarly, the decisions determining the amount of resources that will flow to basic research are not the kind made by rational procedures based on an identifiable body of knowledge. The considerations that influence such decisions are complex. They include such views of the value of basic research in the context of public policy as (a) a means of advancing knowledge or specific fields of knowledge, (b) a precondition of continued long-range capability for the growth of applied knowledge, (c) an ingredient of high-quality education, or (d) a way of gaining the cooperation of top scientists in intellectually unattractive but socially important government projects.

As for the ordering of scientific fields according to internal criteria of scientific merit, the difficulties of arriving at and applying common and specific sets of criteria have clearly contributed to the reluctance of scientists to perform these labors. In the beginning, they tended to argue that the interdependence among the various parts of the scientific enterprise and the difficulties of predicting the future growth of knowledge rendered such tasks difficult, if not impossible (14). During the period of rapid growth in government support for science after World War II—particularly after *Sputnik* and up to the latter part of the 1960s—problems of internal allocation of resources among scientific fields were not acute. In the absence of compelling intellectual standards of scientific merit or external pressures of decreasing support, the various scientific groups were largely free to enhance their wealth in the competition for public funds by using their respective political assets. Often this meant cultivating particular indicators of their achievements and promise (15).

Particular indicators were deployed by representatives of various disciplines in order to exhibit their specific contributions—the physicists to weapons development, the life scientists to the fight against disease, the astronomers to the space program. The social scientists, capitalizing on the public sense that government was not coping effectively with acute social problems, tried to emphasize their actual or potential contribution to improving economic and welfare policies. The very notion of social indicators sprang from the belief that social-science knowledge is *ipso facto* useful for public policy.

It was not that criteria for evaluating the comparative merit of scientific fields had suddenly been discovered which led the National Academy of Sciences and other central organs of the scientific community in the latter 1960s to try to introduce some order in methods of determining the budgetary needs of basic research. This effort resulted, instead, from a growing belief that public policy trends rendered this strategy of evaluation both useful and necessary (16). Decreasing public commitment to scientific research was producing pressure for more selective government support and for the ordering of scientific fields according to scientific merit. Increasingly, leaders of the scientific community have come to recognize that unless internal criteria of comparative scientific merit were integrated into the allocation of public resources, the unrestrained impact of extrascientific social, economic, and political considerations would distort the distribution of resources of science, to the particular detriment of scientifically strong yet politically weak fields.

Such considerations led to a shift from the "free political competition" of fields of science for public support to a "moderately controlled competition in which each [field] is bound to use its political resources economically and in coordination with interests of other sciences and the state of science as a whole" (17).

The National Academy of Sciences (NAS) appointed a special Committee on Science and Public Policy (COSPUP)* to restrain the parochialism of specialized scientific groups that used the collective authority of the Academy to endorse narrowly self-serving reports and policy recommendations. The committee tried to coordinate these field reports with the state and needs of other parts of science and to extend their socioeconomic reference beyond the interests of special social and economic interest groups. The growing interest in comparing various parts of science was accompanied by a growing tendency to develop indicators like the "health of science," and to describe such general states as "the overall balance of scientific effort." A typical evaluation of the overall contribution of science to such national goals as economic growth was included in an NAS report on basic research and national goals (18).

---

*The many NAS-COSPUP field reports include *Chemistry: Opportunities and Needs* (1965); *The Mathematical Services: A Report* (1968); and *Astronomy and Astrophysics for the 1970* (1972).

These developments have clearly reflected the operation of forces both within and outside the scientific community to raise, define and influence the state of science as a social problem (19) deserving serious consideration at the highest levels of public policy making. But this growing political demand for science indicators was reinforced neither by a satisfactory conceptual basis for constructing such overall indicators nor by sufficient political consensus on the goals of science policy.

By comparison with macrolevel science-policy efforts, micro-level attempts—to order fields of research *within* specific disciplines and evaluate the general state of single disciplines—appear to have been somewhat more successful. This may stem from easier consensus on defining worthwhile research and a greater commonality of interests, specifically with regard to the relevance of single disciplines to particular regions of public policy. In at least some of these attempts, for example, *Physics in Perspective* (20), the use of criteria of both scientific and social merit to rate fields has enabled scientists to balance inequalities of scientific and social merit against each other, despite the problems created by mixing hierarchies. This has brought about results that diminished the value of the exercise for policy makers but also eased the internal tensions generated by a "zero sum" gamelike competition of various scientific groups for scarce public support.

Nevertheless, both macro- and micro-level policy decisions ultimately involve allocation of resources among competing groups. Since each group is likely to prefer that important allocation decisions be based on criteria and measurements helpful to its own position, the process of advancing and certifying measurements and shaping allocation procedures becomes an integral part of a political contention. All in all, the process appears to be closer to the political model of decision making—which consists of competing interests arriving at ad hoc arrangements through pressures, bargaining, and compromises—than to the rational model of decisions based on relatively unambiguous, agreed-upon, functionally and technically preferred standards (21). Concepts of fairness are clearly not neutralized in the process by concepts of merit, and factors such as the degree of influence in the public arena and the distribution of other institutional and political resources are not irrelevant to the results.*

This state of affairs may appear regrettable, and, perhaps, short-lived. But it seems clear that those procedures which have shown the greatest effectiveness in bringing about collective decisions involving groups with competing and incommensurable goals have not been the rational and systematic application of criteria but political procedures based, characteristically,

---

*Agreement may more easily be achieved, of course, when indicators are perceived as not being a significant factor in the differential allocation of resources.

upon seldom predictable and often functionally irrelevant balances of op-posing forces. In the absence of another basis for arriving at decisions equally rational from the perspectives of all involved, political procedures—particularly the various forms of compromise—can at least make collective choices possible and ensure their acceptance as the legiti-mate outcome of a gamelike interaction, if only until the next round.

These observations do not mean, of course, that more knowledge about the factors contributing to the welfare of particular fields of science, the use of science in the service of social goals, or the more general welfare of the scientific enterprise cannot eventually represent more weight. But since these decision making and allocation processes are basically political in nature, their time-dimension is relatively short; and typically, their basis is in the social forces and organizational factors that affect the interests and set the boundaries within which such cognitive inputs as valid and accurate indicators can be integrated to improve performance. Hence, information may be technically—as against politically—useful in such group decisions, mostly in the very limited cases where the structure of complementary in-terests among the involved groups is stable enough to allow institutionaliza-tion of compromises. Allocation decisions are therefore "depoliticized" and relegated to technical and administrative agents, who are comparatively free to rely on functional and professional considerations in the selection of standards. (Such situations also exist for limited periods of time, particularly in relatively depoliticized areas of public policy with an effective consensus on specific goals—for example to combat a specific disease or win a war. However, they appear less likely to occur at the synoptic level of the alloca-tion process.)

SCIENCE AS A CULTURAL SYMBOL

It is clear that outside the methodological and theoretical framework of scientific inquiry, such indicators of science as measurements of its financial support, publications, number of students, or contributions to technology are not likely to furnish solid ground for assessing the state of science as a differentiated system of related operations, or for evaluating how various public-policy choices affect it. As we have seen, even with an available body of useful knowledge about science, the largely political nature of the processes through which the levels and the distribution of support are de-termined would impose serious constraints on the use of such knowledge for rendering those processes more rational and systematic.

If these observations are accepted, however, the question remains: What are the consequences of the development and use of indicators of various aspects of science in the political context of science policy? The political visibility of individual science indicators may vary, and may also be subject to different degrees and kinds of interest in different social and political contexts. Furthermore, science indicators, at least in contemporary western societies, seem on the whole to be less central in the context of public policy than several other social indicators. Although one cannot yet rely on the results of research on their comparative social demands and political visibilities, indicators of the state of science, its "growth" or "health," do not appear to be as easily and clearly integrated into the concerns of the individual citizen as are indicators of the state of crime, poverty, health, the economy, and so on. Radical fluctuations in the scope, structure, or direction of scientific activity apparently do not evoke in laymen the strong anxieties or hopes stirred by fluctuations in the volume of crime or the relative size of various parts of the population.

These differences in public perceptions obviously do not stem solely from the fact that indicators about the state of science have not been sufficiently developed and publicized, although that could be one cause. Rather, they stem from the circumstance that the state of science is not perceived as a directly relevant instrument for enhancing individual and social values, and the state of science is not taken as directly related to the state of individual or group well-being. Social attitudes toward the role of science in promoting or reflecting individual and social well-being vary, of course, across historical and cultural contexts. The growth and welfare of science have been perceived, in various times and places, as factors in the achievement of diverse goals such as reduction of material scarcity, promotion of social consensus and rationality in human affairs, moral development (22), achievement of cultural and ideological superiority (23), enhancement of military capability (24), and increase in international cooperation (25). Others have perceived the growth and centrality of science as related also to social and moral decay and political repression (26).

An early advocate of science, Thomas Sprat, associated the advancement of science with the progress of mankind's dominion over things, widening social consensus, closer interaction and cooperation among members of different social classes, increase in the social regard for the value of conscience and truth, the opening of private collections to the wider public, promotion of the glory of national culture, and manifestation of the national character of the British people (23). Despite obvious continuities, even superficial comparisons with contemporary social perceptions of science suggest significant changes of kind and intensity in the associations between science and social values and, consequently, in the cultural bases of sociopolitical justifications of science. Future studies of the relations between changes in the "social significance" of science and kinds and intensity of changes in the social demand for information about science would be greatly aided if the fluctuations in the former would be more adequately mapped out.

In the early 1950s Robert Merton noted the long period of relative neglect of the sociology of science and suggested that interest in that field would develop greatly as the state of science became increasingly defined as a social problem. He observed that the politicization of science in Nazi Germany and the Soviet Union and the science-associated drama of the first atomic bombs awakened public concern with science and contributed to this trend of viewing science as "a social problem, like war or the perennial decline of the family, or the periodic event of economic depressions" (27).

Developments in social perceptions of the relations between science and social values cannot, of course, be induced mechanically by developing and diffusing indicators about the various dimensions of the scientific activity. Instead, political visibility of such indicators and public interest in them appear to depend on the degree to which science is associated socially with matters of public concern. When that association is lacking, science indicators can hardly become central to the political process of public policy, even if they are technically available.

The publication of *SI-72* and the growing interest in various science indicators may have reflected a later stage of response to the evolving social demand for information about science and its relations with society—the demand noted by Merton in connection with the rising interest in the sociology of science. Specifically, at the level where science evolves as a problem for public policy, the demand for indicators seems to be shaped by variations of kind and emphasis in the connection between the state of science and other problems of public policy. Science indicators used for evaluating science as a factor in economic growth obviously differ from those used for evaluating science as a means of winning the space race or securing an American leadership position as a world science center (28). Such variations in the demand for science indicators reveal variations and shifts in the political and ideological premises underlying different definitions of the social role of science and its status as a policy problem.

Although these political definitions cannot be easily affected by technically adequate but politically insensitive indicators about the "real" and "objective" state of science, the technical applicability and functional aspects of such indicators developed in accordance with professional standards may be of interest in the limited forums of decision making and to administrators directly involved in making and carrying out science policy. In the less-exposed forums where decisions are made rather than justified or criticized, it may be easier to separate the technical-functional merits of indicators from their political and ideological uses.

Frequently, however, even policy-making professionals cannot keep these two sets of considerations apart, especially where publicizing indicators is either inevitable or necessary for their use. Consider, for instance, indicators of the social distribution of sickle-cell anemia or venereal disease. Such indicators can, of course, be very useful to public-health officials in designing strategies and programs for coping with the problem. But if publicity cannot be prevented—or, as in at least some cases such as one of those cited above, widespread publicity is an essential factor in limiting the spread of the disease, the development, deployment, and use of such indicators may have a multiplicity of other, largely unintended, uses and effects. Depending on prevailing social beliefs, the same indicators can be used to define the problem of controlling disease as a medical, moral, religious, or political problem.

Indicators can also be employed to "blame" or humiliate parts of the population; justify various forms of social control; or alter patterns of social interaction such as marriage, residence, tourism, and marketing. These effects may influence the social demand for and reception of indicators, as well as their credibility, interpretation, and the freedom to evaluate their technical utility in the making of public policy, without having to take into account their social and political uses.

Reactions to the cholera epidemic in the United States in 1832 furnish one illustration. The epidemic provoked a conflict between the Medical Society of New York and political and economic interests in the city about the desirability of allowing the facts of the situation to be published. When the Medical Society concluded that decisive action required public description of the magnitude of the problem, it was immediately attacked "as a group of private citizens having no authority to make statements affecting the welfare of the entire city." Banker John Pintard asked whether the eager physicians had "any idea of the disaster which such an announcement would bring to the city's business." The city's Board of Health, a public and politically accountable body, was much more equivocal than the Medical Society: The Board certified the diagnosis of the situation but acted on it only when the crisis had reached a higher point (29).*

To say that social interests and beliefs significantly affect the willingness to utilize indicators in the context of public policy does not, of course, deny that indicators may have independent impact on the perceptions and definitions of social problems. Indeed, especially in the long run, if indicators are

---

*Maggie Nunley has called my attention to a film ("Panic in the Streets"), made about 25 years ago, which featured a similar story. Set in modern times, it showed a U.S. Public Health doctor trying to ward off an epidemic of plague in New Orleans and running into similar problems with local officialdom. And, of course, as Robert Merton reminds me, Ibsen immortalized the same syndrome in his An Enemy of the People (1882).

seen to be confirmed by experience—even if the confirmation is insufficient by scientific standards—their users may resist restrictions imposed by social beliefs and political interests. However, this important aspect of indicator use lies outside the limits of the present discussion of the constraints imposed by social and political factors.

Political Constraints on Reception and Integration of Indicators

Once indicators of "states" or processes having social significance come into the political context, they are seen against new frames of reference within which their meanings and functions may change significantly. Thus, the selective social reception and integration of indicators can best be understood in relation to these acquired contextual meanings and functions —and not to those which indicators may originally have for the experts who develop and use them in accordance with technical criteria. Even superficial observation of the attempts to introduce knowledge as a factor in social discourse and action reveals persistent differences between the way in which academics and the rest of the public perceive knowledge; evaluate its relevance and applicability; and judge its meanings, virtues, and limitations. What scientists regard as objective descriptions or explanations of phenomena are often seen by laymen as ideological statements or support for or criticism of values, expectations, and practices of various social groups.

The social "careers" of measurements such as statistics of unemployment and productivity (30), average I.Q. scores in various social groups (31-32), comparative rates of population growth, and estimates of the spread of malnutrition among social classes and ethnic groups have been shaped not only by their technical merits but also by their political and ideological uses.

The reception and integration of indicators about scientific activity into the political context of science policy will probably be subject to similar constraints. The fact that, unlike knowledge about natural phenomena, knowledge about the scientific enterprise is basically knowledge about society and culture only reinforces these constraints. That social scientists cannot dramatize and make their claims of knowledge more socially credible by the generation and manipulation of visible technologies or by the unambiguous confirmation of unique predictions is one cause. Another is that questions about "true" and "desirable" relations between science and society obviously interpenetrate with continuing political and ideological debates.

For this and other reasons, in the political context social definitions of the state of science and choices of actions to influence it cannot be exclusively grounded in knowledge: Factors such as jurisdictional conflicts among institutions, conflicting interests in the allocation of resources, and different sociocultural orientations toward the social rationale and functions of science also play important roles. What may be perceived by the expert as false or incorrect definitions of the state of science can be plausible and perfectly adequate to the political actor. For, as has already been implied, in the context of policy making the political actor does not necessarily use the indicators to serve the same purposes as those that guide scholars or technical experts. For example, his definitions may not be intended to serve as guidelines for more adequate explanations of the state and development of science. They may be designed instead to communicate to the general audience (as opposed to the special audience of science) his value commitments and intentions, with the aim of reinforcing their confidence in him.

Such differences between the uses of indicators are particularly evident in economics, where political leaders often try to back up claims with "the numbers" and expert authority (30). Politically, the particular merits of imprecise or ambiguous descriptions of reality may lie precisely in their capacity to reassure diverse or conflicting audiences. The technical inadequacy of such descriptions of reality as tools for orienting policy makers toward effective actions may be compatible with their adequacy in projecting the appearance that such actions have been taken or are planned. The conditions of public trust and political authority often lead political actors to the peculiar middle ground between apparent and concrete actions where they act just enough to live up to the image by which their constituents judge their performance but not enough to bring about actual results that may alienate some of their supporters. Such considerations of "optimal political benefits" often make of government programs more an expression of some sort of political ritualism than a considered attempt to cope with real problems.

Hence, the dilemma of using indicators in science policy, as in other areas of public policy, is that when science comes to be more a subject of public interest it also becomes more subject to the influence of the political process—which then imposes serious constraints upon the freedom to rely solely on cognitive criteria in selecting indicators for use in public policymaking. Correlatively, the situation in fields such as systematic biology or pure mathematics reveals that the farther science moves away from the center to the periphery of public concerns, the less likely it is to attract the massive support and commitment of large-scale public resources that come with active public acceptance. The result is that science policy requires a trade-off between the outside need—for indicators as a way of achieving political visibility for science as a public-policy problem—and the internal need for indicators to describe the state of science so that science policy can be informed and rationalized. In a politically heterogeneous society, where public visibility of an issue or activity depends upon the extent to which it is perceived as highly relevant to the values and ends of important social groups and institutions, visibility often exacts the price of having the processes whereby such issues are dealt with become subordinated to the logic of political compromises. Thus, the growing visibility and involvement of science in public affairs may contribute to the growing role of political forces, as against substantive considerations, in the selective reception and integration of measures and indicators of the state of science in the policy-making and allocation processes.

Political Profiles of Indicators

That political variables have a central role in defining the state of science in society need not mean, however, that the selective reception and integration of data about the scientific enterprise are dominated by intellectually elusive or undiscernible factors. We can try to develop tools for studying and describing trends in the reception and integration of indicators, of science and other social and natural phenomena, in the context of public policy. Such a study would focus less on the technical relevance of indicators to standing issues or programs of action than on the perceived implications for the parties involved in projecting such indicators into political and policy contexts.

Consider, for example, the injection into the political context of indicators of social distribution of birth-control practices, spread of defense expenditures, or correlations between genetic and behavioral traits. Each of these categories of indicators is likely to activate a different cluster of social groups who expect their vital values to be affected by socially certifying such knowledge and making it available. Religious and ethnic groups may be activated by the social distribution of birth-control practices; business firms, military organizations, environmentalists, and so on may be activated by the spread of defense expenditures; and policemen, psychologists, legal experts, teachers, and others involved in the social definition and control of criminal behavior, education, and so on may be activated by publicized correlations between genetics and behavioral traits. Each of these clusters of groups involves different distributions of values, interests, and power; and in each cluster the groups that expect to be affected will enter into some sort of interaction in an attempt to influence the social use and integration of the relevant indicator in their own interests.

At least two types of criteria can be distinguished by which such groups evaluate the acceptability and the desirability of developing and utilizing indicators. The first criterion is whether given indicators appear to reinforce, challenge, or remain neutral with respect to socially held definitions of the situation, or, more generally, social outlooks on reality. When public attitudes are affected by these considerations of symbolic fit with prevailing systems of ideas and beliefs, the social reception of indicators can be described as dependent, at least in part, on their "communicative load" in the given social context. Criteria of the second type refer to the perceived practical *consequences* of the use and integration of indicators in social action, rather than the meanings ascribed to them. When considerations of consequences affect acceptability of indicators, their social reception can be described as dependent, at least in part, on their "instrumental load."

The communicative load of, say, scientifically established correlations between genetic and behavioral traits would refer, typically, to the perceived ideological implications of publicly certifying and publishing indicators of such correlations. The "career" of these indicators in the policy-making context has been influenced by the fact that they have been widely perceived as challenging cherished concepts of human nature and society (32). Similarly, the social status of measures and indicators linking crime and pathology or deviance and social class have been clearly influenced by their perceived unsettling implications for premises of the social order (33). Also, resistance to the certification and use of average I.Q. scores of various social groups (31) doubtless reflects reactions to their communicative load as politically useful aids to suspected attempts to explain social differences in biological terms.

The instrumental load of the same indicators—of correlations between genetic and behavioral traits, crime and pathology, or social deviance and class—would be their perceived impact on the relative social distribution of public resources and other benefits among the involved groups or on the selection of particular techniques for controlling disease, birth, crime, and so on (34).

Ideally, an analysis of the actual or anticipated communicative and instrumental loads of indicators in the political context should help us to understand or predict the presence and weight of the various sociopolitical mechanisms controlling the reception, rejection, and use of indicators. The communicative and instrumental loads of indicators describe their distinguishable interactions with the symbolic and distributive dimensions of the political process. The relative influence of considerations of technical adequacy and of political acceptability in the social reception of given indicators would of course tend to vary with contexts.

For example, the indifference or resistance of the Roman Catholic Church to the publication of indicators of "population explosion" (in the context of public debates about government encouragement of birth control) were not inspired by evaluating the indicators' validity or utility for public policy but by the inadmissibility of their communicative load: their clash with the concept held by the Church of the meaning and nature of population growth and sexual intercourse (35).* Similarly when Daniel Patrick Moynihan recommended that public policies towards the black problem focus more on remedies for perceived pathologies in the black family structure than on direct manifestations of racial discrimination, social resistance arose from the negative communicative load for groups sensitive to the social image of blacks. Part of their reaction also resulted from the negative instrumental load of Moynihan's statistics and diagnosis, the fear that it would be used to press for psychological help rather than for jobs and other material support for blacks (37).

The effects of differing criteria can also be seen clearly in attempts to define various categories of crime as medical rather than as moral or legal problems. These moves won support or were fought not only with reference to the comparative validity and effectiveness of the differing approaches but also on the grounds of their implications for the legal system's concepts of man and society (communicative load) and the distribution of authority and resources between legal and medical personnel or between law enforcement and health agencies (instrumental load) (38–43[†]; 33; 34, pp. 111–129).

An analysis that could establish whether any given set of indicators has or would have negative, positive, or neutral communicative and instrumental loads for the most important groups involved in public-policy processes should help the effort to understand and anticipate selective reception and integration of these indicators. For instance, indicators perceived by most significant actors as neutral or beneficial seem to have a "history" different from indicators perceived by some actors as costing them money or benefits. Indicators with uses appearing to emphasize benefits for some groups at the expense of others will probably not interact with a given political process in the same way as indicators with uses appearing to involve costs and benefits shared by the same groups.

The merit of this perspective is that it directs attention to the central mediative role of factors that some experts may tend to ignore as irrelevant in judging the merits of indicators. Because the reception of indicators (and other forms of knowledge or information) does not depend simply on their truth value or technical accuracy but also, or in some cases, primarily on such factors as their communicative and instrumental loads, the integration of indicators into the making of collective decisions and choices may often show not that policy making has been depoliticized and rationalized but that in the process of public policy making, certain bodies of information

*The constraints imposed on a Catholic President in initiating birth-control programs were noted by John F. Kennedy and led to the special role of the National Academy of Sciences in bringing the subject to a political focus during his administration (36).
†I have been instructed in this connection by J. A. Pitts, a student in my graduate class, Spring 1974, at the University of Pennsylvania.

and knowledge have come to have more important and acceptable political and organizational functions than others.

Functions of Indicators

What then are the implications of these observations for the roles and functions of measurements and indicators of the state of science in the political context of science policy? Consideration of the various nontechnical uses and functions of indicators in that context should help us to understand why, of the range of possible indicators, only specific ones were adopted, deployed, and used and why different groups both within and outside the scholarly community advocated diverse science indicators such as the volume of research papers, number of patents, number and distribution of students, comparative support for science in different countries or fields, and contributions of science to economic growth.

If the purpose is to enhance the "rationality" and "realism" of science policy, the development and publication of such indicators are useless operations because of the following limitations: in the context of public policy the reception of science indicators depends at least as much on political considerations of their consequences as on their substantive reference to the subject matter, and the political characteristics of the policy-making and allocation processes affecting science are not conducive to the use of criteria of intellectual and technical merit in the selection of measurements of the state of science.

If, however, the above observations concerning the weight of political and organizational considerations in developing and applying indicators are correct, insights into the limited contributions of indicators to the rationality of policy making are unlikely to restrain their use. But fortunately, we need not put this guess to the test. Although bodies of information about science apparently cannot be expected to influence the social perception of science and actions with regard to it simply by virtue of their intellectual adequacy, they can serve a useful purpose so long as they are *not* evaluated simply in terms of the goals of advancing our knowledge or understanding of science and substituting knowledge for politics as a basis of policy decisions. Furthermore, I maintain that these indicators are inadequate for advancing either of those goals precisely because of what makes them useful as *coordinating* and *mediating* devices in the interactions of the various groups and institutions involved in science policy making and resource allocation.

INSTITUTIONALIZATION OF SCIENCE INDICATORS

If the injection and reception of science indicators into the political context of *science policy* cannot be seen as a process by which knowledge is applied to depoliticize and enhance the rationality of science policy, it can perhaps be seen as part of an effort to influence the political definitions of science-as-a-problem for public policy. From a slightly wider perspective, the attempt to construct indicators may be compared with the construction and use of devices like the cost-of-living index, legal rules, standardized weights and measures, money, or traffic signs. The utility of each of these lies primarily in its application as a sort of standardized "social technology" for coordinating and regulating complex systems of social interaction.

When science indicators are advanced in order to influence public policies toward science, they must, of course, be acceptable to the main groups and interests involved. From the perspective of the scientific community, the promotion of such indicators may have obvious merits. They may help to project into the social context of public policy the idea that science is a living system of cooperative social behavior and interrelated institutions vulnerable to the cumulative effects of the decisions and actions of private and public bodies. Like the state of the economy, the state of science may be subject to imbalances harmful to the development and growth of socially important resources, activities, and institutions. To the extent, then, that science indicators can be seen as a way to alert policy makers to the need for considering the relations between their decisions and the "welfare of science," indicators may have a positive communicative load for the scientific community.

If these indicators also appeared to contribute to the strength of the scientific community in the competition for social resources, they would also have a positive instrumental load for that community. Indicators can also facilitate the capacity of the leadership of the scientific community, which perceives the welfare of science generally rather than from the perspectives of specific fields of disciplines, to define the criteria by which the welfare of particular scientific fields can be balanced against the welfare of science as a whole (44).

Science indicators may also be perceived as beneficial from the perspective of the political and the policy-making groups. For one thing, political actors often prefer to deal with interest groups that are articulate and coordinated in defining their preferences: the uncoordinated constituency that gives rise to a hopeless multiplicity of unrelated, unselective and often conflicting demands may increase the political costs and uncertainties of any decision or action. Furthermore, politicians and administrators often prefer to appear to base their decisions (whether increasing or decreasing support for science) on statistical or similarly "impersonal" grounds, thus minimizing the political costs of decisions that alienate potential supporters. A good illustration of this can be seen in attempts by the Israel Department of Education, in the past few years, to develop quantitative indicators in order to base government allocation of support among Israel's universities on "objective criteria." The results are not yet clear, but the Department obviously hoped such indicators would strengthen the government's capacity to present decisions in a sensitive public-policy area as resulting from technical computations rather than discretion—and thus decrease its own vulnerability to political pressures.

If science indicators are accepted by the parties as a kind of language, the availability of standardized vocabulary could increase the efficiency of the systems of communication and interaction behind science-policy and allocation decisions. The utility of such standardized indicators as a coordinating tool may be particularly significant in this case because scientists, by comparison with such social groups as farmers or war veterans, apparently suffer from a relatively weak capacity to organize and act collectively. The very emphasis on unhierarchical collegial values in the organization of research that may be conducive to the inner working of the scientific enterprise may also impose severe constraints on the capacity of scientists to form groupings and organizations likely to generate effective action in the political forums of the larger society (45). Moreover, the policy-makers who can influence the growth and direction of science are also organizationally discrete. Because of the wide relevance of science to diverse social goals—ranging from the promotion of health through agricultural productivity to security needs and military potential—its utilization contributes to the diffusion of science-policy decisions. The effect of this is a scattering of decisions on the distribution of resources for research over the entire spectrum of public and private institutions. Thus, indicators can function not only as a standardized language for describing the overall "health" and "balance" of science, but also—in an especially significant role—as a coordinating tool for articulating and representing the competing preferences and interests of the various groups and institutions involved.

Indicators that measure the movement of financial and human resources into and within science, as well as other aspects of the scientific enterprise, can be used as a subtle mechanism to translate the transactions of the involved groups into a shared, more operational, and "pragmatic" language better suited to the context of policy making. For example, alternative preferences for supporting science are easier to formulate in terms of patterns of financial and human resources than in terms of units of intellectual worth or social merit. The fact that quantitative measures are likely to be more easily defensible in political and bureaucratic contexts may make such indicators more attractive; and their apparent technicality may make the decisions in which they are used seem more impersonal and objective.

Validity versus Regulatory Functions

Although the technical appearance of indicators may thus enhance their political utility, the history of the Cost-of-Living Index, especially in the period immediately following World War II, shows that the coordinative and regulatory social functions of statistics clashed with the steps necessary to improve their validity and technical refinement (46). Following initial publication of the index in 1921, it served not merely as an "important statistical tool for economists, statisticians and administrators in the United States . . . but also as a *criterion in wage negotiations and in the adjustment of salaries*" [emphasis added]. As a result of the latter function, changes in the standards of the index prepared by the Bureau of Labor Statistics (BLS) had important potential consequences for the dynamics of bargaining among labor, business, and government (see Kruskal, this volume). For example, during World War II the index in many countries diminished in validity because of wartime changes in the structure of consumer behavior. But, any changes designed to make the Index more scientifically adequate were seen as destabilizing a delicate system of social arrangements and bringing about undesirable economic and social consequences. Thus, even though updating and modifying the index was advocated by professional statisticians and economists, the suggestion to accommodate to the "state of the art" in the relevant sciences encountered serious resistance (47).

It seems likely, then, that the logic behind the continued use of the dated index was that of economic and political interests rather than of economic or statistical science. The involved parties used the Index as a regulatory or coordinating tool. It was sustained not because of its scientific validity but as a convention with a lifespan that would end when the benefits to the interested parties ceased to outweigh the costs. And indeed, when the trade unions came to believe that the unofficial, scientifically more accurate index was more beneficial to their interests than the official version, they decided to attack the BLS decision to sustain the old index. Their weapon lay in "exposing" its inadequacies by citing standards of economics and statistics.

The case of the Cost-of-Living Index illustrates still another point: When bodies of information function as a data base for collective behavior, demands for revising or reinterpreting information may, in fact, be aimed at adapting operating arrangements to the interests of involved groups. But whatever the specific aim inspired by the ever-changing preferences and power of interest groups, their impulse to avail themselves of the most up-to-date scientific knowledge proves irresistible only when emendations of the data base would improve their own position.

But in fact, indicators would probably lose much of the social and policy-making significance they have if they were subject to continual revision to reflect every change in scientific standards. Like laws or national currencies, indicators must be relatively stable and reliable if they are to prove useful in coordinating social behavior. The likelihood of tensions between these sets of demands (having indicators be both stable and accurate) reveals the impossibility of using knowledge in political contexts without conceding professional standards. For conflict is inevitable when indicators constitute simultaneously "social technology" and a "report on the world" to decision makers. To function as mere coordinating tools, indicators need not be any truer than legal fictions. But if they are only that—if they depart radically from recognized professional standards—not only is their diagnostic and instrumental usefulness impaired, but so too is their regulatory utility and credibility as a kind of authoritative "knowledge."

Nevertheless, because the development and use of indicators are not rigorously disciplined by the internal criteria of scientific inquiry, they can have a flexibility that facilitates their responsiveness to change in political and policymaking needs. For, once outside the specific context of scientific discourse, continual changes in the theoretical and technical components of the indicators' "cognitive environment" could justify initiatives for changing them, even when the changes are actually reflecting the logic of shifts in interests and power structures more than the advancement of knowledge.

Indicators and Political Interests

To understand the development and use of indicators in the political context is thus to understand, at least in part, their role in influencing political and organizational behavior. In his analysis of the phenomenon of "the self-fulfilling prophecy," Robert Merton dealt extensively with the role of cognitive constructs of reality in generating and orienting behavior. He observed that "public definitions of a situation become an integral part of the situation and thus affect subsequent developments" (48). From this perspective, we can understand that as the publication of science indicators constitutes an attempt to fix the parameters of public definitions of the "state of science" in the context of public policy, that act inevitably becomes a part of the political process. With indicators functioning as a means of altering public definitions of "reality" and thereby influencing the priorities and direction of public action, professional considerations of their merits are not likely to be divorced from their behavioral implications.

The political dynamics of developing indicators in the context of public policy is apparent in the case of social indicators. (See Duncan, this volume.) Attempts to institutionalize them as terms of reference for public policies apparently began as a result of the rising social and political power of groups who challenged the centrality of economic indicators in public definitions of "objective states" and "social problems" by attempting to incorporate into the policymaking process various criteria meant to reflect the social costs of economic growth. Thus only as interest groups like consumers, women, blacks, and environmentalists became effectively organized were appropriate indicators institutionalized for describing the side effects of food additives and drugs, assessing resource distribution by race and sex, and reducing pollution levels.

At least in part, the promotion of science indicators, and the publication of *SI-72* and *SI-74*, constitute an attempt to dramatize politically the antici- pated costs should declining trends in public support for science continue. Indicators measuring fluctuation in that support became central when the "health" of science came to be identified largely with a certain rate of growth, and with that alone. No serious attempt was made to develop indicators describing the *costs* of growth in terms of, for instance, the rela- tion between an increasing government share of the economic resources of science and restrictions on its institutional autonomy. No indicators de- scribed the "health" of science with reference to, say, scientists' oppor- tunities to select research projects according to internal professional priorities.

Thus, though cognitive and technical considerations remain relevant for the selection and construction of indicators in the public context, the pro- cess cannot be understood fully without examining the operation of political and organizational factors. To do that, we must look at the forces determin- ing the relative weights assigned to these two sets of considerations in the selection and utilization of indicators for public policy.

INDICATORS AND THE "UTILITY OF TRUTH"

To observe that the reception and use of policy-relevant bodies of informa- tion are influenced by political, social, and organizational factors is not to imply, of course, that the truth value of indicators is entirely irrelevant. Instead, it is simply to ask the following questions.

1. In what way, to what extent, and under what circumstances is the validity of information used as a component of public-policy decisions signifi- cant?
2. How do the cognitive characteristics and political aspects of indicators interrelate?
3. How should efforts to improve indicator accuracy and validity be evaluated?

The following discussion explores only briefly some of the problems and possibilities arising in an attempt to answer these questions.

Although cognitive criteria of validity and accuracy may not be particu- larly relevant for the sociopolitical uses of indicators, these criteria are obviously significant when indicators are used as diagnostic aids for choosing among alternative courses of intervention in the reality to which they refer. The distinction between cognitive and noncognitive criteria for judging indicators suggests the possibility that cognitively and technically superior indicators may be politically unacceptable, whereas those that are politically acceptable may be cognitively and technically deficient.

For any specific indicator, can we distinguish between situations where its cognitive and political merits are compatible and situations where they are not—and represent the spectrum of mixed situations between these two extremes? Such knowledge about the relations between the cognitive and political properties of information in the context of public policy could, of course, be quite useful: Experts like to be able to balance the addition of margins of validity to the informational components of public policy—thus enhancing its "rationality"—against the possible political costs of such changes. Knowledge, based on cognitive and political criteria, about the comparative performance of alternative informational inputs is as relevant to the rationality and efficiency of the actions of indicator experts as the indicators they furnish may be to the rationality and efficiency of the actions of policy makers.

Unfortunately, however, the possible significance of such knowledge is not matched by the level of its availability. The following discussion is therefore limited to only a few strategic aspects of the problem. For this purpose, I outline the elements of a possible approach to study of the interrelations between professional and politico-organizational factors in developing and using indicators in the context of public policy.

Cognitive and Noncognitive Functions

For analytical purposes, let us examine the interrelations considered by experts who operate in the political environment of public policy making. They must weigh the anticipated utilities of improving the validity and accuracy of operating indicators against their—possibly latent—political and organizational functions (49). From this perspective, indicators could be described as simultaneously relatable and potentially contributive to these three components of the public policy process.

1. The cognitive component—indicators representing relevant aspects of reality, thus acting on the *rationality* of public policy.
2. The normative component—indicators appearing to influence the underlying ends and normative parameters, thus determining the *acceptability* of public policy.
3. The behavioral component—indicators affecting the interactions of the

people and institutions involved in making public policy, thus potentially increasing or decreasing the behavioral *feasibility* of public policy

Because the aim is to develop a basis for conceptualizing and systematically describing the relations between indicators and policy making, the three properties of indicators in this scheme cannot be derived from their content or from the separate qualities of the policy-making process. Each one describes a distinctive—and potentially independent—relation between indicators and policymaking.

Using these three types of relations, indicators—and for that matter, other bodies of knowledge or information—may be classified according to their profiles in terms of the three components above. Thus, different indicators applied to the same policy process will relate differently to its cognitive, normative, and behavioral components; the same indicators will also exhibit different profiles in different contexts.

In the context of a given sphere of public policy, an indicator may have high value as a contributor to the cognitive components of rational decisions, even while its effects on the normative parameters of public policy diminish the acceptability of those decisions. Or an indicator may contribute to both the technical rationality and the normative acceptability of a public program, while, because of the circumstances and timing of its publication, it may affect the social interaction of the participants so much, that it in fact undermines the behavioral feasibility of implementing the program.

There is no need to list all the possible combinations of these three characteristics of indicators to recognize that a refined version of this approach could help to classify various indicators according to their actual or anticipated contextual profiles and to compare them in terms of the relative political feasibility of their integration into the policy process. It could perhaps prove useful also in distinguishing between regions of public policy according to the degree of their hospitality to indicators in general or to any given set of indicators. For example, in regions of public policy where the decision-making procedure itself is uninstitutionalized and is as much a subject of political contention as the policy content, or where selecting means is inseparable from setting goals, the weight of the scores of an indicator in the cognitive dimension would be likely to diminish in relation to the normative and behavioral scores. By contrast, in regions of public policy where goal-setting and decision-making procedures are already supported by effective consensus, the consideration of the cognitive merits of indicators could be expected to be more weighty and more easily separable from considerations of their political and administrative uses. Such differences may be anticipated also between those regions of public policy where political compromise has generated a mandate for advancing internally consistent goals and those where compromise is based on a mandate to advance, at least in part, conflicting goals (50). One might reasonably expect that the weight of cognitive scores would be greater in the former instance.

Nevertheless, only substantial empirical research, furnishing a basis for juxtaposing the selective utilization of indicators and the profiles of their comparative relations to the three components of public policy, could furnish a basis for testing such hypotheses and enhancing our understanding of decisions to develop and use indicators in the context of public policy. But with the diverse and heterogeneous goals public policy must satisfy, to define a "rational policy," based upon the most adequate cognitive and technical sets of indicators may well be beyond expertise.

In most cases the balance of diverse preferences underlying a collective action means that some policy ends are to be carried out more fully than others. Altering such tacit agreements by making public action more effective and rational with reference to some of the less preferred ends may violate the level of performance sanctioned by the normative parameters of collective action.

Normative Parameters

In public policy and collective action, the compromises among competing participants are complex and subtle. Thus, it cannot be taken for granted that better information would be more beneficial to all the interested parties. Acts of introducing new indicators or improving indicators already in use always involve the risk of violating or altering the normative parameters of action. Thus, experts devising indicators for use in public policy-making often must operate in a narrow zone beyond which neutrality becomes advocacy and professional accountability becomes bureaucratic or political cooperation.

In the sociopolitical setting of public policy-making, arguments concerning the adequacy or utility of a body of information are rarely separable from debates on the normative parameters of the decisions involved. An illustration of this may be seen in the controversy in the United States over a government proposal to build an antiballistic missile (ABM) system. After one of the participants in the debate questioned the validity of calculations made by other participants, the Operations Research Society of America (ORSA) tried to evaluate the professional performance of the experts involved and to define professional standards for objective and reliable calculations (51). Paul Doty, a critic of ORSA's recommendations and findings, observed that they gave too much attention to measurement criteria for determining the need for an ABM system and not enough to the adequacy of the solution. He pointed out that "this limitation of scope not only imposes severe restrictions on what is being examined but prejudices the outcome because it cuts out most of the grounds on which anti-ABM arguments were made and gives undue emphasis to the narrow terrain on which most pro-ABM arguments rest" (52).

Doty seems to interpret the demand for accuracy in that instance as a one-sided attempt to narrow the normative grounds for taking an action. Whatever one's evaluation of Doty's position on the ABM question, his opinion that "the fine tuning of a calculation involving a number of parameters having substantial uncertainities is seldom justified, particularly if it obscures larger issues" at least illustrates a case where unqualified adherence to higher professional standards was found unacceptable. In the complex interaction systems underlying collective actions, the reception of an indicator tends to depend on its compatibility with a wide set of norms; thus, if attempts to enhance its validity and accuracy by scientific standards or in any other way alter the normative references of a problem, the indicator's status as an acceptable term of reference may easily be damaged.

With respect to any field of social action, available knowledge may not be symmetrically instrumental for achieving all relevant ends. This naturally complicates the situation by increasing the possibility that upgrading the validity and accuracy of information by applying the most advanced available techniques may imply unacceptable de facto changes in the balance of acceptable ends and preferences. In some cases, then, improving the information can be like dosing a patient with a more effective disease-specific drug at the cost of side effects that may damage his or her general health.

In the context of public policy-making, the relative autonomy of professional judgment in guiding selection and use of indicators also relates to the extent to which any given set of indicators lends itself to other purposes in the political-bureaucratic environment. The range of actual or potential uses of indicators beyond their areas of specific technical relevance can be termed their *political versatility*. Indicators of the social distribution of a certain physical handicap may be used both to improve public health or welfare services and to stigmatize parts of the population. Crime indicators can be used both as a tool for improving the control of crime and as a means of affecting the value of real estate in certain parts of the city. And when indicators about the state of the economy are used in partisan political evaluations of government performance, their application to regions where they are relevant for improving the technical components of public policy may then become limited. All these are examples of indicators whose high political versatility may restrict their use, even if their cognitive merits as inputs into policy making are granted.

It would be interesting, of course, to find out whether the political versatility of indicators affects the patterns of their deployment in the context of public policy and under what circumstances higher degrees of political versatility—representing wider ranges of interaction between indicators and the political process—inevitably diminish the freedom of experts to manipulate indicators according to professional criteria. (Often, the phenomenon of "secrecy" in government operations may be seen as a reflection of efforts to restrict by fiat the actual political versatility of information. However, these efforts may not necessarily be intended to protect the autonomy of professional standards.)

Behavioral Components

On examination of the relations between indicators and the behavioral components of public policy, the particular timing, forms, and forums in which indicators are introduced take on special significance. The context and form in which indicators are socially introduced, though possibly external or irrelevant to their cognitive "substance" or their normative implications, are significant in determining the potential role of indicators in influencing social behavior. This can be easily illustrated by the situation in the historic decision of the U.S. Supreme Court posing, for purposes of defining the dimensions of individual freedom of speech, the example of a man shouting "Fire!" in a crowded theatre. That cry, in those circumstances, may instantly improve information relevant to the audience's vital interests (assuming, of course, that the fire is real). But if, because of the circumstances of the information's release, the conditions of well-coordinated group behavior are undermined by panic, the information, however valid, has failed to promote the goal of public safety. Thus, aside from the cognitive qualities of any given piece of information, the act of communicating it at a particular time and in a specific context may play an important role.

The behavioral matrix of medical services can serve as another illustration. Consider a situation in which the enhancement of health depends principally on the extent to which curable patients seek help in medical institutions and on the quality of the treatment they receive. Assume further that experts who seek to rationalize medical services and enhance health have obtained information that 40% of the drugs prescribed by physicians are either without any significant effects or have adverse effects. Finally, let us stipulate the condition that, given the structure of the communications system and the prevalent attitudes of patients, this information cannot be communicated to the doctors without also sharing it with the wider public—thus risking a radical reduction of the social demand for medical help. In this situation, deploying information to rationalize medical treatment may undermine the behavioral conditions of an effective system of medical service. Here again, the fact that a body of information is cognitively relevant for rationalizing an action is not sufficient to ensure its actual utilization for that purpose.

Sometimes, however, the links between the cognitive and behavioral aspects of informational inputs may be regarded as beneficial rather than costly. For example, public agencies may furnish producers with economic incentives to improve the quality of their products by exploiting the effects on consumer behavior of publishing information about undesirable side effects of certain consumer products. This technique can sometimes achieve desirable results without recourse to administrative regulatory measures.

INDICATORS AND POLITICS IN GROUP INTERACTION

In view of the special constraints on the use of indicators to rationalize action in the political-bureaucratic context, it is not surprising that government experts who supply information for use in collective actions tend to be more sensitive than their professional peers outside (especially in academe) to the need to balance the cognitive, normative, and behavioral components of action (53). Government data specialists act as brokers of the stock of available knowledge for political planners of collective action. They often develop a unique skill in assessing trade-offs between an increase in the cognitive value of information and its functions with reference to the normative and behavioral dimensions of action. In this role they are often vulnerable to criticism from both sides: For outside professionals, the losses in cognitive value and rationality are more visible than the gains in the acceptability and feasibility of action; for political and bureaucratic groups, the socio-behavioral requirements of collective action are more visible than losses in cognitive value and rationality.

Although admittedly presented in simplified form, the scheme in Figure 1 conceives of expert teams developing and deploying information for collective action, as if they were the focus of the countervailing pressures of three groups: outside (usually academic) professionals pressing for increasing cognitive rationality, politicians tending to emphasize normative adequacy of the information used to ensure the acceptability and legitimacy of collective action, and administrators or bureaucrats concerned primarily with the role of information in relation to behavioral mechanics and feasibility.

This scheme represents the three types of indicator effects on public-policy decisions or programs—that is, on their rationality, acceptability, and behavioral feasibility—and it also represents the three groups to whose perspective each of these effects is central. Such differing orientations and interests naturally contribute to each group's tendency to assign differing communicative and instrumental loads to given indicators.

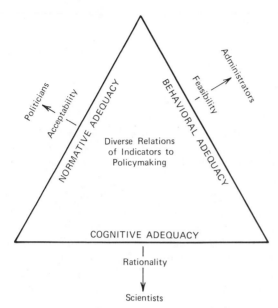

Figure 1   Heuristic scheme for assessing impact of indicators on policymaking.

The experts who develop and diffuse these indicators in the policy-making context depend on political, bureaucratic, and outside (usually academic) experts for cooperation. Thus, it is natural for them to compare indicators with reference to the actual or potential responses of these groups. The experts may have to consider both the profile of a new or modified indicator in relation to the cognitive, normative, and behavioral components of public policy and how its integration into the policy-making process may affect their image as professionals, accountable to their peers inside and outside the government. And they must also take into account the image and status needs of the politicians as representatives of certain interests and values and the needs of the administrators as conscientious operators.

Once the decisions of indicator experts are viewed this way, it is easier to recognize that the weight they give to professional and technical considerations may reflect the significance they assign to the comparative reactions of their professional peers outside the government and the other involved groups. Such factors as strength of the experts' professional allegiance, visibility to outside professionals of experts' deviance from professional standards, availability of credible professional sanctions against such deviations, and degree of mobility of professionals between government and academic institutions may be significant for fixing the limits of their ability to resist pressures for becoming responsive to nonprofessional considerations.

Professional versus Political Accountability

In the case of *science* indicators, however, a special complication stems from the double role of the scientists as the experts who describe reality and as the social group whose activities and institutions are being described. As professionals, they have a special stake in adhering to objective technical standards of description; and as an interest group, they have a particular political stake in the kind of indicators used to describe science and the way they are used in the policy process. So far, it is difficult to assess whether or how this peculiarity affects the interaction and representation of cognitive and political criteria in the case of science indicators. However, even though from an analytical point of view the development of public-policy indicators and statistics in nonscientific institutions seems to be a different kind of activity from the same work pursued in the context of research, both academic scientists and government experts often appear to ignore this difference. The academics often judge and evaluate the activities of government experts by rigorous professional standards, and the latter contribute to this by seeking to cultivate credibility as objective and independent professionals bound by scientific standards.

Because government experts are usually much less free than their peers in academe to ignore the normative and behavioral implications of statistics, the expectations that both groups of scientists will be bound by the same professional code inevitably generates tensions. These tensions become particularly acute when the measurements developed by government experts are used by their academic counterparts as input into research or as a basis for professional assessments of public policy and government performance (see Kruskal, this volume). In these instances, even minor deviations of the former from strict professional standards, no matter how necessary in the light of situational demands, appear to their academic peers to be an intolerable threat to the integrity of the role of the scientist as researcher, critic, and social agent of knowledge and rationality.

As critics of government experts, the influence of academic professionals stems from the relevance of their skills and knowledge for validating information used as the cognitive basis of effective actions and from their social authority to discredit publicly, in the name of science, those policies they do not sanction.

Ironically, however, the presumed power of nongovernment (usually academic) scientists to affect both the actual cognitive basis of rationality and effectiveness in public policy, and the layman's belief that such a sound basis for public policy does or does not exist can become a liability. Although relevant for improving substantive validity and reliability, open discussion of errors and repetitious demands for revisions may have the effect of undermining the status of and the trust in certified information as objective and reliable. Such trade-offs between actions contributing to the *substance* instead of the *appearance* of validity recall the problem confronted by Michaelangelo when he had to sacrifice some of the engineering requirements of structural strength in order to give the dome of St. Peter's the appearance of a massive and powerful structure (54).

In the social context, the appearance of truth does not necessarily depend on adherence to scientific standards of truth any more than aesthetic experience in architecture necessarily depends on adherence to standards of engineering. Surely, scientific truths, like principles of engineering, limit the extent to which appearance may successfully deviate from substance. But within such constraints the psychology of perception is as real a factor in determining which definitions of reality become socially acceptable and underlie cooperative behavior as it is in shaping aesthetic experience. Government bodies, particularly those producing and certifying information for regulative use, may be better able to afford a decline in the *actual* validity of the information than damage to its appearance of validity. Therefore, well-intended attempts by outside scientists to upgrade the cognitive qualities of information used by government experts may sometimes involve too great a loss of the trust and credibility that render such information useful in regulating and orienting collective behavior. (See the rest of this volume.) In such cases, government experts may be led to reject outside recommendations, even when they are valid, because whatever gains in rationality they are intended to produce seem to be offset by their behavioral and normative effects.

Professional Authority as Political Power

Nongovernment professionals who insist on integrity and purity of professional standards may be tempted to back up their criticisms and demands with their power to influence the professional reputation of the government experts. Even though the scientific community lacks the organizational capability of some other pressure groups, it has proved on occasion that it has the capacity both to discredit experts who appear to bend professional standards in order to accommodate political, economic, or administrative interests and to back up those who do resist pressures to make such expedient accommodations (55).

However, employing this power may result in a downward spiral of diminishing effectuality. For government experts need a degree of credibility as professionals if they are to resist pressures for the sacrifice of professional standards to political and administrative considerations. Therefore, damaging their professional status, no matter how justified in the light of their performance, may have the effect of decreasing their ability to resist future pressures.

Some of these complexities in the relations between government and academic scientists can be illustrated in the debate about the validity of the consumer price index (CPI).* Two University of Chicago statisticians who criticized the Bureau of Labor Statistics (BLS) for failing "to publish studies of accuracy so that outside scholars can press their own judgments about the quality of the Consumer Price Index" charged that "it flies in the face of long established scientific standards to expect outside professionals to accept on faith the Bureau's product without having the means of forming their own judgments based on published quantitative evidence." The Bureau publications, they went on to assert, "give a misleading picture of the CPI's accuracy and help stifle scientific criticism" (56).

Nevertheless, from the perspective of the BLS, such demands to facilitate the constructive role of scientific criticism and make the CPI more acceptable to scientists could not be fully satisfied without hurting the utility and acceptability of the index for its other users. The Commissioner of Labor Statistics, in responding to the Chicago scientists, alluded to this when he stated that the Bureau must be responsive not only to "economists and statisticians making economic analyses" but also to the needs of "businessmen and labor leaders using the index for collective bargaining and escalation . . . congressional committees, government officials and the general public" (57). The Commissioner especially emphasized the active role played by both business and labor councils as advisers of the Bureau.

The need to be responsive to such diverse constituencies illustrates the simple fact that where experts furnish the data base for complex interaction systems, the weight of scientific criteria in guiding choices from among alternative informational policies tends to be restrained by considerations of their comparative political implications—including the effects on the organizational resources and influence of the experts themselves.

In weighing possible compromises that may involve a degree of sacrifice in standards of validity and accuracy, as defined by academic professionals, government professionals are therefore likely to consider together with the substantive merits and demerits of alternative methodologies and alternative categories of information, the power and propensity of their outside peers to penalize deviance—by using social prestige and professional authority to hurt their professional status and undermine the social credibility of their output. The Chicago statisticians alluded to this power when they warned the BLS that "measured in numbers, the professional public is small but all

*I am grateful to William Kruskal for calling my attention to this debate. His kindness, however, should not be taken to imply responsibility for the interpretation and use of this material in the following discussion.

other users of the Index rely on their judgments" (58). Yet the significance of the power of the professional public is only relative to the power that can be exercised by interest groups like business and labor.

The cognitive, normative, and behavioral parameters of public policy and collective action are, then, socially represented by groups that do not hesitate to use their power to enhance their interests and discourage noncompliance with their policy intents. In view of the multiple functions of indicators in such complex systems of social interaction, indicators probably cannot be improved in one or more of these functions without being less useful or even detrimental to others. Thus, the structural and dynamic characteristics of the system of social interaction underlying the policymaking process are just as significant as the cognitive and technical characteristics of indicators for understanding the acceptance or rejection of the indicators.

What are the consequences of indications of the significance of sociopolitical and organizational factors in mediating the roles and functions of knowledge and information for the premise that more valid and accurate indicators tend to contribute to more rational decisions and more effective actions? Clearly, the premise, standing alone, serves neither to explain nor to guide the process of developing and deploying indicators in the policy-making context. Therefore, it is at least worth considering whether, in devising strategies for enhancing rationality, one should not explore the possibility of increasing the impact of knowledge by anticipating and operating indirectly on the sociopolitical and organizational components of the policy-making process. Taking into account the role of these mediating variables in filtering cognitive and informational inputs would probably produce better results than concentrating exclusively on the cognitive and technical bases of policy making.

TOWARD A POLITICAL ECONOMY OF INDICATORS

Perhaps what is needed to evaluate alternative cognitive inputs into policy making is a sort of political economy of indicators for public policy that would supply an intellectual basis for evaluating the cognitive, normative, and behavioral functions of indicators. Ideally, such a comparative approach to the assessment of indicators should systematically synthesize the following knowledge about them: their technical instrumentality for specific decisions and actions, their socioaxiological relations to the normative parameters of those decisions and actions, and their impact on the patterns of social interaction underlying policy making.

A political economy of indicators could systematically consider the possible "external costs" of using indicators to enhance rationality. Ideally, it could help identify indicator strategies that could yield "net rationality" (i.e., performance improvement resulting from changes in the cognitive basis of action not outweighed by costs to the other components). Perhaps such a systematic approach would better equip information experts for avoiding the pitfalls of rashness and timidity by, respectively, inducing improvements that lack net rationality and by failing to implement improvements because of unwarranted predictions of external costs.

Since experts on indicators who work within interaction systems do, in fact, use such judgments, a more analytical basis could introduce more systematic assessments of alternative criteria and courses of action. By helping to define the limits beyond which cognitive, normative, and behavioral components of group decisions and actions lose their elasticity, a political economy of indicators could delineate the zone within which various combinations of these components constitute practical alternatives. Furthermore, it could aid in identifying the direct or indirect measures that would overcome noncognitive constraints on assimilating cognitive inputs. Those measures which can improve the overall technical performance of the interaction system, regardless of the choice among politically competing alternatives of action, would then be more easily assimilated.

Such a comprehensive perspective on the use of indicators in sociopolitical contexts should provide some guidance for experts on indicators. It could aid them in recognizing situations where it is reasonable to expect short-run losses in net rationality to be offset by long-run gains, or situations in which short-run gains expected from informational inputs may involve unreasonable risks to the vitality of the system of social interactions upon which a given public benefit depends. For example, accurate indicators about the situation on a battlefield may demoralize an army and undermine its capacity to generate the coordinated group behavior necessary for its ability to fight. Yet, they may be vital for the capacity of the officers to adapt the actions of the army to the "realities" of the situation. In each such case to identify the point at which these two functions of information can be most adequately balanced requires careful consideration of the cognitive and behavioral profiles of various categories of information and also of the merits of alternative strategies for delivering it in the specific context.

A political economy of science indicators for science policy may or may not be in the making. But the kinds of considerations conducive to such a practical discipline reveal the inadequacy of evaluating strategies of the National Science Board, the National Science Foundation, and other such bodies simply by reference to the truth value or professional standards of the measurements and techniques they use. Without an analysis of the axiological premises of science policy and of the behavioral effects of various types of science indicators in the political organizational context of public policy, no comprehensive evaluation could be made.

The experts involved in the development and deployment of science indicators have not been indifferent to these various dimensions of utilizing information in the context of science policy. Therefore, the choices (conscious or not) of categories of information to be developed and deployed are as open to such evaluation as is the professional adequacy of the techniques and measurements used by the experts who did the choosing. The validity of such choices would largely depend on the quality of the presumptions underlying the experts' expectations that, in a given time and context, particular indicators may serve particular purposes.

In assessing the decisions and actions of the committees and expert teams that developed and deployed *SI-72*, it is, therefore, pertinent to examine the wide range of mostly nontechnical decisions affecting the functions of indicators in relation to science policy. Among these decisions are: the use of statistics labeled the "patent balance of the United States" and of the "Science Citation Index" as references for evaluating the state and role of science; the breakdown of federal expenditures for research and development according to eight functional areas, which include for instance, health, advancement of science and technology, and transportation—but omit communications; the passing of early drafts of the science indicators report to the Office of Management and Budget (OMB), the Treasury Department, the President's Science Advisory Council, and other bodies; the inclusion in the report of a study of public attitudes toward science and technology; the setting of limits beyond which OMB pressures for adjustments and changes in the report were to be resisted; and the compliance with OMB's desire that the report it found disagreeable not be released with the fanfare of a press conference. Taken as a whole, these decisions represent a complex fusion of considerations of scientific-technical as well as political and organizational "sense." As a sort of mixed strategy for developing and utilizing indicators, they cannot be evaluated without considering the merits and demerits of alternative mixes of cognitive, political, and organizational factors.

Concern with the operation of political and organizational constraints on the autonomy of cognitive and technical standards should not be taken as denying the existence of cognitive constraints on the use of indicators to achieve political and policy goals. Consider, for example, the report's evident inability to satisfy OMB's interest in "output indicators," which are particularly useful for justifying allocations of public money to science in terms of "what the taxpayers get for their money." The professional literature on measurements of research outputs may indicate that this omission was largely the product of technical difficulties or cognitive constraints rather than of preference by the sponsors of *SI-72* (59). This also illustrates my suggestion that, because measurable aspects of reality need not be symmetrically instrumental for satisfying the ends of all the participants, a cognitive constraint may become an obstacle to arriving at a necessary compromise. For, apparently, the cognitive constraints on developing output indicators were responsible in large measure for the resistance of the OMB and other government agencies to official recognition of *SI-72* as providing balanced, objective, and therefore "obliging" statistics.

This example and others further support the need for considering and comparing not only the cognitive merits of indicators but also the various relations between their cognitive and noncognitive characteristics in the policy-making context. A political economy of indicators for public policy encompassing such a comprehensive perspective should illuminate the way indicators can be used in political contexts to enhance various goals, including the rationality of public choices.

CRITICS AND ADVISERS

The foregoing observations are not meant to suggest that non-government experts reviewing the validity and accuracy of measures and information developed by government experts should necessarily temper criticism to accommodate the latter's needs. Considering their different perspectives, some conflicts between government and non-government experts concerning these criteria and their application are not only inevitable but desirable; they may represent a division of labor between scientists' roles as advisers and as critics.

The adviser's need to mix information with decisions and action is precisely what lends a special significance to the critic's role in keeping professional standards for the evaluation and certification of information "pure" and separate. Scientists who evaluate the adequacy of information by narrow professional standards help to sustain its cognitive value as a basis of rationality in public action. Comparative evaluations of the trade-offs between the functions of information in relation to the rationality, acceptability, and feasibility of collective action would be impossible without this invaluable defining and maintaining of socially visible standards of cognitive rationality. By articulating and bringing standards of cognitive rationality to bear on the evaluation of public policies and programs, the scientific critics complement the contributions of the cultural, political, and legal critics who judge public policies by other standards. Jointly, these groups help delineate the area remaining for practical decisions and actions between the sets of constraints imposed by "reality" and by prevailing social values and cultural codes.

REFERENCES

1. Bertram N. Gross, "A Historical Note on Social Indicators," in R.A. Bauer, Ed., *Social Indicators*, M.I.T. Press, Cambridge, 1969, pp. ix, xvii.
2. Bertram N. Gross, "Social Systems Accounting," in *ibid.*, p. 255.
3. In letter of transmittal to the President of the United States, *Science Indicators 1972: Report of the National Science Board, 1973*, National Science Board, National Science Foundation, Government Printing Office, Washington, D.C., 1973, pp. iii–iv.
4. National Academy of Sciences, *The Behavioral and Social Sciences: Outlook and Needs*, Government Printing Office, Washington, D.C., 1969, p. 102.
5. Joseph Ben-David, "Scientific Productivity and Academic Organization in Nineteenth Century Medicine," in B. Barber and W. Hirsch, Eds., *The Sociology of Science*, Free Press, New York, 1962.
6. Robert K. Merton, "Science and the Social Order," in Merton, *Sociology of Science*, The University of Chicago Press, Chicago, 1973.
7. Warren D. Hagstrom, *The Scientific Community*, Basic Books, New York, 1965.
8. Alma Wittlin, *The Museum, Its History and Its Tasks in Education*. Routledge & Kegan Paul, London, 1949, p. 58.
9. Florian Znaniecki, *The Social Role of the Man of Knowledge*, Harper & Row, New York, 1940.
10. Aaron Wildavsky, *The Politics of the Budgetary Process*, Little, Brown, Boston, 1964.
11. *New York Times Magazine*, May 12, 1974, p. 59.
12. Kenneth J. Arrow, *The Limits of Organization*, Norton, New York, 1974, pp. 24, 25.
13. Carl F. Christ, "The 1973 Report of the President's Council of Economic Advisers: A Review," *The American Economic Review, LXIII*, **4**:56 (September 1973).
14. Private Communications from members of *COSPUP*, April 1970.
15. Yaron Ezrahi, "The Political Resources of American Science," *Science Studies*, **I**, No. 2, 112–133 (1971).
16. *Ibid.*, pp. 225–229.
17. *Ibid.*, p. 227.
18. National Academy of Sciences, NAS-COSPUP, *Basic Research and National Goals*, Government Printing Office, Washington, D.C., 1965.
19. Herbert Blumer, "Social Problems as Collective Behavior," *Social Problems*, **18**, No. 3, 298–396 (Winter, 1971); Merton, *The Sociology of Science*, pp. 219–20.
20. NAS, *Physics in Perspective*, Government Printing Office, Washington, D.C., 1972.
21. Robert A. Dahl, *Pluralistic Democracy in the United States*, Rand McNally, Chicago, 1967.
22. Antoine-Nicholas de Condorcet, *Sketch for a Historical Picture of the Progress of the Human Mind*, trans. by Barraclough, Weidenfeld and Nicolson, London, 1955; Auguste Comte, *Positive Philosophy*, trans. by Harriet Martineau, Blanchard, New York, 1955.
23. Thomas Sprat, *History of the Royal Society*, [1667] I. Cope and H. W. Jones, Eds., Washington University Press, St. Louis, 1958; J. D. Bernal, *The Social Functions of Science*, M.I.T. Press, Cambridge, 1964; Michael Polanyi, *Science, Faith and Society*, Phoenix Books, Chicago, 1964.
24. Don K. Price, *Government and Science*, Oxford University Press, New York, 1962.

25. Andrei D. Sakharov, *Progress, Coexistence and Intellectual Freedom,* Penguin Books, Harmondsworth, 1968.

26. J. J. Rousseau, "The First Discourse," in *The First and Second Discourses,* St. Martin's Press, New York, 1964; Theodore Roszak, *The Making of a Counter-Culture,* Anchor Books, Garden City, N.Y., 1969.

27. Robert K. Merton, *The Sociology of Science,* p. 218.

28. Daniel S. Greenberg, *The Politics of Pure Science,* New American Library, New York, 1967.

29. Charles E. Rosenberg, *The Cholera Years,* The University of Chicago Press, Chicago, 1962, p. 27.

30. Edward S. Flash, *Economic Advice and Presidential Leadership,* Columbia University Press, New York, 1965.

31. R. J. Herrnstein, *I.Q. in the Meritocracy,* Little, Brown, Boston, 1973.

32. "Environment, Heredity, and Intelligence," *Harvard Educational Review, Reprint Series* No. 2, 1969.

33. Ian Taylor et al., *The New Criminology,* Harper & Row, New York, 1974; E. Rubingtor and N. S. Weinberg, *The Study of Social Problems,* Oxford University Press, New York 1971, pp. 15–46.

34. A. B. Biderman, "Social Indicators and Goals," in R. A. Bauer, Ed., *Social Indicators,* M.I.T. Press, Cambridge, 1968.

35. John T. Noonan, *Contraception,* New American Library, New York, 1965.

36. Private communication from Robert Green of NAS-COSPUP, August, 1969.

37. Lee Rainwater and William L. Yancey, *The Moynihan Report and the Politics of Controversy,* M.I.T. Press, Cambridge, 1967.

38. U.S. Senate Committee on the Judiciary, Subcommittee on Constitutional Rights, Hearing I, *Constitutional Rights of the Mentally Ill,* 87th Cong., May 2–5, 1961, Governmen Printing Office, Washington, D.C., 1961.

39. U.S. Senate Committee on the Judiciary, Subcommittee on Constitutional Rights, Hearing II, *To Protect the Constitutional Rights of the Mentally Ill.,* 88th Cong., May 28–31, 1961 Government Printing Office, Washington, D.C., 1961.

40. U.S. Senate Committee on the Judiciary, Subcommittee on Constitutional Rights, Hearing on S935, May 2–8, 1963, Government Printing Office, Washington, D.C., 1963.

41. Edward J. Sachar, "Behavioral Science and Criminal Law," *Scientific American,* **209** 39–45 (1963).

42. A. R. Louch, "Scientific Discovery and Legal Change," *The Monist,* **49,** 485–503 (1965).

43. Sheldon Glueck, *Law and Psychiatry,* Johns Hopkins University Press, Baltimore, 1962.

44. Ezrahi, *op. cit.,* p. 227.

45. Alice Kimball Smith, *A Peril and a Hope: The Scientists' Movement in America 1945–1947,* The University of Chicago Press, 1965.

46. K. S. Arnow, "The Attack on the Cost of Living Index," in H. Stein, Ed., *Public Administration and Policy Development,* Harcourt & Brace, New York, 1952, pp. 778–853.

47. *Ibid.*

48. Robert K. Merton, *Social Theory and Social Structure,* The Free Press, New York, 1968, p 477.

49. On the concept of latent functions, see Merton, *ibid.,* pp. 73–138.

50. On the structure of policy ends and the problem of rationality in policymaking, see, for example, Charles E. Lindblom, "The Science of Muddling Through," *Public Administration Review,* **19,** 79–88 (Spring, 1959).

51. P. Doty, "Can Investigators Improve Scientific Advice? The Case of the ABM," *Minerva,* **10,** 289–294 (1972).

52. P. Doty, "Science Advising and the ABM Debate," p. 19, in Charles Frankel, Ed., *Controversies and Decisions: The Social Sciences and Public Policy,* Russell Sage Foundation, New York, 1976.

53. See Merton's discussion of the role of the intellectual in bureaucracy, *Social Theory and Social Structure,* pp. 261–278.

54. Geoffrey Scott, *The Architecture of Humanism,* Scribners, New York, 1924, p. 113.

55. S. A. Lawrence, "The Battery Additive Controversy," in E. A. Bock and A. K. Campbell, Eds., *Case Studies in American Government,* Prentice-Hall, Englewood Cliffs, N.J., 1962, pp. 354–356, 364.

56. W. H. Kruskal and L. G. Telser, "Food Prices and the Bureau of Labor Statistics," *Journal of Business,* University of Chicago, **33,** No. 3, pp. 258–279 (July 1960).

57. Ewan Clague, "Comment," *ibid.,* p. 280.

58. W. H. Kruskal's reply, *ibid.,* p. 285.

59. H. G. Johnson, "Federal Support of Basic Research: Some Economic Issues," in NAS, *Basic Research and National Goals;* Edward Shils, Ed., *Criteria for Scientific Development: Public Policy and National Goals,* M.I.T. Press, Cambridge, 1968.

APPENDIX

 Participants in the Conference on Science Indicators

The following persons participated in the Conference on Science Indicators held at the Center for Advanced Study in the Behavioral Sciences, Stanford, California, on June 13–15, 1974:

Robert Brainard, National Science Foundation
Jonathan R. Cole, Columbia University
Stephen Cole, State University of New York, Stony Brook
Lloyd Cook, National Science Board
Emilio Daddario, Office of Technology Assessment
Otis Dudley Duncan, University of Arizona
Yehuda Elkana, Van Leer Jerusalem Foundation
Yaron Ezrahi, Hebrew University
Eugene Garfield, Institute for Scientific Information
Zvi Griliches, Harvard University
Hubert Hefner, National Science Board
Gerald Holton, Harvard University
Thomas Kennedy, National Institutes of Health
Daniel Kevles, California Institute of Technology
William Kruskal, University of Chicago
Joshua Lederberg, Stanford University
Robert K. Merton, Columbia University
Ronald Overmann, National Science Foundation
Robert Parke, Social Science Research Council
Donald Ploch, National Science Foundation
Steven Shapin, University of Edinburgh

Henry Small, Institute for Scientific Information
Arnold Thackray, University of Pennsylvania
Hans Zeisel, University of Chicago
John Ziman, University of Bristol
Harriet Zuckerman, Columbia University

APPENDIX

B

Notes on Contributors

JONATHAN R. COLE is Professor of Sociology at Columbia University and Associate Director of the University's Center for the Social Sciences. The coauthor with Stephen Cole of *Social Stratification in Science* (1973), he has completed a monograph on the changing place of women in science.

STEPHEN COLE is Professor of Sociology at the State University of New York, Stony Brook. In collaboration with Jonathan R. Cole, he has completed a monograph on the operation of the peer-review process in science, and with others, a case study of the relations between theory and research in the development of sociology and solid state physics. His books include *The Unionization of Teachers* (1969) and *Social Stratification in Science* (with Jonathan R. Cole, 1973).

LORRAINE DIETRICH is an instructor in the Department of Sociology at the University of Arizona, and is completing her doctoral dissertation on measurement of scientific growth.

OTIS DUDLEY DUNCAN is Professor of Sociology at the University of Arizona. His long-term interest in research on social indicators has focussed particularly on the measurement of social change, using replicated surveys. His books include *Introduction to Structural Equation Models* (1975) and, with Peter Blau, *The American Occupational Structure* (1967).

YEHUDA ELKANA holds the Henya Sharef Chair in Humanities, and is Associate Professor of History of Science at The Hebrew University of Jerusalem. The author of *The Discovery of the Conservation of Energy* (1974), he is at work on a multivolume study of *Images of Scientific Knowledge*. He is also Director of the Van Leer Jerusalem Foundation.

YARON EZRAHI, Senior Lecturer in Political Science at The Hebrew University of Jerusalem, took his doctorate at Harvard University with Professors Don K. Price and Judith Shklar. His research focuses on the uses of scientific knowledge in public policy decisions and the interactions of science and politics as cultural systems.

EUGENE GARFIELD is President of the Institute for Scientific Information in Philadelphia, publisher of the *Science Citation Index*. He took his doctorate on chemical linguistics with Zellig Harris at the University of Pennsylvania and pursues a continuing interest in the uses of citation and bibliometric analysis for studying scientific and scholarly literatures.

ZVI GRILICHES, Professor of Economics at Harvard University, is the author of *Price Indexes and Quality Change* (1972). His early research examined the economics of technological change, and he continues work on the economics of education and returns to investment in industrial research.

GERALD HOLTON, a student of P. W. Bridgman, is Professor of Physics at Harvard University. He works on the thermodynamics of proteins at high pressure and the history of recent physics. He is the author of *Thematic Origins of Scientific Thought: Kepler to Einstein* (1973) and *Introduction to Concepts and Theories in Physical Science* (1973).

MANFRED KOCHEN, whose doctorate is in applied mathematics, is Professor of Information Science and Urban and Regional Planning at the University of Michigan. He is the author of *Principles of Information Retrieval* (1974) and editor of *The Growth of Knowledge* (1967). His current research centers on the use of information systems in decision making and on algorithms for cognitive learning.

WILLIAM KRUSKAL, Ernest de Witt Burton Distinguished Service Professor of Statistics at the University of Chicago, is coauthor with Leo Goodman of a series of papers on measurement of association. His research interests include the concept of the representative sample and the uses of statistics in public policy.

JOSHUA LEDERBERG is currently at work on problems in molecular genetics, the applications of artificial intelligence, health science and public policy, and the sources of problem differentiation in science. Recipient of the Nobel Prize in Medicine for his studies of genetics in bacteria, he is Professor of Genetics at Stanford University and the author of "A View of Genetics" (*Prix Nobel,* 1958).

MORTON MALIN, Vice President of the Institute for Scientific Information at Philadelphia, took his doctoral degree in diplomatic history at the University of Maryland. He is the author of a series of papers on citation indexing.

ROBERT K. MERTON, University Professor at Columbia University and Resident Scholar at the Russell Sage Foundation, is the author of *Social Theory and Social Structure* (1968) and *The Sociology of Science* (1973). His current research centers on the historical sociology of scientific knowledge and the self-fulfilling prophecy.

DEREK J. de SOLLA PRICE, Avalon Professor of History of Science at Yale University, took doctorates in physics at University of London and in history of science at Cambridge University. The author of *Science Since Babylon* (1975) and *Little Science, Big Science* (1964), Price focuses his research on quantitative studies in science policy.

HENRY G. SMALL holds a doctorate in history of science from the University of Wisconsin and is Senior Research Scientist at the Institute for Scientific Information. In collaboration with Belver Griffith, he has been developing the method of co-citation analysis for studying the structure of scientific literatures and is also concerned with other applications of citation analysis to the history and sociology of science.

ARNOLD THACKRAY is the author of *Atoms and Powers* (1970) and *John Dalton* (1972). He is Professor of History and Sociology of Science at the University of Pennsylvania and is currently engaged in studies of the interactions of science, technology and society, the historiography of science, and quantitative history of science.

HANS ZEISEL, Emeritus Professor of Law and Sociology at the University of Chicago Law School, continues his empirical studies of legal institutions. He holds doctoral degrees in jurisprudence and political science from the University of Vienna and is the author of *Say It With Figures* (1968, 5th edition) and coauthor of *The American Jury* (1971).

JOHN ZIMAN, author of *Public Knowledge: The Social Dimension of Science* (1967) and *The Force of Knowledge* (1976), is Melville Wills Professor of Physics at the University of Bristol. He works on the theory of disordered systems and examines unquantifiable ingredients of scientific knowledge and scientific life.

HARRIET ZUCKERMAN, Associate Professor of Sociology at Columbia University, is the author of *Scientific Elite: Nobel Laureates in the United States* (1977). Her research focuses on the cultural structure of science and scholarship, and cognitive consensus in scientific specialties.

Index of Names

335

Index of Subjects